EurographicSeminars

Tutorials and Perspectives in Computer Graphics

Edited by W. T. Hewitt, R. Gnatz, and D. A. Duce

W. T. Hewitt M. Grave M. Roch (Eds.)

Advances in Computer Graphics IV

With Contributions by
E. Fiume, I. Herman, R. J. Hubbold and W. T. Hewitt,
A. Gagalowicz, C. Bouville and K. Bouatouch,
T. Nadas and A. Fellous

With 138 Figures, Including 28 in Colour

Springer-Verlag
Berlin Heidelberg New York
London Paris Tokyo
Hong Kong Barcelona
Budapest

EurographicSeminars

Edited by W. T. Hewitt, R. Gnatz, and D. A. Duce
for EUROGRAPHICS –
The European Association for Computer Graphics
P. O. Box 16, CH-1288 Aire-la-Ville, Switzerland

Volume Editors

W. T. Hewitt

Manchester Computing Centre
Computer Graphics Unit, Computer Building
University of Manchester
Manchester M13 9PL, U.K.

Michel Grave

35 Rue Lauriston
F-75116 Paris, France

Michel Roch

Sun Microsystems
12, route des Avouillons
CH-1196 Gland, Switzerland

ISBN-13: 978-3-642-84062-3 e-ISBN-13: 978-3-642-84060-9
DOI: 10.1007/ 978-3-642-84060-9

45/3140-543210 – Printed on acid-free paper

Preface

This fourth volume of Advances in Computer Graphics gathers together a selection of the tutorials presented at the EUROGRAPHICS annual conference in Nice, France, September 1988. The six contributions cover various disciplines in Computer Graphics, giving either an in-depth view of a specific topic or an updated overview of a large area.

Chapter 1, Object-oriented Computer Graphics, introduces the concepts of object oriented programming and shows how they can be applied in different fields of Computer Graphics, such as modelling, animation and user interface design. Finally, it provides an extensive bibliography for those who want to know more about this fast growing subject.

Chapter 2, Projective Geometry and Computer Graphics, is a detailed presentation of the mathematics of projective geometry, which serves as the mathematical background for all graphic packages, including GKS, GKS-3D and PHIGS. This useful paper gives in a single document information formerly scattered throughout the literature and can be used as a reference for those who have to implement graphics and CAD systems.

Chapter 3, GKS-3D and PHIGS: Theory and Practice, describes both standards for 3D graphics, and shows how each of them is better adapted in different typical applications. It provides answers to those who have to choose a basic 3D graphics library for their developments, or to people who have to define their future policy for graphics.

Chapter 4, Special Modellings, is an extensive description of all methods used for modelling non-geometrical objects for computer graphics. It covers the fields of texture synthesis, solid texturing, fractals and graftals, as well as the use of botanistic models for describing plants and trees. All these techniques are presented in a synthetic document, each section of which gives references for more detailed approaches.

Chapter 5, Developments in Ray-Tracing, provides much useful information to those who are deeply involved in the development of ray-tracing software. Most of the new techniques applied to enhance the quality of the images produced or to accelerate their computation are described in detail and compared together, and this tutorial will help developers in building high performance software for producing highly realistic images.

Chapter 6, Rendering Techniques, finally, is an up-to-date general presentation of the various techniques used in the visualization of 3D objects. It defines all the basic notions and vocabulary, in order to permit an understanding of the processes involved and the problems encountered. It will help all users of such system to really appreciate their tools.

This collection of tutorials covers topics which are presently very important in Computer Graphics, and we would like to thank the contributors for their high quality work.

Finally, thanks to the women (and men) of Manchester who did so much, so quickly: Jan, Mary, Paula, Maria, Sheila, Jo, Julie, Andy and Steve.

Terry Hewitt
Manchester

Michel Grave
Paris

Michel Roch
Geneva

Contents

Colour Plates

The following 28 plates all refer to Chapter 4.

1

2

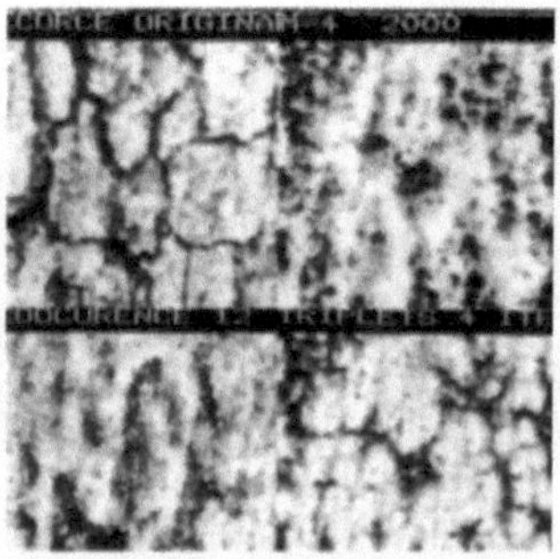

3

4

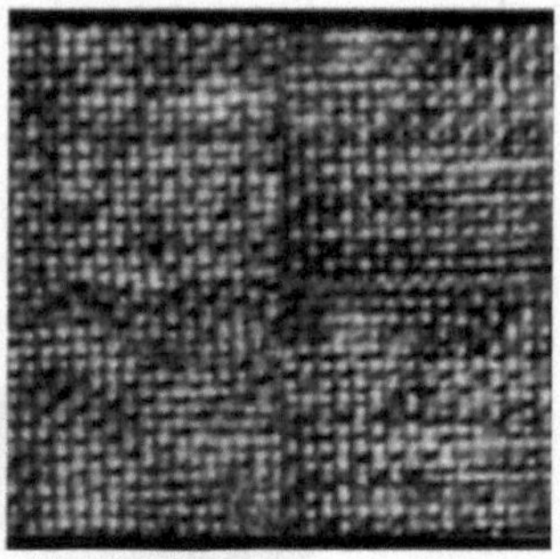

5

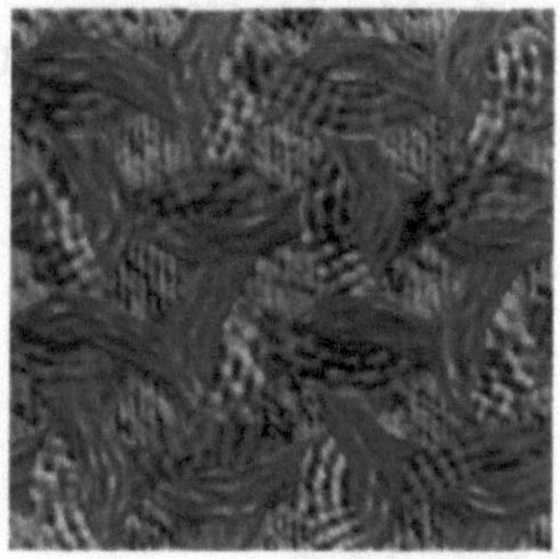

6

1 Natural and synthetic bark using various statistical models (model 1, 3 and 4)
2 Natural and synthetic bark using various statistical models (model 1, 3 and 4)
3 Synthesis of a pullover pattern using our macroscopic texture model
4 Natural and computerized formica displayed with the format of table 4.2
5 Natural and artificial woven string wall covering displayed as in table 4.2
6 Artistical hierarchical texture produced with the synthesis algorithm in [12]

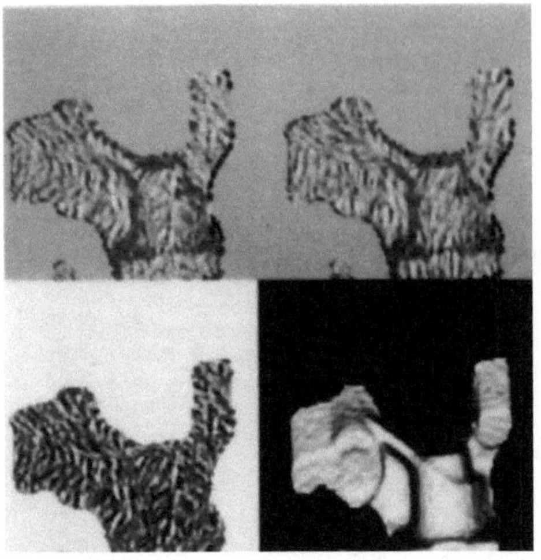

Plate 7
Use of model 2 to map seismic texture on a cast-iron part of a RENAULT car engine

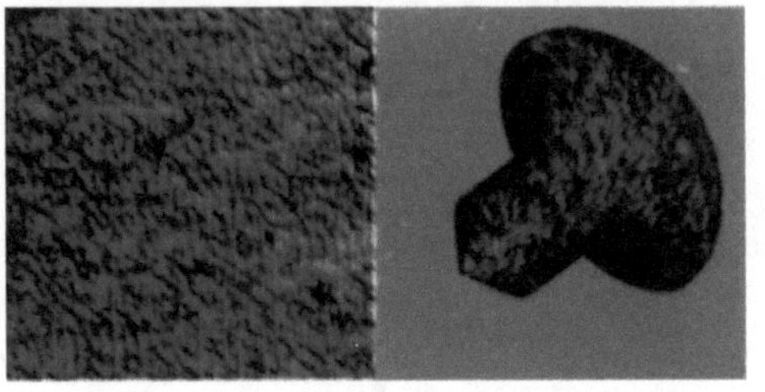

Plate 8
Synthetic nail covered with real colour wool

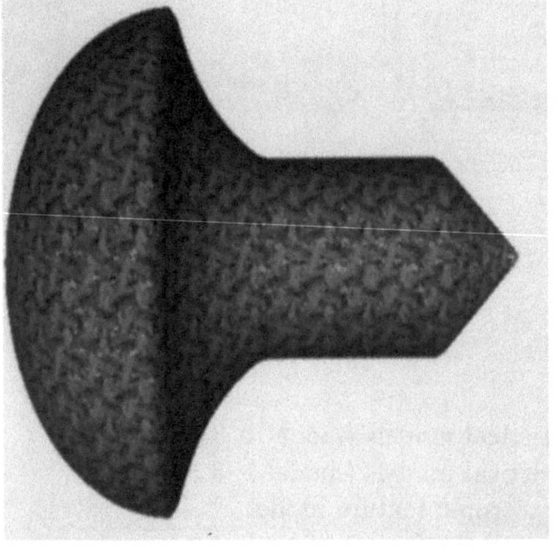

Plate 9
Hierarchical colour texture synthesis on a 3D surface (from [12])

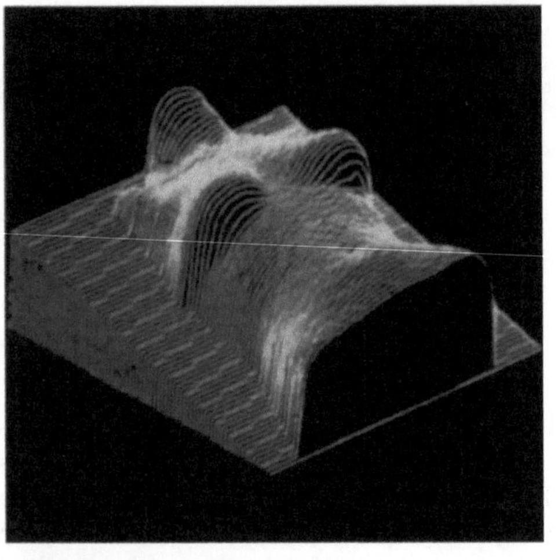

Plate 10
Body shape of a mannequin measured by a 3D laser sensor

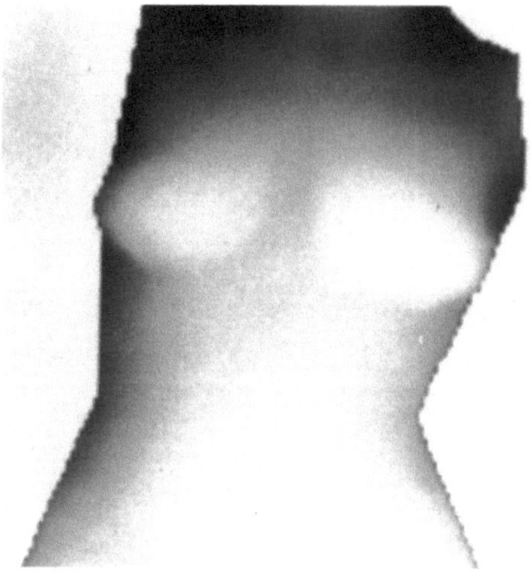

Plate 11
Same body shape in an image format

Plate 12
Picture of a naked mannequin

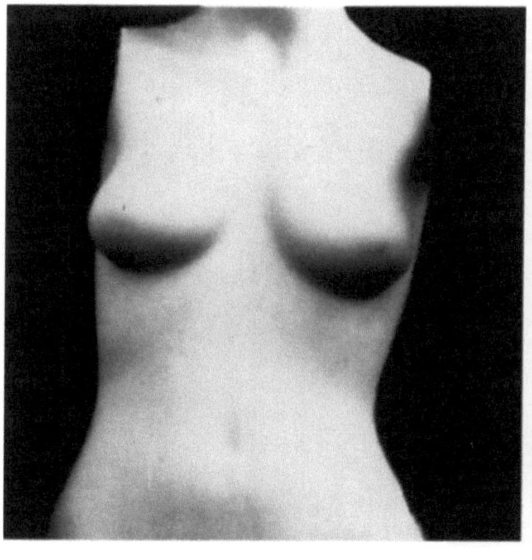

Plate 13
Body of the mannequinn extracted with a threshold technique

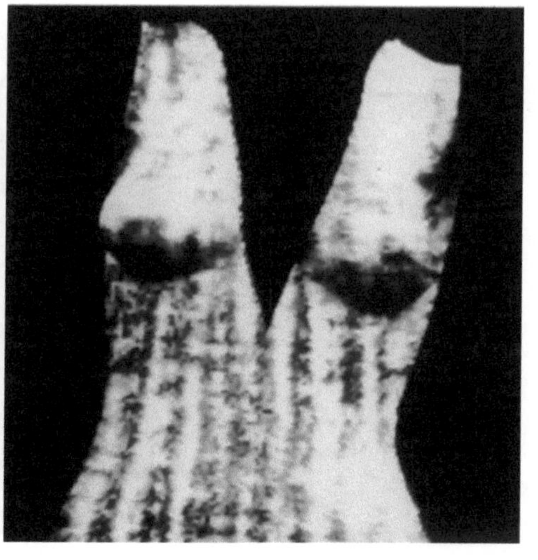

Plate 14
(Textured) dressed mannequinn body

Plate 15
Mannequinn wearing a bathing costume textured by a microscopic model

Plate 16
Natural maple leaf and synthesized fractal approximations using a mean square error criterion

Plate 17
Composite image (created by J. Lévy Véhel)

Plate 18
CARTOON.TREE . A 2D rendering of a context-free grammar phenotype (from [80])

Plate 19
GARDEN. Several context-sensitive graftal species showing the variety obtained (from Smith [80])

Plate 20
Forest scene from the ADVENTURES OF ANDRE AND WALLY B. using particle systems (from [77],[62])

Plate 21
Frame from STAR TREK II: THE WRATH OF KHAN obtained by the use of particle systems (from [77],[71])

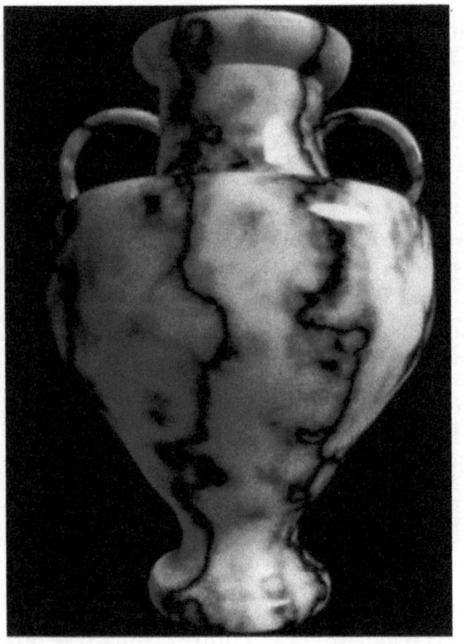

Plate 22
Marvellous marble vase produced by K. Perlin [74] with the use of solid texture model

Plate 23
Bumpy donut created by K. Perlin [74] using a normal perturbation technique

Plate 24
Clouds created by G.Y. Gardner [44] using a similar model (sinewave perturbations)

Plate 25
"Natural" palm tree reconstructed from a botanical model (from [22])

Plate 26
Weeping willow from the same program [22]

Plate 27
Evolution of a maple leaf using the free surface evolution technique (from Lienhardt [56])

Plate 28
Bell flower obtained by Lienhardt's technique (from [56])

1 Object-Oriented Computer Graphics

Eugene Fiume

ABSTRACT
Object-orientation and computer graphics from a natural, if occasionally uneasy alliance. Tenets of object-orientation, such as data abstraction, instantiation, inheritance, and concurrency, also appear in the design and implementation of graphics design. We explore the relationship between object-orientation and computer graphics, and consider the structuring of various kinds of graphics systems in an object-oriented manner.

1.1 Introduction

The term "object" has become one of the most popular computer buzzwords of the day. Interpretations vary as to what object-oriented systems are, and in fact their very diversity may legitimately cause one to wonder if there is anything that these systems share. As in other areas of computer science, "object-oriented" techniques have been employed in computer graphics since well before the term was coined. We shall examine the ways in which computer graphics is inherently object-oriented, and how object-oriented techniques may be of help in structuring graphics systems. We shall also see that object-orientation, despite being a generally useful approach, is not a panacea. Throughout this paper, the issues underlying object-orientated graphics will be pursued from two not necessarily distinct viewpoints: that of how a programmer might construct such a system from a (preferably object-oriented) programming language. In this manner, we hope to gain some insight into the programming and use of programming and use of object-oriented graphics systems. The fact that few such systems currently exist should not deter us from understanding the principles underlying them, for their arrival in large numbers is imminent.

An *object* is an encapsulation of activities and data. Its basic properties are inherited from a prototype, of which the object is an instance. An object may execute independently of other objects, and communicates with them according to some protocol. Rather than viewing an application as a single large piece of code, the object-oriented approach favours viewing it as a set of communicating capsules of activity, perhaps executing concurrently. The general belief is that this approach facilitates structuring a system in a manner that clearly reflects its conceptual design, and that enhances reusability of its components. Of course, not all applications require all of these facilities, but it is generally felt that they are sufficient to handle most of the practical applications one might wish to program or use. Applications from several areas of computer science, including database systems, office information systems, and simulation, have been successfully modelled by object-oriented approaches. Surprisingly, computer graphics has yielded comparatively slowly to object-orientation. This is due in part to the great diversity of graphical applications. Some, such as computer animation, have greatly benefited from object-orientation. On the other hand, graphical applications such as modelling and rendering systems have not.

This chapter has several goals:

- to introduce the basic principles of object-orientation
- to show how object-oriented techniques can be applied to computer graphics

- to give examples of object-oriented graphics systems
- to point to current research areas and future directions
- to provide categorised references for further information.

1.2 Basic Principles of Object-Orientation

1.2.1 Motivation

Object-oriented programming represents the evolution and synthesis of many well-known and important concepts in software engineering and programming methodology. The conventional history credits Simula as the major inspiration for early object-oriented languages such a Smalltalk. However, equally important to the evolution of today's systems are the more conventional "modular" programming languages, beginning with Algol-60, and progressing to Pascal, Modula, Euclid, Ada, Mesa, and CLU. In the addition of greater functionality such as concurrency, communication, dynamic binding, and strong typing, object-oriented systems owe a great deal to process-based systems such as CSP, Ada, Mesa, and Actors, and systems based on Lisp. Current, well-known object-oriented languages include Smalltalk, Objective C, C++, Flavors, and Cedar. The number of experimental systems that have been or are being developed is also increasing dramatically. The most important practical problem to which object-oriented is applied is that of *re-usability*. Software systems often exhibit an excessive amount of redundancy and object-oriented systems contain a number of facilities to allow one to optimise the usability of an object in other domains. In traditional programming environments, reusability has been enhanced by grouping useful modules into libraries which could be accessed by a wide community of programmers. The difficulties often encountered in such libraries is the lack of uniformity in the access and use of various modules in the library, poor naming practices, and the lack of structure within the libraries themselves. To counteract this, a large programming community often establishes standards or conventions for the naming and structuring of programs, for the use of libraries, and for the linkage of programs to library modules. Object-oriented environments tend to relax strict conventions, while attempting to maintain the consistency of interfaces between modules.

In [84], Wegner gives a taxonomy of object-oriented systems according to the following categories:

- objects
- classes
- inheritance
- data abstraction
- strong typing
- concurrency
- persistence

Very few systems contain all of these properties and features, and Wegner attempts to name and to draw distinctions among systems having various subsets of them. It is not clear that these dimensions exhaustively characterise all object-oriented systems.

Nevertheless, they form a good basis from which to begin our discussion. Also missing is the fact that *environmental* support tools greatly facilitate the use of an object-oriented system, and one must be cautious about separating pragmatics (environmental issues) from semantics (the actual language features). After all, it is difficult to imagine writing or using a Smalltalk program without using a fast bit-mapped workstation and other facilities. This support could include:

- graphical interaction
- change and version management
- structure editors
- computer-aided software engineering tool
- object browsing
- behavioural access methods
- object visualisation
- object debuggers.

The fact that it may not be correct to separate pragmatics from semantics will unfortunately not stop us from doing so. In the interests of space and time, in this document, we shall only deal with identifiable object-oriented language features.

It is important to note that object-oriented programming is largely a programming *style*. As indicated earlier, a large number of languages have been developed that enforce a particular style, but it is possible to adopt a personal programming style within a traditional programming language such as C, Pascal, or Fortran, which approximates fairly well many of the issues discussed in this document, particularly if support tools such as preprocessors and macros are available.

1.2.2 Object-Oriented Systems

To illustrate the concepts of object-orientation, we shall work with a hypothetical object-oriented language using graphical examples. Our discussion of the concepts listed above will be somewhat informal. See the bibliography for references to papers that develop these concepts more carefully. More sophisticated examples relating to specific differences between graphics systems and other applications can be found in section 1.3.

Objects and Classes

Any application structured in an object-oriented manner is composed, naturally enough, of a set of *objects*. An object has a name, it can take on some domain of values, and, depending on the preferred model of computation, either it can perform operations, or operations can be performed on it. Some objects never change their state (i.e., value), as in the case of constants. In this case such an object is said to be *immutable*. Otherwise, an object can potentially change its state and is said to be *mutable*. To help one organise and define objects easily, several language features are required over and above this very basic notion of an object.

Classes or *object schemata* offer a way of defining a prototype from which objects can be instantiated. The idea of defining and instantiating basic data types was borrowed by computer science long ago from Russell's type theory, and can be found in languages such as Pascal and C. Classes provide a straightforward extension to these type mechanisms. For example,

class Quadrilateral $(P_1, P_2, P_3, P_4 : \mathbf{R}^2)$

may define a quadrilateral object with initial vertices $P_1, \ldots, P_4$. Then the declaration

quad1: **Quadrilateral** ((0,0),(1,0),(1,1),(0,1))

defines an object named quad1, which is an instance of the class **Quadrilateral**, with specific *instance variables* (0,0) for P_1, (1,0) for P_2, (1,1) for P_3, (0,1) for P_4. This class can be used in a manner analogous to the use of other data types in the system. One can, for example, make arrays or lists of quadrilaterals. The difficulty so far is that it is not at all clear what can be done with quadrilateral objects. We have no idea what is inside them, nor what operations we can perform on them, which is not a particularly pleasant state of affairs. To rectify this, data abstraction facilities are required.

Data Abstraction

By *data abstraction*, we mean (at least) two things. First, the set of operations that may be performed on an object are known and stated publicly. Second, it is only through this set of operations that the state of the object may be discovered. Let us return to our quadrilateral and refine it to reflect these notions.

class Quadrilateral $(P_1, P_2, P_3, P_4 : \mathbf{R}^2)$
Operations
{ Translate: **Quadrilateral** $\times \mathbf{R}^2 \rightarrow$ [**Quadrilateral**]
Rotate: **Quadrilateral** $\times \mathbf{R} \rightarrow$ [**Quadrilateral**]
Scale: **Quadrilateral** $\times \mathbf{R}^2 \rightarrow$ [**Quadrilateral**]
Get Vertices: **Quadrilateral** $\rightarrow$ **list of** $\mathbf{R}^2$
}

We shall assume that we have access to primitive data types such as **R** (or some computable subsets) and **N**, as well as aggregation operators such as **list of** and **array of**. Elements of a list will be enclosed in parentheses. The use of square brackets in the above definition denotes that the result is performed on the same object as the argument. Thus if quad1 is a quadrilateral, then as defined above, quad1.Translate(1,3.3) affects quad1. Otherwise, quad1.Translate would return a new object of class **Quadrilateral** with, one would expect, appropriately translated vertices. A useful way to visualise an object's operations is as a Pascal record, except that the elements addressed in the record are activities. For example, if object O supports operation *op*, then the statement $O.op$ requests that O perform operation *op*.

Before going on, it is worth making one observation, at the risk of dwelling too long on an obvious fact. As long as we have access to an object's name, we can potentially ask it to perform an operation. Since it is an independent entity, in that it does not share its address space with other components of the application, it can in principle "reside" elsewhere in the system. Exactly *how* the system handles object invocation is usually irrelevant information to a programmer or user, except in highly-concurrent or time-critical applications. Even then, as we shall see when we discuss active objects, the mechanisms for distribution and communication can be made fairly abstract and convenient to use (which does not necessarily make the job of concurrent programming simple). In distributed systems, the implementation of object operation invocation is generally accomplished using a *remote procedure call* or at a lower level by a *message passing* protocol. In a uniprocessor system, the implementation is often simply a procedure call, though there are many exceptions to this rule.

One of the difficulties with object-orientation is that, especially in computer graphics, object-orientation appears to take something away, namely the ability to "look inside" objects, without giving something back, namely a precise characterisation, in an implementation nonspecific manner, of what object operations are. One can argue that the above example is guilty of exactly this, in that the set of operations that one may perform on a quadrilateral is defined only syntactically, without any sense of what the operations actually do. For example, apart from the difference in their names, the Translate and Scale operations are syntactically indistinguishable. In the main, this is a problem that most object-oriented systems share. However, these problems have existed since well before the dawn of object-orientation. The approach simply highlights the need for more precise characterisations of objects and their supporting mechanisms. A *formal semantics* of an object class would characterise the behaviour of the operations of an object in terms of an *abstract representation* of the state of the object. This approach is useful for capturing the semantics of passive objects (i.e. objects that are acted on by another agent). To illustrate this notion, consider the naive semantics of our quadrilateral in figure 1.1.

The semantics in figure 1.1 point us in the right direction regarding how all instances of **Quadrilateral** behave. However, the semantics is still somewhat incomplete (e.g., what order are the vertices in, what is the mathematical definition of $Rotate_z$, or is θ specified in degrees or radians?). Even so, the specification is much richer than one would normally see in a typical class definition, which points to an ongoing problem in the usability of object-oriented systems.

What goes on "inside" and object? Essentially, that is a secret and usually life is simpler that way. Presumably, an object's inside realises the advertised operations, but we do not know how it does so, nor whether or not that is *all* it is doing. For all we know, it might be computing π to arbitrary precision for its own enjoyment while otherwise inactive.

Inheritance

Inheritance provides a way of defining new classes from others by allowing a new class to take on and refine some of the characteristics of one or more existing object classes. This facilitates the reusability of entire object classes, allowing uniform sets of operations to pervade a hierarchy of classes. For example, suppose that instead of having defined the class **Quadrilateral**, we had first defined a class called **Polygon** as follows.

class Polygon (**list of** $\mathbf{R}^2$)
Operations
{ Translate: **Polygon** $\times$ $\mathbf{R}^2$ $\rightarrow$ [**Polygon**]
Rotate: **Polygon** $\times$ $\mathbf{R}$ $\rightarrow$ [**Polygon**]
Scale: **Polygon** $\times$ $\mathbf{R}^2$ $\rightarrow$ [**Polygon**]
Get Vertices: **Polygon** $\rightarrow$ **list of** $\mathbf{R}^2$
}

The formal semantics for the above operations would be as given earlier. Next, we could define a class of quadrilaterals as follows.

class Quadrilateral ($P_1, P_2, P_3, P_4 : \mathbf{R}^2$)
inherits Polygon(P_1, P_2, P_3, P_4)

In object-oriented systems supporting inheritance, the implementations of the operations for the class **Polygon** would automatically become those for the operations of the class **Quadrilateral**. However, it is often the case that these operations could be implemented

class Quadrilateral $(P_1, P_2, P_3, P_4 : \mathbf{R}^2)$

Operation Syntax:

{ Translate: **Quadrilateral** $\times$ $\mathbf{R}^2$ $\rightarrow$ [**Quadrilateral**]

Rotate: **Quadrilateral** $\times$ $\mathbf{R}$ $\rightarrow$ [**Quadrilateral**]

Scale: **Quadrilateral** $\times$ $\mathbf{R}^2$ $\rightarrow$ [**Quadrilateral**]

Get Vertices: **Quadrilateral** $\rightarrow$ **list of** $\mathbf{R}^2$

}

Operation Semantics:

{ **Abstract Instance Variables:**

$P_1(x_1, y_1,), P_2(x_2, y_2), P_3(x_3, y_3), P_4(x_4, y_4) : \mathbf{R}^2$,

L_1:**Line** $(P_1, P_2), L_2$:**Line**(P_2, P_3),

L_3:**Line** $(P_3, P_4), L_4$:**Line**(P_4, P_1).

where **Line** (P,Q) $=_{df} \{(1-t)P + tQ : t \in [0,1]\}, P, Q \in \mathbf{R}^2$

Translate$(t_x, t_y) =_{df}$

$P_1 \leftarrow (t_x + x_1, t_y + y_1)$,

$P_2 \leftarrow (t_x + x_2, t_y + y_2)$,

$P_3 \leftarrow (t_x + x_3, t_y + y_3)$,

$P_4 \leftarrow (t_x + x_4, t_y + y_4)$.

Scale$(s_x, s_y) =_{df}\ s_x\ s_y \neq 0 \Rightarrow$

$P_1 \leftarrow (s_x\ x_1, s_y\ y_1)$,

$P_2 \leftarrow (s_x\ x_2, s_y\ y_2)$,

$P_3 \leftarrow (s_x\ x_3, s_y\ y_3)$,

$P_4 \leftarrow (s_x\ x_4, s_y\ y_4)$.

Rotate $(\theta)\ =_{df}$

$P_1 \leftarrow$ *Rotate*$_z(P_1,\ \theta)$,

$P_2 \leftarrow$ *Rotate*$_z(P_2,\ \theta)$,

$P_3 \leftarrow$ *Rotate*$_z(P_3,\ \theta)$,

$P_4 \leftarrow$ *Rotate*$_z(P_4,\ \theta)$.

Get Vertices $=_{df} (P_1, P_2, P_3, P_4)$.

}

FIGURE 1.1. Naive semantics of a quadrilateral

more efficiently for the subclass. That is, we may wish the subclass to have the same operational syntax and semantics, but to have a different implementation. For example, if we had defined the predicate *Inside* for general polygons, then the implementation of *Inside* would also work for, say, the class **Rectangle**, but clearly a more efficient check for insideness can be written for rectangles than for general polygons. It is generally possible for a programmer to make such optimisations, sometimes called *tailoring*, in many object-oriented systems. It is also possible to add more operations to the definition of a subclass.

As stated earlier, inheritance provides a way of defining a hierarchy of classes in which subclasses inherit all the operations of their superclass, and perhaps contain additional operations. While this is a useful way of defining new classes from old, it can also lead to difficulty, because the higher-level objects tend to be "cast in stone". For example, suppose we have defined a class hierarchy in which the reals, **R**, are a subclass of the natural numbers, **N**, in which class **R** inherits the (syntax) of the operations $+, -, \times$ from **N**, and to which the operation $\div$ is added. If it is observed at a later date that a *successor* operation is desired for **N**, we are in trouble, because *successor* is not well defined for the subclass **R**. Of course, at play here is the fact that **N** and **R** are fundamentally different mathematical structures, and to view either one as a "subclass" of the other is simplistic (although often practical).

Occasionally it is desirable to synthesise the operations of two or more object classes into a new class. Such a synthesis is called *multiple inheritance.* Suppose, for example, their exists a class called **AnalyticTexture**, which provides an operation giving the colour for arbitrary point $(x, y) \in \mathbf{R}^2$. It might be defined as follows:

class AnalyticTexture
Operations
$\{ \quad \tau : \mathbf{R}^2 \rightarrow C$
$\}$

where C is some colour space (which is also and object class). Then we could define a "textured quadrilateral" as follows:

class TexturedQuadrilateral (P_1, P_2, P_3, P_4: **list of** $\mathbf{R}^2$)
inherits Polygon(P_1, P_2, P_3, P_4) **and AnalyticTexture**

The problem with this definition is that the texture would be defined outside the quadrilateral as well. It is straightforward in principle to parameterise τ to subsets of $\mathbf{R}^n$. Multiple inheritance mechanisms can be quite useful when one wishes to bring together object classes with disjoint sets of operations. Dealing with non-disjoint sets of operations is also possible, but more tricky (see the references for further information).

Strong Typing

Object-oriented languages equipped with *strong typing* provide policies for detecting and resolving conflicts that occur in expressions involving several objects (or types) of different classes. Detecting when objects can be mixed in expressions can be difficult, mainly because the semantics of object classes is usually not well-enough understood by the compiler. A compiler should allow the mixture of objects that have the same *behaviour* in an expression, but this notion is difficult to convey. As it stands, programmers are often required to write "casting" operations that convert one object class to another for the purposes of expression evaluation or assignment. As well, the "type" of an object is often simply a syntatic thing, and type checking often just reduces to ensuring that objects are of identical class. As a result, it can sometimes be problematic to do very simple things, such as adding a value of type $\mathbf{N}$ to one of type $\mathbf{R}$. See [23] for a careful analysis of the notions of *overloading* and *polymorphism.*

Persistence

Some object-oriented systems are now beginning to incorporate *persistent objects.* This means that an object remains active and accessible even after the application that has created it is no longer active. Almost every computer user already knows that things like files and documents have this property. Persistent objects essentially generalise this notion. Many users of graphics systems are familiar with this notion, in that graphics systems often contain facilities to maintain graphical workspaces, for example, that persist over multiple invocations of a modelling and rendering system.

Concurrency

For some graphical applications such as user interface construction, the use of *concurrency* is vital. A systems may for example support multiple input devices that can be manipulated concurrently, or one may wish to capitalise on a distributed architecture in order to implement a computationally complex graphics algorithm. In such cases, it is beneficial to allow a programmer direct access to some kind of concurrency mechanism within the language. Several recent object-oriented programming languages support concurrency mechanisms [3, 87, 2, 59].

1.3 Object-Orientation in Computer Graphics

Object-orientation is most successful in an application in which the interface between components is (to use undefinable and unmeasurable terms) simple, uniform, and consistent. In these cases, new objects or operations can be more easily "snapped" in or out, and new connections can be more easily made. The varying success of object-orientation in areas of computer graphics bears this out. Some areas of computer graphics have exploited object-orientation for some time, whereas other areas are more resistant to it. Computer animation and, more recently, user interfaces, rather naturally embrace it, whereas graphics systems based on the traditional "pipeline" view are only slowly accommodating object-orientation. At the outset, it is important to mention that while many graphics systems have object-oriented features, few existing graphics systems, if any actually are completely object-oriented.

1.3.1 Modelling and Rendering

The Structure of Graphics Systems

It is assumed that the reader is familiar with traditional graphics systems and the so-called graphics pipeline [33]. To support an open-ended object-oriented graphics systems, we shall argue that a *server-based* approach is particularly convenient.

A graphics system can be embedded into a computing environment in various ways, and there are several possible views that one may have of it. Three such views are:

- *The user view.* A user interactively creates and manipulates graphical objects and displays them on a variety of output devices
- *The application program view.* An application requires some graphical support and makes use of a library of objects and operations
- *The graphics package view.* Software that supports user and application program requests.

Object-orientation is not simply a new kind of programming. It is more correctly viewed as a general structuring mechanism. Consequently, non-programmers and programmers alike can use object-oriented techniques in their work[1]. Our examples will generally try to capture this fact. Figure 1.2 is a highly-idealised depiction of one way in which an object-oriented graphics system may be incorporated into a computing environment (interaction components omitted) and in which the above views, among others, are possible.

The goal of object-orientation is to increase uniformity of graphic objects and operations, to maximise their re-usability, and to minimise their redundancy. One would hope, for example, that a graphical object *Sphere* is available both to programmers and to interactive users of a graphics system, and moreover that the *same* object is used by both communities. To do so requires some care, since difficulties may arise as to the separation of graphic objects from the operations that may be performed on them. Every graphic

[1] One can of course argue that any interaction with the computer is a form of programming

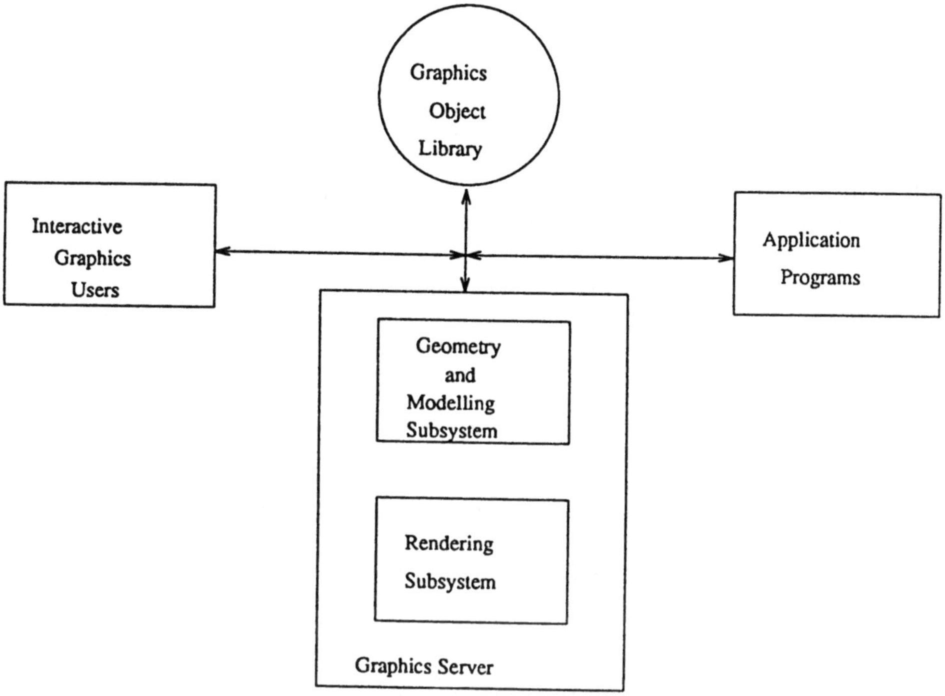

FIGURE 1.2. Schema of a server-based graphics system

object should, naturally enough, contain a "display" operation. Where should the support for this operation be placed? It would be inefficient to embed a rendering subsystem within every object that could conceivably be displayed. Window management systems face an analogous dilemma, and the possible solutions are instructive. In the windowing environments supported by Sun Microsystems, for example, if an application uses windows within the Sunview system, the necessary window display support is internally provided by linking it into the application program, which results in large (but efficient) executable modules; on the other hand, in Sun's NeWS and MITs X window system [71], display support is provided externally by a distributed *server*[2]. The design of an object-oriented graphics system must be similarly sensitive to such design issues. A server-based design as illustrated in Figure 1.2 is more easily extensible, and is in many ways analogous to the old notion of a "graphics pipeline". On the other hand, it is probably less efficient and less simple to customise. Under this model, a graphic object need not do its own displaying (or visibility determination, or shading, etc); rather, it must have a way of describing itself to a display process.

Hierarchies and Parts

From an early stage, many graphics modelling systems have incorporated some notion of structured graphic objects and class-instancing mechanisms [25, 33]. Indeed, one can argue that such mechanisms existed in computer graphics coincidentally with the advent of Simula [79]. There are several reasons for supporting the use of these facilities.

[2]SUN and NeWS are registered trademarks of Sun Microsystems

First, assuming the object to be represented is hierarchical, a complex object may be created as the composition of many independent *parts* (or instances of simpler objects). Parts are "glued" together in a hierarchical fashion that reflects the designer's overall view of the structured object. While it is possible to design pictorially equivalent non-hierarchical objects using linear data structures (such as *segments*) found in GKS or ACM SIGGRAPH's CORE, hierarchical mechanisms provide convenient structuring mechanisms that are comparable to those in programming languages.

Second, a hierarchy (or indeed any aggregation of parts, including segments) can be transformed as if it were a single object. This assumes the transformation distributes over the composition operation. For example, suppose object O is composed of a union of parts $\{O_1, O_2, \ldots, O_n\}$, with each $O_i \in \mathbf{R}^3$ and that $\tau : \mathbf{R}^3 \rightarrow \mathbf{R}^3$ is an arbitrary transformation.[3] Then a fundamental assumption to all graphical systems (including non-hierarchical ones) for τ to be admissible as a transformation on graphical objects is that

$$\tau O = \{\tau O_1, \tau O_2, \ldots, \tau O_n\}.$$

Generally this is true if τ is a geometric (or linear) transformation, and if the graphical primitives that are manipulated are not space-dependent. See below for a further discussion on primitives.

Third, a carefully-defined hierarchy can be used to reflect spatial relationships of the constituent parts, such as inclusion or bounding volume, and can sometimes be used to speed up complex operations on graphic objects, such as visibility. See Clark's classic paper on this topic [25].

Fourth, nodes in an object hirearchy need not be passive data structures. It is certainly possible in principle to embed so-called "procedural" or active graphical objects into the hierarchy. The ability to embed active objects as a part of an overall graphical object definition is critical to modern systems in which one can construct models of natural phenomena, stochastic objects, and adaptive objects such as those that dynamically determine an appropriate level of representative detail. Later we shall discuss other applications of active objects.

Figure 1.3 gives a schema of a structured object using the object-oriented approach. In this figure an object O consists of three sub-objects (or parts) *O.1, O.2,* and *O.3*, which in turn consist of other parts. Geometric operations performed on O percolate down into the hierarchy. In principle, it would be possible for some of the parts to be active objects. As a refining example, consider Figure 1.4, which contains two scripts: an abbreviated script from a graphics package, as well as code that a programmer might write in an object-oriented programming language to do the same thing. Scenes are constructed from instances of basic modelling primitives such as lines, polygons, surfaces, and so on.

The scripts in Figure 1.4 each define a "master" object O_1, which is analogous to our notion of a class. Each script specifies several modelling primitives, which are themselves instances of classes. Once a class has been defined, it can either be used to construct other classes, or it can be instantiated to create a scene as illustrated in the script in Figure 1.5. Note that the programmer version of the script makes the assumption that part P_3 of instance I_1 is visible beyond the scope of I_1. In some object-oriented systems, this is not possible. Observe that objects are instances of classes, and that operations may be performed on them. While our example considers only modelling and geometry, it is easy to imagine that similar approaches can be taken with the encapsulation of a

[3] The article "a" in "a union" was carefully chosen, because there are many ways of defining a union operation on objects; this is due to the fact that the semantics of an object consists of something more than simply its set-theoretic volume [29].

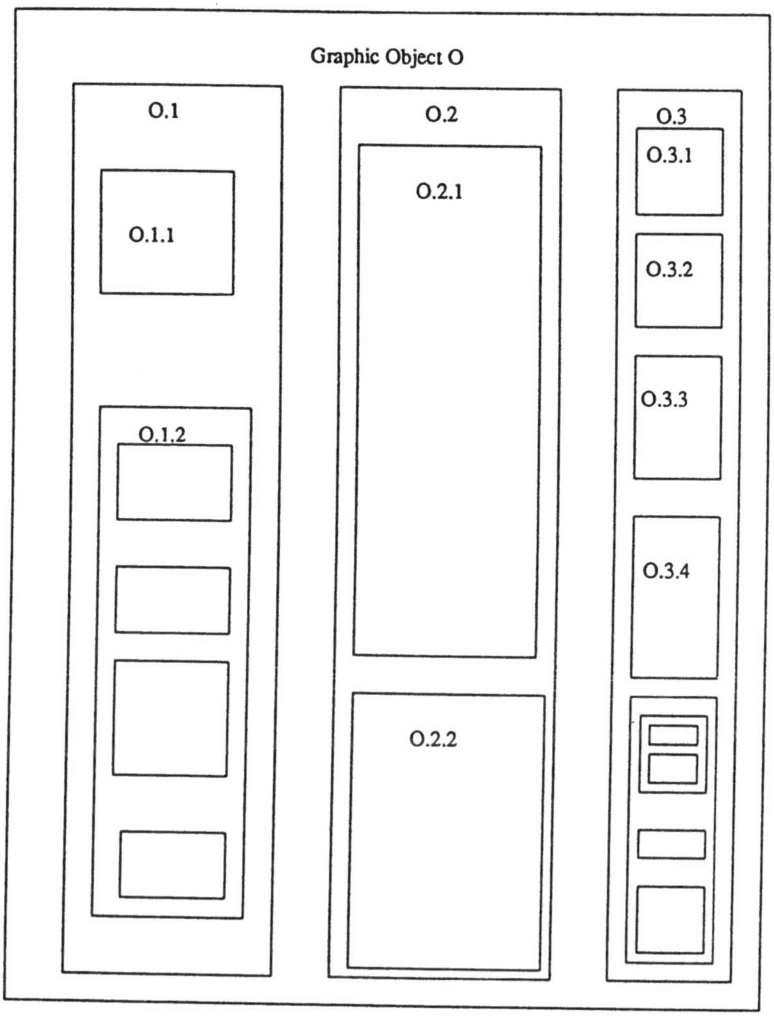

FIGURE 1.3. A structured graphical object

rendering subsystem. For example, if the graphic primitives have a uniform interface, different renderers can be used as desired. An object-oriented characterisation of cameras (i.e. viewing parameters) is also an obvious extension [82]. There has also been some work in generic object-oriented rendering systems [17], as well as modelling objects at varying levels of spatial detail [12].

Graphic Primitives are Elusive Objects

In many ways, graphical objects are "worst-case" objects insofar as fitting them into traditional hierarchical class definitions. Moreover, the strict autonomy of objects can be overly restrictive in graphics systems. We shall motivate these thoughts in this section. There are some obvious differences between the notion of objects presented in paragraph 1.2 and the graphical objects illustrated in Figures 1.4 and 1.5, which we now describe.

First, the operations that can be performed on graphical objects are typically *implicit.* That is, the system allows a set of operations such as the linear transformations to be

Interactive Script

```
define object class O1
  polygon P1
    vertex V1 normal N1
    .
    .
    .
    vertex Vn normal Nn
  end P1
    .
    .
    .
  polygon Pm
    .
    .
    .
  end Pm

end O1
```

Application Program

```
O1: Graphic Object
{   P1: Polygon ((V1,N1),..., (Vn,Nn))
      .
      .
      .
    Pm: Polygon...)
}
```

FIGURE 1.4. A structured object definition (user view above, programmer view below)

Interactive Script

```
add instance I1 O1            {Make I1, I2 an instance of O1}
add instance I2 O1
rotate I1 90 degrees about z-axis
translate I2 2.0 4.3 6
scale I1.P3 0.5 0.5 0.5       {Only affects P3 within I1}
```

Application Program

```
I1, I2 : O1
I1.Rotate(z,90)
I2.Translate(2.0,4.3,6)
I1.P3.Scale(0.5,0.5,0.5)
```

FIGURE 1.5. Scripts manipulating a structured graphic object (user view above, programmer view below)

performed uniformly on modelling primitives. A transformation on an object is assumed to distribute over its parts. The actual semantics of these operations is not normally defined. This can be a difficulty in some cases. For example, if one has defined a surface such as a Bézier surface in terms of control points, it is important to know whether the perspective transformation will be performed on the control points or on the actual surface.[4]

Second, the parts of an object are often visible and accessible to the user. Observe, for example, the part P_3 of object I_1 in the script above is named and transformed. While it is possible to hide object definitions or to keep their names a secret, it is quite common and acceptable not to hide the inside of an object. Some object-oriented languages allow internal parts of an object to be explicitly "exported". This helps the situation somewhat, but it tends to cause rather long-winded object definitions.

Third, as was discussed earlier, there are a great many aspects of a graphics system that are not easily captured solely in terms of individual operations on objects. Because objects interact in space, visibility, shading, clipping, and rendering can require a context that is much richer than that of any individual object. Worse still, a class of objects may not even be closed under an operation. Consider as a trivial example the fact that the output of an object-space visibility algorithm can contain concave polygons (and indeed polygons with holes) even if the input is only a set of convex polygons.

Fourth, and related to the third point, a graphic object may have many related representations at different stages in the graphics pipeline. For example, at the modelling stage (e.g., to a user manipulating a graphical model), a surface may be viewed as a set of bicubic parametric patches: at an intermediate geometric stage (e.g., to a programmer writing a shader), the object may be represented as a polygonal mesh; at the rendering stage (e.g., to a programmer writing a renderer), the object may be transformed into a wire frame drawing or a shaded image. Once again, it is not object orientation that causes the problem, but rather that the problem of multiple representation is made more apparent by object-orientation. One possible solution is to define the different possible views of a surface, say, as different classes, each with a separate set of operations. The link between representations would then be established by special transformations, namely *metamorphoses*, which would map one representation to the other. To continue our example with Bézier patches, as a modelling primitive, a Bézier surface should probably support a *change control point* operation. This operation would not make sense for the polygonal mesh representation. On the other hand, the geometric transformations would make sense to both representations. In fact, it would be useful to allow a user to orient the polygonal-mesh view of a surface as desired and then to have the control-point view inherit the user-specified geometry. A metamorphosis from a control-point representation to a polygonal mesh would be accomplished in one of the standard ways, using a technique such as forward differencing or direct evaluation[33].

All of the above points can be combined into one main conclusion: the definition of graphical object in terms of traditional object or data type hierarchies is not particularly easy. Consider, for example, a simple portion of a plausible hierarchy of graphical objects as illustrated in Figure 1.6. In this figure, a graphic primitive is one of a modelling, geometric, or display primitive, and examples are given of each. The straight lines indicate inheritance relationships, which can be adequately captured using standard object-oriented systems. The difficulties are in the metamorphic relationships expressed by the arcs, which can be virtually impossible to express. For example, the one metamorphic re-

[4]More formally, if P is a perspective transformation, C is a set of control points, and $S(C)$ is the surface induced by the control points, then it is not always true the $PS(C) = S(PC)$. It is thus important to know when P can be applied.

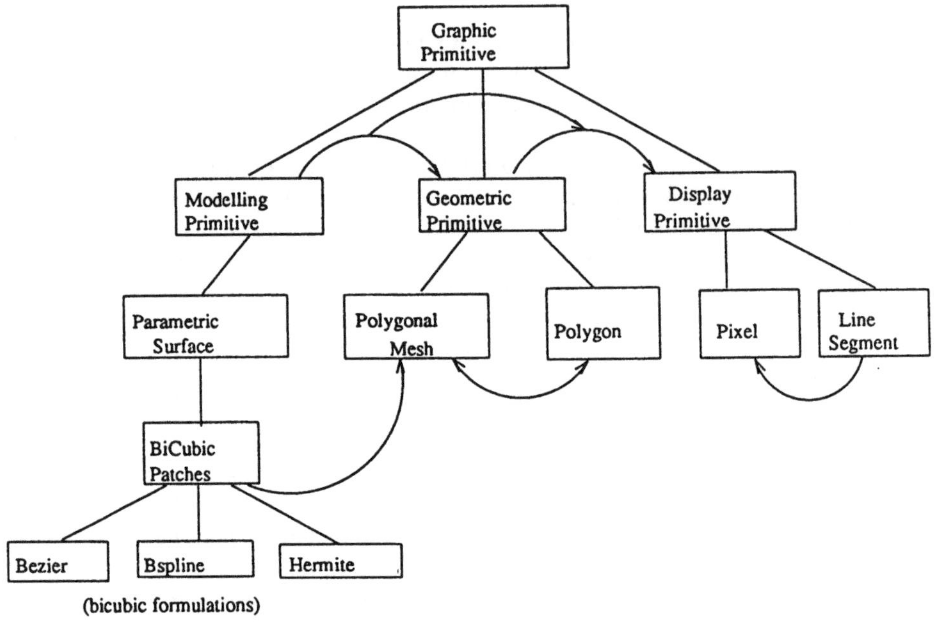

FIGURE 1.6. A partial graphical object "hierarchy"

lationship would be the arc between a geometric primitive and a display primitive, which is otherwise known as *rendering*. There are much more subtle metamorphoses that are not expressed in the diagram. For example, *clipping* is a transformation from a geometric or modelling primitive back to one of these classes. However, a clipping operation must properly preserve properties of object primitives such as normals and colours at the clipping boundaries. Must a clipper be aware of the inside of objects in order to do this? How can the several types of visibility be incorporated? How about the many illumination models, rendering and filtering techniques, and texture mapping? We indicated in paragraph 1.2 that the precise semantics of internal object operations is not easy to specify. It is clear from this discussion that a semantics of extra-object operations is at least as difficult. The first step in resolving these problems is in the clarification of precisely what a "graphical primitive" is, and the operations one can perform on and across them [29, 30].

As it currently stands, metamorphoses and general operations on aggregations of objects are defined in a system-dependent, *ad hoc* manner. For individual systems, of course, this may be satisfactory. Difficulties arise when systems must be portable to other environments, or when a need for the standardisation of graphics systems becomes important.

1.3.2 Computer Animation

Of all the areas of computer graphics, computer animation has been the one to embrace object-orientation most enthusiastically. This is hardly surprising, for the correspondence between the "actors" in an animation and objects is an obvious one, and one that has allowed animations to be conveniently structured. As stated earlier, the applications that most successfully utilise object orientation are those in which interaction between components is uniform, simple, and consistent. In animation, the object interface is typically very uniform, and is usually of two varieties:

1. Each object is a static model, much like the graphic objects described earlier. The operations supported would be incremental geometric transformations as well as operations to change attributes of the object. For a given "script" of actions that are desired of an object, an animation scheduler would compute the incremental changes in all objects and invoke the appropriate object operations

2. Objects are active and have more information regarding the actions they should be performing. They perform their actions when they are told to by a scheduler or clock.

Of course, hybrids of the two approaches exist, as we shall see. In either case, the simplicity of the interface has allowed for the creation of special-purpose languages that help animators to construct *scripts* to co-ordinate the temporal behaviour of objects. A similar trend is occurring in user interaction and in the animation of programs and algorithms. Moreover, the uniformity of the interface often allows one to reuse interesting objects and operations. For example, a trajectory may be defined that operates on arbitrary objects of variety 1 above. In fact, the same trajectory may be interpreted non-spatially to alter dynamically other attributes of an object, such as its surface normal, its colour, and so on.

Temporal Scripting

As an example of the degree to which uniformity of interaction can be exploited in an object-oriented system, we shall consider a simple application in which the author is involved [31]. The goal of this project was not to produce Disney-quality animation, but rather to provide mechanisms that make it simple to introduce animation to a wide variety of applications such as electronic documents, mail systems, program illustration, and user interfaces. Suppose we have a set of graphical objects that "know" their spatiotemporal behaviour. More specifically, each object O has a duration $O.d$, and for any time $t \in [0, O.d)$, the object can be asked to produce its "behaviour" at that time (or since the last time it was asked). For a graphic object, a behaviour is simply a list of graphics commands that are pipelined into a graphics subsystem, the details of which are irrelevant here. Another possibility, however, is that an object could be of an entirely different medium such as a sound or video recording. For simplicity, we shall stick to graphic objects. From a set of these primitive animated graphic objects, interesting composite animations can be synthesised by means of a simple but surprisingly powerful temporal scripting language. For example, suppose E_1 and E_2 are scripts (e.g., animated objects). Then

$$E1\&E2$$

states that the animation specified by E_1 is to *execute simultaneously* with that of E_2, and the duration of the composite animation is the maximum of the durations of E_1 and E_2. Similarly,

$$E_1; E_2$$

states that animation E_1 is followed by animation E_2, and that the overall duration is that of E_1 plus that of E_2. The analogue of simultaneous execution is *simultaneous termination*, which is expressed by

$$E_1 \wr E_2$$

The expression

$$\textbf{delay}\ t$$

simply introduces a delay of t time units. Time units themselves can be scaled arbitrarily. A general synchronisation operator is of particular interest, in that it can be used to express the above operations as special cases. The expression

$$E_1[t_1] \asymp E_2[t_2]$$

states that the animation of E_1 at (local) time t_1 must coincide with that of E_2 at time t_2. The times t_1 and t_2 can be arbitrarily arithmetic operations based on object durations. For example,

$$E_1[\$/2] \asymp E_2[\$/2]$$

states that the half-way points of animations E_1 and E_2 must coincide (the "$" in the above expressions stands for the duration of the largest sub-expression to the left of the square brackets). Several other operators exist, but this suffices for our example, since some nontrivial combinations can be concisely expressed. For example,

$$(\mathrm{A}\ \&\ (\mathbf{delay}\ 4\ \mathrm{B};\ \mathrm{C}))[\$/2] \asymp (D \wr E)[\$/4].$$

We do not claim that this is a particularly user-friendly specification language, but it is quite precise, it can be used to express interesting animations, and it is particularly handy for putting together animations from and existing animation library. The ease of reuse of animated objects is strongly facilitated by the temporal scripting approach.

Another fact that we discovered (more by accident than by design), is that the scripting language could be used to define the motion of individual objects. That is, motion also has a duration and a "behaviour" at each time instant. Therefore, the scripting language can be used to orchestrate motion, as the following example illustrates.

tea: **Teapot** {Motion1;Motion2 & Motion3}

In this case, tea is an instance of static object **Teapot**, which is animated by a series of motions. These motions are themselves objects which are instances of motion classes, and which include motions such as continuous geometric transformations and trajectories. It was satisfying to see that the language could be used at many levels in the animation system.

The first implementation of the above language was modest: we simply took advantage of the basic UNIX[5] tools such as the C-shell, and the compiler aids, lex and yacc. Herein lies a valuable lesson: to a user of our system, the language is entirely object-oriented. Beneath the object-oriented veneer, the language was a collection of basic UNIX tools. Subsequent versions of the implementation became more sophisticated, incorporating the use of true (and pseudo) concurrency and C++[78]. While the outward appearance of the language changed to meet different needs, it is important to remark that the language itself need not have changed at all, despite the completely different underlying representation of animated objects. In this respect, the encapsulation implicit in object-orientation is very helpful.

The idea of scripting independent animated objects is now a fairly well established practice [68, 49, 51, 31]. Recent work has been concentrating on the problem of specifying interactions among otherwise independent objects [69, 52, 66]. Among the popular new approaches are those based on physics, which model object interactions using physical analogies such as fields and forces, and constraint-based approaches, in which constraints between objects are defined and maintained by the animation system.

---

[5]UNIX is a trademark of AT&T Bell Laboratories

1.3.3 User Interface Construction

The first object-oriented approaches to user interface construction were developed at Xerox PARC. The Xerox Star[6] was an early commercial product that was developed using an object-oriented methodology [46], and it demonstrated the utility of facilities such as multiple inheritance (based on traits), and classes (and subclasses). The approach is particularly successful at the presentation level, namely at the level seen by the user, for there always appears to be a direct correspondence between icons and images the user sees, and an underlying object representation. This results in a clean and uncluttered system design. There is no need to reproduce here the very good and instructive discussion found in [46].

Recent work has considered more specialised user interface construction techniques for handling concurrent input and adaptive user interfaces [39]. It is conceptually straightforward to imagine devices and dialogues as classes that can be initiated just as if they were data types. Indeed the object-oriented approach can be used to extend process-based approaches to user interface construction, especially concurrent implementations [8, 22]. We shall consider concurrency in a broader context in the next section. In general, however, real practical progress has been fairly slow in coming. It is not easy to pinpoint a single reason for this problem, but one problem certainly is that a formulation of a precise semantics of interaction is currently nonexistent. The semantics of traditional passive data types is well advanced. However, the semantics of concurrency and input-output is not, and without it a precise characterisation of what interactive systems do is simply not possible. In practice, however, there are two classes of user interfaces that appear to be particularly well-suited to object-oriented approaches: user interface management systems (UIMS) and direct manipulation user interfaces (DMUI).

In a DMUI, various elements of an application have a visual screen representation with which a user may directly interact using an input device. In this sense, the lesson of the Star user interface provides ample justification for an object-oriented approach to DMUI construction. Research is now gaining maturity in this area (see [7]). In a UIMS, the goal among others is to allow entire dialogues to be snapped into or out of a user interface. Once again, the argument for object-orientation is compelling. Both of these areas are attracting active research interest. We continue our discussion of interaction in the next section, where it shall become clear that object-orientation is useful to both of these classes of user interfaces.

1.3.4 Active Objects

Concurrency is essential to real-time, interactive graphics applications. A wide variety of notations and languages expressing concurrency have been proposed [4]. Although object-oriented systems have been slow in accommodating it, concurrency is now becoming increasingly popular in experimental languages [3, 87, 2, 59]. In this section the notion of an active object, that is, an object that is a self-contained computational unit, will be motivated. It will be seen that it provides great flexibility in putting together nontrivial interactive graphic applications.

To make issues concrete, we shall outline a language for active objects containing a minimum of concurrency features. First and foremost, we make the assumption that *every* object is active. An object such as a data structure can operate in a passive mode simply by executing only when it is asked to by an external agent. On the other hand, an object will be able to take a more active role than this. As before, an active object

[6]XEROX and Star are trademarks of XEROX Corporation.

will support a set of operations. Some operations will be inaccessible to external objects. We call such operations *hidden operations*. At any time, at most one operation may be active within an object. It is entirely permissible for an object to invoke one of its own operations. However, it would compete with all other objects requesting these operations. Since many independent activities are possible, there may be several objects that simultaneously request an operation of the same object. We shall assume that the system decides on some (presumably fair) ordering of these requests.

When an object is instantiated, its *Initially* operation, which is one example of a hidden operation, is invoked. An object instantiation creates another "thread of control" in the system. Within the *Initially* code, an object can set itself up and commence execution. If it chooses to, it can return to a passive state by ceasing to invoke operations within itself or on other objects. An object A can invoke object operation op_B of object B in one of two ways.

1. *Synchronous operation invocation.* The execution of A is suspended until B indicates completion of the requested operation by issuing a **reply** command, or by the termination of the operation. A specifies a synchronous operation request by: $B.op_B$

2. *Asynchronous operation invocation.* Unlike the synchronous case, the execution of A is not suspended. Instead, A asks B to perform an operation and passes to the system the name of one of its own operations that is invoked to return the value of op_B's **reply**. A specifies an asynchronous operation request by: $B.op_B@op_A$, where op_A is the optional operation of A that is to be invoked when B issues a **reply**. The domain of op_A must be consistent with that of the value generated by the original $B.op_B$. If op_A is not specified, no value is returned. In fact, a **reply** itself may be viewed as an asynchronous operation invocation (of op_A) in which no return value is expected of A by B. That is, if A has issued the command $B.op_B@op_A$, then **reply** $= A.op_A@$.

Asynchronous operation invocation allows multiple operation requests to be active simultaneously, and it is also useful for performing event-driven behaviour.

We now illustrate the use of this simple operation-oriented concurrency mechanism by means of an example. We shall construct and interactive application consisting of a button, a dial, a graphic object, and a dialogue manager. The idea is that a user will be permitted to use the dial to rotate the graphic object until a button is pushed. We must therefore cope with multiple concurrent input devices. We shall focus our attention on the object called **Dialogue**, but first we shall summarise the operations that other objects will (externally) support.

We begin with the button.

```
class Button
{  Operations:
       ButtonDown: Null →  Boolean
}
```

The operation *ButtonDown* returns the next time the button is depressed. The format of dial is similar.

```
class Dial
{  Operations:
       NextValue: Null →  Z
}
```

NextValue simply returns the displacement since the operation was last invoked.

```
class GraphicObject
{  Operations:
       RotateZ: Z →  Boolean
}
```

In our example, the only operation of interest for a graphic object is a rotation about the z-axis.

The dialogue will support three operations, but only one of them will be visible (i.e. invokable).

```
class Dialogue
{  Operations:
       *Initially: Null →  Null
       *Update:     Null →  Null
        Finish:    Null →  Null
}
```

Both *Initially* and *Update* are hidden operations. The *Finish* operation will be passed to the button object as the operation to be invoked when the button is depressed. The code for the dialogue follows.

```
Dialogue
{ b: Button         { Instance Variables}
  d: Dial
  o: GraphicObject
  done: Boolean

  *Initially
   { b.ButtonDown @ Finish
     done = FALSE
      self.Update @
   }

  *Update
  {  if not done
      o.RotateZ(d.NextValue)
       self.Update @
  }

  Finish
  { done = TRUE
  }
}
```

A few words of explanation are in order. The asterisks in front of the operation denote that the operation is hidden. We begin with the *Initially* operation within the dialogue. It requests that the *Finish* operation be invoked when the button is depressed. It then initialises a variable and asynchronously invokes its own *Update* operation.

Note that there is no possibility of a race condition with the update of the *done* variable because the *Finish* operation cannot commence execution (even if the button has been depressed) until the *Initially* operation is complete.

When the *Initially* completes, one of two operations may be executed, depending on whether or not the button has been pressed. If it has not, then the *Update* operation outstanding from the *Initially* begins execution. In this case, a value is synchronously requested from the dial, the graphic object is updated, and a new update request is asynchronously generated, just as the current update operation terminates. If at any time the button is depressed, the *done* flag is set, and no further updates are possible.

Even for a small example such as the one above, it is clear that a large number of plausible configurations of a concurrent system are possible. It is not always easy to program in a concurrent environment, but it is certainly true that concurrent programming structures are necessary to interact with real-world parallelism. Note that the above dialogue is a class that can be instantiated just like any other class. To make the example more realistic, some parameterisation of the button, dial and graphic object is required to bind them to real devices and structures at run time. Observe that it is straightforward to "snap" a new dialogue in to change the interaction style or the devices used. Furthermore, note that the style of programming very much supports the use of direct manipulation interfaces.

1.3.5 Other Applications

Several additional application areas are gaining in popularity: constraint-based systems, algorithm animation, and graphical programming systems.

The notion of graphical constraints was developed long ago by Sutherland in his landmark Sketchpad system [79]. Loosely speaking, a constraint is a predicate relating several aspects of a system whose truth is maintained by a *constraint-satisfaction* mechanism. Two trivial examples of graphical constraints are: defining a rectangle as a suitably-constrained quadrilateral, and constraining the lines drawn on a display to be either horizontal or vertical. Sutherland demonstrated the value of constraints in a graphical setting. In ThingLab, Borning has since demonstrated the power of constrained objects in an object-oriented programming environment [14]. Furthermore, Borning and Duisberg have suggested several ideas for constructing object-oriented user interfaces using constraints [15]. Graphical constraints are also at the heart of Nelson's Juno system [57].

The advent of new workstations has seen the development of radically different programming environments. Some environments such as ThingLab suggest novel graphical ways in which to program computers. The research area of graphical programming is becoming increasingly active. See, for example, the work of Reiss [67]. Similarly, the graphical depiction of programs, objects, and algorithms is also becoming popular [18].

1.4 Conclusions

The use of object-oriented techniques in computer graphics is at once both old and new. We have seen that many notions quite related to object-orientation, such as classes, instances and hierarchies, have existed in computer graphics virtually since its inception in the early 1960's. On the other hand, the emergence of modern object-oriented systems is beginning to have a renewed effect on the design and implementation of graphics systems. This paper has introduced the basic notions of object-orientation from the perspective of computer graphics, and has outlined some of the areas of computer graphics that are good targets for object-orientation, including modelling and rendering systems, computer animation, and user interface construction.

1.5 A Note on the References

After each reference in this bibliography, a number of abbreviated keywords may appear which categorise the reference:

OOS - presents an object-oriented system

OOI - an introduction to object-orientation

OOA - an object-oriented application

OOCG - relates to object-oriented computer graphics

OOCA - relates to object-oriented computer animation

OOUI - relates to object-oriented user interface

PL - supplementary or supporting programming language issues

Specification - specification and verification of objects, data types and programs.

Acknowledgements:

The financial assistance of an operating grant and University Research Fellowship from the Natural Sciences and Engineering Research Council of Canada is gratefully acknowledged. The assistance of Oscar Nierstrasz of the University of Geneva in the preparation of this bibliography is gratefully appreciated.

1.6 References

[1] G A Agha. *ACTORS: A model of Concurrent Computation in Distributed Systems.* MIT Press, Cambridge, Massachusetts, 1986. PL.

[2] P America. POOL-T: A Parallel Object-Oriented Language. In A Yonezawa and M Tokoro, editors, *Object-Oriented Concurrent Programming*, pages 199–220. MIT Press, Cambridge, Massachusetts, 1986. OOS.

[3] D B Anderson. Experience with Flamingo: A Distributed, Object-Oriented User Interface System. *ACM SIGPLAN Notices*, 21(11):177–185, November 1986. OOS.

[4] G R Andrews and F B Schneider. Concepts and Notations for Concurrent Programming. *ACM Computing Surveys*, 15(1):3–43, March 1983. PL.

[5] J G P Barnes. An Overview of Ada. *Software, Practice and Experience*, 10:851–887, 1980. PL.

[6] A J Baroody and D J de Witt. An Object-Oriented Approach to Database System Implementation. *ACM Transactions on Database Systems*, 6(4), December 1981. OOA.

[7] P S Barth. An Object-Oriented Approach to Graphical Interfaces. *ACM Transaction on Graphics*, 5(2):142–172, April 1986. OOUI.

[8] R J Beach, J C Beatty, K S Booth, D A Plebon, and E L Fiume. The Message is the Medium: Multiprocess Structuring of an Interactive Paint Program. *Computer Graphics (Proc. Siggraph 82)*, 16(3):277–287, 1982. OOUI, PL.

[9] S Bergman and A Kaufman. BGRAF2: A real-time graphics language with modular objects and implicit dynamics. *Computer Graphics (Proc. Siggraph 76)*, 10(3):133–138, 1976. OOCA.

[10] K S Bhaskar, J K Peckol, and J L Beug. Virtual Instruments: Object-Oriented Program Synthesis. *ACM SIGPLAN Notices*, 21(11):303–314, November 1986. OOA.

[11] G Birtwistle, O Dahl, B Myhrtag, and K Nygaard. *Simula Begin.* Auerbach Press, Philadelphia, 1973. PL.

[12] E H Blake. A Metric for Comuting Adaptive Detail in Animated Scenes Using Object-Oriented Programming. In *Proceedings of the Eurographics-87.* North-Holland, August 1987. OOCG, OOCA.

[13] D G Bobrow, K Khan, G Kixzales, L Masinter, M Stefik, and F Zdybel. Common-Loops: Merging Lisp and Object-Oriented Programming. *ACM SIGPLAN Notices*, 21(11):17–29, November 1986. OOS.

[14] A Borning. The Programming Language Aspects of ThingLab, a Constraint-Oriented Simulation Laboratory. *ACM Transactions on Programming Languages and Systems*, 3(4):353–387, October 1981. OOA, OOUI.

[15] A Borning and R Duisberg. Constraint-Based Tools for Building User Interfaces. *ACM Transactions on Graphics*, 5(4):345–374, October 1986. OOA, OOUI.

[16] A Borning and D H H Ingalls. Multiple Inheritance in Smalltalk80. In *Proceedings of the National Conferece on AI, Pittsburgh*, 1982. PL, OOS.

[17] D E Breen, P H Getto, A A Apodaca, D G Schmidt, and B D Sarachan. The Clockworks: An object-oriented Computer Animation System. In *Proceedings of Eurographics-87 Conference*. North-Holland, August 1987. OOCA, OOCG.

[18] M H Brown. *Algorithm Animation.* MIT Press, Cambridge, Massachusetts, 1988.

[19] K B Bruce and P Wegner. An Algebraic Model of Subtypes in Object-Oriented Languages. *ACM SIGPLAN Notices*, 21:163–172, October 1986. Specification.

[20] Special issue on Smalltalk. *Byte*, 6(8), August 1981. OOI,OOS.

[21] Special issue on Object-Oriented Systems. *Byte*, 11(8), August 1986. OOI, OOS.

[22] L Cardelli and R Pike. Squeak: a Language for Communicating with Mice. *Computer Graphics (Proc. Siggraph 85)*, 19(3):199–204, July 1985. OOUI, PL.

[23] L Cardelli and P Wegner. On Understanding Types, Data Abstraction and Polymorphism. *ACM Computing Surveys*, 17(4):471–522, December 1985. Specification.

[24] T A Cargill. Pi: A Case Study in Object-Oriented Programming. *ACM SIGPLAN Notices*, 21(11):30–360, November 1986. OOA, OOS, PL.

[25] J H Clark. Hierarchical Geometric Models for Visible Surface Algorithms. *Communications of the ACM*, 19(10):547–554, October 1976.

[26] B J Cox. *Object-Oriented Programming - An Evolutionary Approach.* Addison-Wesley, 1986. OOI, OOS.

[27] G Curry and R Ayers. Experiences with TRAITS in the XEROX STAR Workstation. *IEEE Transactions on Software Engineering*, 10(5), September 1984. OOS, PL, OOA.

[28] G Curry, L Baer, D Lipkie, and B Lee. TRAITS: an Approach for Multiple Inheritance Subclassing. *SIGOA Newsletter, (Proceedings ACM SIGOA)*, 3(12), June 1982. OOS, PL.

[29] E Fiume. *A Mathematical Semantics and Theory of Raster Graphics.* PhD thesis, Department of Computer Science, University of Toronto, Toronto, Canada, M5S 1A, 1986. available as CSRI Technical Report CSRI-185.

[30] E Fiume and A Fournier. Toward a Precise Characterisation of Graphic Primitives. In preparation.

[31] E Fiume, D C Tsichritzis, and L Dami. A Temporal Scripting Language for Object-Oriented Animation. In *Proceedings of Eurographics-87*. North-Holland, August 1987. OOCA, OOCG.

[32] J D Foley and C F McMath. Dynamic Process Visualization. *IEEE Computer Graphics and Applications*, 6(2):16–25, March 1986. OOA, OOCG.

[33] J D Foley and A van Dam. *Fundamentals of Interactive Computer Graphics.* Addison-Wesley, 1982.

[34] C M Geschke, J H Morris Jr, and E H Satterthwaite. Early Experience with Mesa. *Communications of the ACM*, 20(8):540–553, August 1977. PL, OOS.

[35] A Goldberg. *Smalltalk 80: the Interactive Programming Environment.* Addison-Wesley, 1984. OOS, OOI, OOUI.

[36] A Goldberg and D Robson. *Smalltalk 80: the Language and its Implementation.* Addison-Wesley, 1983. OOS, OOI.

[37] J Guttag. Abstract Data Types and the Development of Data Structures. *Communications of the ACM*, 20(6):396–404, June 1977. Specification.

[38] C Hewitt. Viewing Control Structures as Patterns of Passing Messages. *Artificial Intelligence*, 8(3):323–364, June 1977. PL.

[39] R D Hill. Supporting Concurrency, Communication and Synchronization in Human-Computer Interaction - The Sassafras UIMS. *ACM Transactions on Graphics*, 5(3):179–210, July 1986. OOUI.

[40] C A R Hoare. Monitors: An Operating System Structuring Concept. *Communications of the ACM*, 17(10):549–557, October 1974.

[41] C A R Hoare. Communicating Sequential Processes. *Communications of the ACM*, 21(8):666–677, August 1978. PL.

[42] C A R Hoare. *Communicating Sequential Processes.* Prentice-Hall, 1985. PL, Specification.

[43] G Krasner. *Smalltalk-80: Bits of History, Words of Advice.* Addison-Wesley, 1983. OOA, OOI, OOUI, OOS.

[44] L Lamport. Specifying Concurrent Program Modules. *ACM Transactions on Programming Languages and Systems*, 5(2):190–222, April 1983. Specification.

[45] B W Lampson and D D Redell. Experience with Processes and Monitors in Mesa. *Communications of the ACM*, 23(2):105–117, February 1980. PL, OOS.

[46] D E Lipkie, S R Evans, J K Newlin, and R L Weissman. Star Graphics: An Object-Oriented Implementation. *Computer Graphics (Proc. Siggraph 82)*, 16(3):115–124, July 1982.

[47] B Liskov and J Guttag. *Abstraction and Specification in Program Development.* MIT Press/McGraw-Hill, 1986. PL, Specification.

[48] B Liskov, A Snyder, Atkinson, and C Schaffert. Abstraction Mechanisms in CLU. *Communications of the ACM*, 20(8):564–576, August 1977. PL, Specification.

[49] N Magnenat-Thalmann and D Thalmann. Subactor Data Types as Hierarchical Procedural Models for Computer Animation. In *Proceedings of Eurographics-85.* North-Holland, August 1985. OOCG, OOCA.

[50] N Magnenat-Thalmann, D Thalmann, and M Fortin. Miranim: An extensible director-oriented system for the animation of realistic images. *IEEE Computer Graphics and Applications*, 4(3), March 1985. OOCG, OOCA.

[51] G Marino, P Morasso, and R Zaccaria. NEM: A Language for Animation of Actors and Objects. In *Proceedings of Eurographics-85*. North-Holland, August 1985. OOCG, OOCA.

[52] T Maruichi, T Uchiki, and M Tokoro. Behavioural Simulation Based on Knowledge Objects. In *Proceedings of the European Conference on Object-Oriented Programming, Paris, France*, June 1987. OOA.

[53] N Meyrowitz. Intermedia: The Architecture and Construction of an Object-Oriented Hypermedia System and Applications Framework. *ACM SIGMAPLAN Notices*, 21(11):186–201, November 1986. OOA.

[54] D A Moon. Object-Oriented Programming with Flavors. *ACM SIGPLAN Notices*, 21(11):1–8, November 1986. OOS.

[55] J E B Moss and W H Kohler. Concurrency Features for the Trellis/Owl Language. In *Proceedings of the European Conference on Object-Oriented Programming, Paris*, pages 223–232, June 1987. OOS, PL.

[56] J Mylopoulos, P A Bernstein, and J K T Wong. TAXIS: A Language Facility for Designing Database-intensive Applications. *ACM Transactions on Database Systems*, 5(2):185–207, June 1980. PL.

[57] G Nelson. Juno, A Constraint-Based Graphics System. *Computer Graphics (Proc. Siggraph 85)*, 19(3):235–243, July 1985. OOCG.

[58] O M Nierstrasz. Hybrid: A Unified Object-Oriented System. *IEEE Database Engineering*, 8(4):49–57, December 1985. OOS, PL.

[59] O M Nierstrasz. Active Objects in Hybrid. *ACM SIGPLAN Notices (Proc. OOPSLA-87)*, 22(12):243–253, December 1987. OOS, PL.

[60] K Nygaard. Basic Concepts in Object-Oriented Programming. *ACM SIGPLAN Notices*, 21(10):128–132, October 1986. OOI.

[61] P D O'Brien, D C Halbert, and M F Kilian. The Trellis Programming Environment. *ACM SIGPLAN Notices (Proc. OOPSLA-87)*, 22(12):91–102, December 1987. OOS, OOA.

[62] OOPSLA '86 Conference Proceedings, Portland, Oregan. *ACM SIGPLAN Notices*, 21(11), 1986. OOS, OOA, OOUI, OOCG, PL, Specification.

[63] OOPSLA '87 Conference Proceedings, Orlando Florida. *ACM SIGPLAN Notices*, 22(11), 1987. OOS, OOA, OOUI, OOCG, PL, Specification.

[64] D L Parnas. A Technique for Software Module Specification with Examples. *Communications of the ACM*, 15(5):330–336, May 1972. PL, Specification.

[65] K W Piersoll. Object-Oriented Spreadsheets: The Analytic Spreadsheet Package. *ACM SIGPLAN Notices*, 21(11):385–390, November 1986. OOA.

[66] X Pintado and E Fiume. Grafields: Field-Directed Dynamic Splines for Interactive Motion Control. In *Proceedings of Eurographics-88*. North-Holland, September 1988. OOA,OOCG.

[67] S P Reiss. An Object-Oriented Framework for Graphical Programming. *ACM SIGPLAN Notices*, 21(10):49–57, October 1986. OOA, OOCG, OOUI.

[68] C Reynolds. Computer Animation with Scripts and Actors. *Computer Graphics (Proc. Siggraph 82)*, 16(3), 1982. OOCA, OOA, OOCG.

[69] C Reynolds. Flocks, Herds and Schools: A Distributed Behavioral Model. *Computer Graphics (Proc. Siggraph 87)*, 21(4), 1987. OOCA, OOA, OOCG.

[70] C Schaffert, T Cooper, B Bullis, M Killian, and C Wilpolt. An Introduction to Trellis/Owl. *ACM SIGPLAN Notices*, 21(11):9–16, November 1986. OOS, PL.

[71] R W Scheifler and J Gettys. The X-Window System. *ACM Transactions on Graphics*, 5(2), April 1986.

[72] M Shaw and W Wulf. Abstraction and Verification in Alphard: Defining and Specifying Iteraction and Generators. *Communications of the ACM*, 20(8):553–564, August 1977. PL, Specification.

[73] A H Skarra and S B Zdonik. The Management of Changing Types in an Object-Oriented Database. *ACM SIGPLAN Notices*, 21(11):483–495, November 1986. OOA.

[74] D C S Smith, C Irby, R Kimball, B Verplank, and E Harlem. Designing the Star User Interface. *Byte*, 7(4):242–282, April 1982. OOUI, OOA, OOS.

[75] A Snyder. Encapsulation and Inheritance in Object-Oriented Programming Languages. *ACM SIGPLAN Notices*, 21(11):38–45, November 1986. PL, OOS.

[76] Issue Special. Object-Oriented Programming Workshop. *ACM SIGPLAN Notices*, 21(10), October 1986. OOS, OOA, OOUI, OOCG, PL, Specification.

[77] M Stefik and D G Bobrow. Object-Oriented Progamming: Themes and Variations. *The AI Magazine*, December 1985. OOI.

[78] B Stroustrup. *The C++ Programming Language*. Addison-Wesley, 1986. OOI, OOS.

[79] I E Sutherland. *Sketchpad, A Man-Machine Graphical Communication System*. PhD thesis, MIT, January 1963.

[80] D Swinehart, P Zwellweger, and R Beach. A Structural View of the Cedar Programming Environment. *ACM Transactions on Programming Languages and Systems*, 8(4):419–490, October 1986. OOS, OOA.

[81] L Tesler. The Smalltalk Environment. *Byte*, 6(8), August 1981. OOS.

[82] D Thalmann and N Magnenat-Thalmann. Actor and Camera Data Types in Computer Animation. In *Proceedings of Graphics Interface 1983*, pages 203–210, May 1983. OOCG, OOCA, OOS.

[83] D C Tsichritzis, E Fiume, S Gibbs, and O M Nierstrasz. KNOs: KNowledge Acquisition, Dissemination and Manipulation Objects. *ACM Transactions on Office Information Systems*, 5(1):96–112, January 1987. OOS, OOA, PL.

[84] P Wegner. Dimensions of Object-Based Language Design. *ACM SIGPLAN Notices (Proc. OOPSLA-87)*, 22(12):168–182, December 1987. OOI.

[85] G Williams. The Lisa Computer System. *Byte*, 8(2):33–50, February 1983. OOA.

[86] N Wirth. *Programming in Modula-2*. Springer-Verlag, 1983. PL.

[87] A Yonezawa, J-P Briot, and E Shibayama. Object-Oriented Concurrent Programming in ABCL/1. *ACM SIGPLAN Notices*, 21(11):258–268, November 1986. OOS, PL.

[88] S B Zdonik. Maintaining Consistency in a Database with Changing Types. *ACM SIGPLAN Notices*, 21(10):120–127, October 1986. OOA.

2 Projective Geometry and Computer Graphics

Ivan Herman

ABSTRACT
Projective geometry is the basic mathematical tool to visualize three dimensional objects on a two dimensional surface. As a consequence, it is also the mathematical background for the output pipeline of all three dimensional graphics packages, whether this is explicitly stated or not (usually not). This chapter tries to present some of these mathematical tools to give a deeper insight into these systems, and, at the same time, to assist in the creation of new algorithms and methods to improve them or to elaborate new ones.

2.1 Introduction

This chapter makes an attempt to present some basic elements of projective geometry, and of axiomatic geometry in general, for the purpose of computer graphics. It is not our aim to provide new algorithms, or even to make a detailed presentation of the already existing ones. There are a number of excellent surveys and tutorials for that purpose, and it would not be of much interest to repeat all these here again (see e.g., [18]). However, we have the hope that by using the exact notions of projective geometry, a better understanding may be achieved as far as the mathematical background of computer graphics is concerned, which may also be helpful for the elaboration of new methods, approaches and algorithms. The mathematics presented here serve as the mathematical foundation for graphic packages like for example GKS, GKS-3D or PHIGS, and for most of the application programs running on the top of these. As a rule, however, the mathematics involved is not presented in these documents and also hard to find in the usual textbooks and introductions to computer graphics (e.g., [8], [17], [19], [23]).

2.2 Mathematical Preliminaries

2.2.1 Axiomatic Systems

To really understand the mathematical constructions leading to projective geometry, we start by some (at first glance) elementary problems. We seek an answer to the following question: what is, in reality, mathematics? What is its basic approach in trying to model reality? What is the background of its (sometimes contended) success?

The fact is that an exhaustive answer to these questions is almost impossible. There are libraries full of books, studies and papers dealing with these problems, there are different schools of mathematicians and philosophers who seek the final answer to these questions. It is of course not our aim to solve this problem here (we probably could not do so).

It is however widely accepted that the basic method of mathematics in this respect is the so called *axiomatic method.* Axiomatic systems (as they are called) may be found at the very foundations of all mathematical theories and branches, even if in their everyday practice mathematicians do not always deal with them. The method may be summarized (in a very simplified form) in the following way.

To form a new theory, a set of so called *primitive notions* and a set of *axioms* are accepted. Primitive notions are, in a certain way, just names for some entities which are to be examined via the given theory. As an example, the notions of points and lines are primitive notions of axiomatic geometry; *inside* the theory the mathematician is not really interested in what these notions are in the surrounding world. Of course, everybody has an intuitive feeling about what a point or a line is, but this intuition is just helpful in working in geometry, not more.

Axioms are a set of logical statements which describe the properties and basic relationships among the primitive notions. It is of course not absolutely clear what a "logical statement" is; in fact, there is a separate field of mathematics which tries to make this much more precise (called formal or mathematical logic). Again, everybody has a certain intuitive feeling about what a logical statement is, and this is just enough for our actual purposes.

Having the primitive notions and the set of axioms in hand, the mathematician tries to construct a number of additional logical statements about the primitive notions; the main point is that these new statements, called *theorems*, should be deducible with the help of formal logic from the set of axioms. A theorem is accepted if and only if it may be deduced logically; there is no other way to declare (within mathematics) that the theorem is true or not. In short, the role of a mathematician may be reduced to the task of setting up appropriate axiomatic systems to model a given problem, and trying to deduce as many interesting theorems as possible within the axiomatic system which has been defined. The basic, somewhat philosophical, idea behind this whole method is the assumption that the rules of formal logic reflect some natural phenomena, and, consequently, the theorems deduced with the help of these rules do reflect some properties of the real world.

It is of course required that the whole system should not contain contradictions. This requirement means that it should not be possible to deduce a given theorem *as well as* its logical negation. This is one of the basic properties an axiomatic system should fulfill. Another natural requirement is that the axiomatic system should be powerful enough to deduce really interesting new properties, and not only trivialities; otherwise the whole system is just useless.

Whether a given axiomatic system fulfills these two requirements is among the most difficult questions arising in mathematics. In fact, one of the most fantastic results of our century is the fact that, in a certain way, you may never know whether a given axiomatic system is contradictory or not. A particularly exciting and intellectually challenging presentation of this whole problem may be found for example in [10].

Up to now, we have always spoken about axiomatic systems which try to give an appropriate model describing a part of the surrounding world. In fact, the same approach may be used, and is used, within mathematics as well. Namely, in the course of its developments a given mathematical theory may reach very high complexity levels; therefore, to reduce this complexity and to make it easier to concentrate on a given question, some (mathematical) notions are sometimes defined as being new primitive notions, and some theorems describing the properties of these notions are defined to be axioms. This means that a "derived" axiomatic system is created within the larger one, which gives a way to concentrate on some dedicated problems. In practice, mathematics is full of such hierarchies of axiomatic systems giving birth to new mathematical theories. The so defined "sub-systems" may sometimes prove to be so rich by themselves that they become, after a time, a separate field. In fact, almost all branches of modern mathematics (e.g., topology or functional analysis) have been created this way.

All these things may seem to be a bit too abstract at a first glance, but, hopefully, they will become clearer in the coming sections. The fact is, as we shall see, that projective geometry is typically such a "sub-system" which has proved to be extremely useful by itself, for example in computer graphics.

2.2.2 Euclidean Geometry

The idea of the axiomatic method is far from being new. In fact, this is also one of those fantastic cultural achievements which we have inherited from ancient Greece.

The first axiomatic system in mathematics comes from the well known cultural centre of ancient Alexandria: this is Euclidean geometry[6]. It is of course, as usual, very unjust to connect this theory exclusively with the name of Euclid (365 B.C.(?)–300 B.C.(?)). In fact, his work, called "The Elements", is a collection of the geometrical results proved by a number of outstanding Greek mathematicians who had preceded Euclid by, eventually, several hundreds of years, like Thales, Pythagoras and others. "The Elements" is probably not the first such collection either; unfortunately all others have been lost in the course of history.

However, if we regard "The Elements" as a collection of Greek mathematics, it is hard to find in the history of science an achievement which has had a deeper influence than this. According to some estimates, the number of translations and publications of "The Elements" is the second highest in the whole of history; it is preceded only by the Bible. During the European Middle Ages, geometry, which at that time was more or less equal to Euclidean geometry, was one of the "free sciences" which were in the centre of the intellectual activities in European universities from Oxford to Padova. In spite of that, we could say that not very much had been added to the whole theory up to the 17th-18th centuries; this means roughly 2000 years!

The axiomatic system of Euclid is of course not so clean as in the axiomatic geometry of today. However, we find the primitive notions as well as the axioms as we have stated above; primitive notions are the points, lines and planes; axioms are usually very simple statements about these notions, like "two different points determine one and only one line intersecting both points" or "two lines do not determine a closed area" and the like.

2.2.3 Hyperbolic Geometry

It is not without interest to make a little detour toward what is called hyperbolic geometry. It is of course not our aim to apply hyperbolic geometry in computer graphics. The reason for this detour is that the development of hyperbolic geometry may give a better insight into the problems of axiomatic geometry and of axiomatic systems in general; it will also focus our attention on the problem of parallel lines. All this may hopefully be helpful for us to understand the basis of projective geometry as well.

The axioms of Euclid were usually very simple and clear. There was, however, one exception: the so called "fifth postulate". This axiom makes a statement about the unicity of parallel lines, roughly as follows.

> **The fifth postulate.** If we have a line on the plane and a point outside of it, then there is only one line which intersects the given point and which has no intersection point with the original line.

(This is not the original form of the fifth postulate, but an equivalent form which has proved to be more useful than Euclid's original one.) The fact that such lines *exist* can be proved out of the other axioms alone; this axiom states the *unicity* of such line.

We have to realize that this axiom is really different from the other ones. All other axioms (like the one we have already cited) are easily checkable in practice and use only finite notions. It is therefore acceptable to consider them as modelling reality. However, the axiom about parallels is much different. It is, in reality, impossible to check it; what difference does it make whether two lines do not intersect at all or intersect at a distance of, say, 300,000 km? In other words, this axiom brings somehow the notion of infinity into the whole system of axioms, which gives it a very different flavour.

The fact that this axiom is so different meant that mathematicians always felt the necessity to *prove* it, to consider it as a *theorem* and not as an axiom. This was already clear to Euclid as well; in "The Elements" he tried to prove as many theorems as he could without using the fifth postulate to make the whole structure clearer.

As we have already said, geometry played an essential role in the Middle Ages. One of the greatest challenges for all mathematicians of that time was to prove the fifth postulate, so as to "clean" the axiomatic system of geometry. And they did not succeed.

The breakthrough came only in the 19th century, when some mathematicians realized that this theorem *could not be proved* out of the other theorems. To be more precise, they showed that if you take the axiomatic system of Euclid, you take the fifth postulate out of the system and you add the *negation* of it (that is that there *may* be more than one parallel line), you will get an absolutely consistent geometrical structure as well. This new geometry is equal in its potential power to Euclid's geometry; it is just different.

It is hard to understand today what a revolutionary idea this whole approach was. Up to that time, everybody considered Euclidean geometry as being the most adequate description of geometrical reality; in other words that reality *is* Euclidean. As a result of the appearance of this new mathematical structure it became clear that Euclidean geometry is just one of many possible geometries, which may be a very good modelling tool to describe the surrounding world, but is not necessarily the only one. This idea was so difficult to accept, that for example the "princeps mathematicae", Gauss (1777–1855), who was undeniably one of the greatest mathematicians of all time, and who was one of those who had realized this new idea, did not dare to publish his results. He was afraid to be confronted by the great intellects of his time. Independently of Gauss, two other mathematicians had arrived at the same conclusions, but in contrast to him they went ahead, and really created and published their ideas about the new theory, which was later given the name of hyperbolic geometry. It is worthwhile to cite the name of these mathematicians, as their role in the history of mathematics is enormous: one was the Russian Lobatschevski (1792–1856), the other was the Hungarian Bòlyai (1802–1860).

The birth of hyperbolic geometry has started a proliferation of different "geometries", that is, of mathematical structures similar to Euclid's, modelling some part of the surrounding world. New geometries were defined by creating somewhat similar but different sets of axioms; these geometries were able to give a very adequate and easy description of, for example, the geometry on the surface of a sphere or on some other, more complicated surfaces. In such cases, the primitive notions of the new geometrical system still exist but they aim at modelling other, sometimes more complicated, traditional geometrical notions. Just to mention a practical case: a geometrical structure describing the geometry of a given surface may be defined by choosing the notion of "points" to coincide with the original (Euclidean) notion, while the "lines" within this geometry are the geodesic curves of the surface. In this case the local set of axioms should be defined in a way that they describe the properties of the geodesic curves; once this is done, we get a special geometry which may describe the geometrical behaviour of the given surface (in fact, there exists a complicated surface where hyperbolic geometry is the adequate tool for description). What we have to understand is that these geometries all have their own internal world, just like the Euclidean one; they are just different. As we will see, projective geometry is just one of these new geometrical structures.

2.2.4 The Coordinate System

As we have already seen, not too much had been added to Euclidean geometry up to the 17th century. The most important change occurred with the introduction of the cartesian coordinate system of Descartes (1560–1650).

The use of a coordinate system has become part of our basic mathematical education; consequently we consider it as one of the most natural notions in mathematics. You have to become a student in mathematics to realize that the existence of a coordinate system is a *theorem*, which has to be proved, which you have to learn for your exams. Furthermore, it turns out that the proof itself is not even trivial; it is a long and tedious process.

The theorem about the existence of cartesian coordinate systems may be stated in the following way.

In Euclidean geometry, the notion of the *distance* of two points is introduced (in fact, some of the axioms themselves deal with this notion). Use of this notion may lead to the following theorem:

Theorem 2.1 *If there are three non-collinear points (that is, points not determining a line)* **O**, **A1**, *and* **A2** *on the plane, then there exists one and only one distance keeping, one-to-one correspondence between the plane and the set of two-element real vectors, such that the point* **O** *corresponds to the vector (0,0), the point* **A1** *corresponds to the vector (1,0) and the point* **A2** *corresponds to the vector (0,1).*

"Distance keeping" in this case means that the distance of two points on the Euclidean plane may be determined with the well-known formula $\sqrt{(x_1 - x_2)^2 + (y_1 - y_2)^2}$. In the case of a Euclidean space, an additional **A3** point is also necessary, which should correspond to the vector (0,0,1).

This theorem creates a bridge between two basically different mathematical structures, namely Euclidean geometry and the structure of the two dimensional real vectors. We have to stress the fact that this theorem is not necessarily true for all geometric structures; in fact it will not be true for projective geometry either! This shows that the theorem itself is far from being trivial.

We are of course not interested in the actual proof here. The only thing which is important is to be aware of the fact that the existence of the coordinate system is not a trivial fact; we will need this awareness in what follows.

An interesting consequence of the cartesian coordinate system is the fact that it gave birth to the so called "multi-dimensional" geometries. The usual planar and spatial geometries were identified with the linear structure of the two and three dimensional vectors; it became possible to describe lines, planes, spheres, etc. with the help of real numbers and equations. All these equations are, however, applicable to four or five dimensional vectors as well; for example the well known equation

$$T = t.P + (l - t).Q$$

describing a line crossing the points **P** and **Q** is independent of whether the points are described by two or three element vectors. In other words, all these notions could be transferred into higher dimensions, giving a "geometrical" structure for higher dimensional vector spaces as well. These geometrical analogies may be very helpful to design algorithms for, let us say, four dimensional vectors. In fact, a number of algorithms developed originally for three dimensional geometry, which used the coordinate representation, could be generalized without problems to higher dimensions.

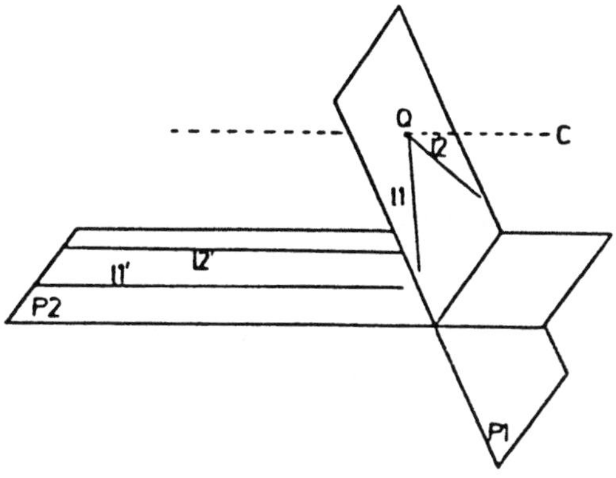

FIGURE 2.1. Projecting Parallel Lines

2.3 Basic Elements of Projective Geometry

2.3.1 Background

As we have already seen, the appearance of parallel lines, which is an inherent fact in Euclidean geometry, has created a number of problems. Parallel lines appear very frequently in different theorems as some kind of special case, which makes the description of some problems clumsy and difficult to understand.

A typical case of such a situation is when the properties of projections are investigated. Figure 2.1 shows such a case: the plane **P1** is projected onto the plane **P2**; the projection is central, with centrum at **C**. The two lines **l1** and **l2** have an intersection point on **P1**, denoted by **Q**. However, the line **CQ** is parallel to the plane **P2**; as a consequence, the lines **l1**′ and **l2**′ in plane **P2** are parallel. In this case, what is the image of **Q**?

Such situations may appear in all systems (mathematical or others) which deal with projections. Let us take an example: painting. One of the main differences (from a crude technical point of view of course) between medieval European painting and Renaissance painting is the fact that in the 15th and 16th centuries painters had acquired a much deeper understanding of the properties of projections; consequently, they were able to create much more realistic pictures. At a time when the gap between art and science was not so desperately deep as it is today, a great German artist, Albrecht Dürer, even wrote a book about projective geometry in 1525 [5].

To come closer to our era, three dimensional computer graphics systems are, of course, typically such environments as well. It is therefore necessary to have a mathematically clean and clear method to handle problems such as the one cited above. Of course, ad hoc methods may always be found to solve a given problem, but if an implementor of a graphic system has not acquired a really consistent model in his/her mind about such situations, the ad hoc methods may easily lead to difficult problems and/or even errors. This is why projective geometry is so important for such cases: it provides such a model, and, consequently, it provides a good framework to handle these problems.

In projective geometry, an alternative geometrical system is created. This geometrical system has two main properties, which are very important for us:

- parallel lines do not exist in this geometry
- the new system contains somehow as a sub-set the classical Euclidean system.

The second requirement is of course very important. In the use of projective geometry, we never forget our ultimate goal: to describe some events in our environment, which are modelled basically by Euclidean geometry. We will see the details of all these in the coming chapters.

2.3.2 The Basic Construction

Two intersecting lines on a plane have something in common, namely their intersection point. This is exactly the missing property of two parallel lines.

Is there anything which two parallel lines have in common, something which two intersecting lines do *not* have in common? In a certain way yes: this is the *direction* of the lines. Two parallel lines have the same direction; and vice versa, if two lines share the same direction, they are either identical or parallel.

Of course, a direction is not a normal Euclidean point; it is an abstract notion which may be defined very precisely within mathematics (directions are so called *equivalence classes*). The exact definition is not really important here; the main point is that if we have a Euclidean plane, denoted by **P**, then a well defined set may also be created, which is the set of all possible directions of this plane **P**. Let us denote this set by **D**. Clearly, **P** and **D** are disjoint sets. Consequently, if we define the set $\mathbf{P}' = \mathbf{P} \cup \mathbf{D}$ (the union of **P** and **D**), this new set will be a super set of our original Euclidean plane.

The set $\mathbf{P}'$ will be used as a basis for projective geometry. In other words, a geometrical structure is defined on this set with primitive notions and axioms, to create a new axiomatic system.

In geometry, the primitive notions are points, lines and the notion of intersection of two lines or the intersection of a point and a line. These are the primitive notions for Euclidean geometry; we would like to extend these notions for our enlarged set. This is done as follows (to make things more understandable, the new notions will be put into quotation marks for a while, to make them distinguishable from the original notions).

Definition 2.1 *A "point" is an element of the set* $\mathbf{P}'$. *In other words, the original Euclidean points and the directions are considered as "points". In case we want to remind ourselves of the original Euclidean environment, the Euclidean points will also be called "affine points", while the directions are also called "ideal points". The set of affine points (that is the original Euclidean plane) is also called the "affine plane".*

Definition 2.2 *A "line" is either a Euclidean line or the set* **D**. *(To be more exact, each Euclidean line is enlarged with an additional "point": the direction corresponding to the given line.) That means we have practically added one new "line" to the original ones. The Euclidean lines may also be called "affine lines", while the collection of ideal points is called an "ideal line".*

Definition 2.3 *Intersection of a "point" and a "line". If both are affine, the notion of intersection is the same as in the Euclidean case. If both are ideal, we* agree *that the "point" is on the "line". If the "line" is affine and the "point" is ideal, we define the intersection to take place if and only if the direction of the line corresponds to the given "ideal point". Finally, if the "line" is ideal and the "point" is affine, there is no intersection.*

Definition 2.4 *Intersection of two "lines" means that there exists a "point" which intersects both "lines".*

A set of theorems may be now proved for these notions, which describe their basic properties. We do not want to enumerate all of them here, since they contain a number of technical details which are of no real interest for us. The main ones are the following:

Theorem 2.2 *For any two non-identical "lines" there exists one and only one intersection point (i.e. there are no parallel "lines").*

Theorem 2.3 *Any two non-identical "points" determine one and only one "line" which intersects both of them.*

In fact, these theorems are almost trivial statements, and may be derived very easily from the basic properties of the Euclidean plane and our definitions. Let us see, for example, a more formal proof of Theorem 2.1.

Let us denote the two lines by **l1** and **l2**. If both these lines are affine, there are two possibilities. Either they have an affine intersection point or not. If yes, this intersection point will also be a point on the projective plane, as defined in Definition 2.3. If not, the two lines are parallel in the Euclidean sense, hence their direction is the same. According to our definitions, this common direction is the intersecting (ideal) point for **l1** and **l2**.

If, say, **l1** is affine and **l2** is the ideal line, the direction of **l1** determines an ideal point. However, by definition, **l2** is the collection of all ideal points; consequently, the intersection of **l1** and **l2** is not void.

With all these definitions and theorems in hand, we now have the possibility to define an absolutely new mathematical structure. This structure contains primitive notions like "points", "lines" and intersection; additionally, there are a number of statements about these notions (the ones we have cited and a number of additional ones) which may be accepted as axioms. With these in hand we get a new geometry; this is called projective geometry, or the geometry of the projective plane.

The fact that we have arrived at this new geometry via a construction starting from a Euclidean plane gives us the possibility to say that projective geometry may contain Euclidean geometry as a kind of sub-structure. Of course, it is perfectly possible to regard projective geometry as a mathematical structure by itself, forgetting about its origin. In fact, most of the mathematical investigations about the properties of the projective plane are done that way.

However, computer graphics is just an application of abstract geometry; that is, we will never forget the origin of projective geometry. For us, Euclidean geometry should always appear as a sub-structure of projective geometry; this approach gives us a consistent model of what may happen within a graphic system.

We have spoken, up to now, about projective planes only. Let us see now the three dimensional case! The same kind of construction may be achieved. We have the directions as well; we can create the corresponding $\mathbf{P}'$ set as well. The extended notions will be as follows:

Definition 2.5 *A "point" is an element of the set* $\mathbf{P}'$.

Definition 2.6 *A "line" is either a Euclidean line or the set of directions* belonging to one Euclidean plane. *That means we have added an "ideal line" to each Euclidean plane.*

Definition 2.7 *A "plane" (which is also a primitive notion for three dimensional geometry) is either a Euclidean plane together with the directions belonging to it or the set of all directions ("ideal points"). In other words, we have added one new plane to the system, which is also called an "ideal plane".*

Definition 2.8 *Intersection of a "point" and a "line". This is similar to the two dimensional case.*

Definition 2.9 *Intersection of a "point" and a "plane". This means that the "point" belongs to the point set of the "plane".*

Definition 2.10 *Intersection of a "line" and a "plane". This means that the intersection of the two point-sets is not void.*

Definition 2.11 *Intersection of two "lines" means that there exists a "point" which intersects both "lines".*

Definition 2.12 *Intersection of two "planes" is similar: for "affine planes", the usual intersection is accepted. Intersection of an "affine plane" and the "ideal plane" is the set of all directions which belong to the "affine plane" (that means an "ideal line").*

The basic theorems are basically identical. These are as follows.

Theorem 2.4 *For any two non-identical but co-planar "lines" there exists one and only one intersection point (i.e. there are no parallel "lines").*

Theorem 2.5 *Any two non-identical "points" determine one and only one "line" which intersects both of them.*

Theorem 2.6 *For any two non-identical "planes" there exist one and only one intersection "line" (that means that there are no parallel planes).*

Theorem 2.7 *Any three non-collinear "points" determine one and only one "plane" which contains all of them (collinearity for three "ideal points" means that they belong to a same "ideal line").*

The resulting structure is the projective space. As far as computer graphics is concerned, this space has great importance; this is the structure which helps us to describe the mathematical background of such systems as GKS-3D or PHIGS.

As a first "application" of these notions, the problem we have presented in figure 2.1 may now be consistently described; if we regard the whole projection as a transformation between two *projective* planes, then, quite simply, the image of **Q** (which is an affine point) is an ideal point of the plane **P2**. It is of course a question of how such a point may be handled in practical cases; we will come back to this in the following sections. However, it should be clear by now that if we regard the whole process as embedded into the projective environment, there is no real exception any more.

2.3.3 Collinearities

From the axiomatic system of projective geometry, a lot of interesting geometrical properties may be derived. However, we should concentrate only on those facts and theorems which have a direct or an indirect consequence on computer graphics; unfortunately, we cannot go into the details of projective geometry here.

All geometrical systems have a deep interest with regard to different transformations. There are different classes of transformations, like rotations, translations, scalings, etc. which are of a great importance in computer graphics as well. It is therefore necessary to find a consistent way to describe these transformations, like for example the projection which has already caused us some problems.

A very general class of transformations is the class of *collinearities* or *projective transformations.* These transformations may be defined in the following way:

Definition 2.13 *A transformation of a plane into another one (or a space into another one) is said to be a* collinearity *if any arbitrary three collinear points are transformed into collinear points.*

The definition does not state whether such a transformation is defined for Euclidean or projective geometry. In fact, the definition itself may be applied to both of them, that is, we may have collinearities both in the classical Euclidean and in the projective environment. Furthermore, it may be proved that if a collinearity is defined between two Euclidean planes, this collinearity may be extended in a unique way into a collinearity of the generated projective planes.

Clearly, the definition includes a very large number of transformations. All transformations which are usually used in computer graphics, like translations, scalings, etc., are collinearities.

What is much more interesting is the fact, that e.g., the projections among planes are also collinearities (only in the projective sense; as we have already seen, a projection may turn an affine point into an ideal one!). On the other hand, the projections cannot be described as the concatenation of simple transformations like rotations, scalings and translations, which means that this class of transformations is very general. To those who have already had a glance into GKS-3D or PHIGS, we can also add the fact that the basic transformations of viewing, as they are defined in these documents, are also collinearities of the projective space (we will come back to this later). In other words, collinearities seem to be the appropriate class of transformations to describe both the well known classical transformations and also the projections. This fact is very important; that is the reason why these transformations are also called "projective" transformations.

To differentiate within this class of transformations, a sub-class is also defined, namely the class of *affine transformations*. The definition is as follows:

Definition 2.14 *A collinearity is said to be an* affine transformation *if the images of all affine points are affine.*

This sub-class is also very important. It contains the "usual" transformations, which we have already cited (translations, scalings, rotations) and all possible concatenations of these. In other words, the basic transformations of computer graphics, except the projections (e.g., all transformations which are usually used in a 2D system) belong to this class.

We have to present now a theorem which will have an importance in the following. This theorem shows also the "power" of the notion of collinearities.

For the purpose of the theorem, we have to accept the following definition:

Definition 2.15 *A finite point set is said to be of* general position, *if any arbitrary three points of the set are not collinear. Furthermore, if the point set is defined in the projective space, it is also required that any arbitrary four points of the set are not coplanar.*

With this definition in hand we can state the following theorem:

Theorem 2.8 *If two sets of four points ([*$\mathbf{A}$,$\mathbf{B}$,$\mathbf{C}$,$\mathbf{D}$*] and [*$\mathbf{A}'$,$\mathbf{B}'$,$\mathbf{C}'$,$\mathbf{D}'$*]) are given on a (projective) plane, and both sets are of general position, there is one and only one collinearity which turns* $\mathbf{A}$ *into* $\mathbf{A}'$, $\mathbf{B}$ *into* $\mathbf{B}'$, $\mathbf{C}$ *into* $\mathbf{C}'$ *and finally* $\mathbf{D}$ *into* $\mathbf{D}'$.

The same theorem is true for projective spaces; the difference is that instead of four points, five points are necessary.

It is of course not possible to prove this theorem here. The proof, by the way, is very far from being trivial. The role of the theorem itself in projective geometry is extremely important; it is one of those theorems which are in a certain way behind a number of additional theorems, even if its role is not visible at a first glance.

2.3.4 Homogeneous Coordinates

The techniques available in computers require some kind of numerical representation of the geometry to make it accessible for computing purposes. That is why coordinate systems play an essential role for us as well. The present section will present the way of creating a numerical representation for projective geometry, called the homogeneous coordinate system.

At first, we have to define what a homogeneous vector is; this definition may be well known, but let us have it here for completeness:

Definition 2.16 *Two non-zero n-dimensional vectors* $(a_1, a_2, \ldots, a_n)$ *and* $(b_1, b_2, \ldots, b_n)$ *are said to be equal in the homogenous sense, if there exists a number* α *such that:*

$$\begin{aligned} a_1 &= \alpha.b_1 \\ a_2 &= \alpha.b_2 \\ &\vdots \\ a_n &= \alpha.b_n \end{aligned} \tag{2.1}$$

When we speak about n-dimensional homogeneous vectors, this means that if we have two vectors which are equal in the homogeneous sense, we do not consider these two vectors to be different. In other words if we use the non-zero vector **a**, we could at the same time use all the vectors of the form $\alpha.\mathbf{a}$ where α is a non-zero number.

It is clear that to define some kind of a coordinate system for a projective plane/space, the classical cartesian coordinate system cannot be used. Instead, homogeneous coordinates should be used to give a possible numerical representation of projective points, lines and planes. We have presented the theorem about the existence of cartesian coordinates in section 2.2.4. The analogous theorem in the projective case is the following.

Theorem 2.9 *If there are four points* **O, A1, A2, E** *of a general position on the projective plane, then there exists one and only linear one-to-one correspondence between the projective plane and the set of three-dimensional homogeneous vectors, so that the point* **O** *corresponds to the vector (0,0,1), the point* **A1** *corresponds to the vector (1,0,0), the point* **A2** *corresponds to the vector (0,1,0), and, finally, the point* **E** *corresponds to the point (1,1,1).*

In case of a projective space, one more point, say **A3**, is also necessary, and instead of three dimensional homogeneous vectors four dimensional ones are to be used. The points correspond to (0,0,0,1), (1,0,0,0), (0,1,0,0), (0,0,1,0) and (1,1,1,1) respectively.

The somewhat unusual approach in this coordinate system definition is that a projective *plane* is described by *three* dimensional (homogeneous) vectors, and the projective *space* is described by *four* dimensional ones, in contrast to the usual Euclidean case. Anyhow, the fact that such a correspondence exists is an extremely important fact.

The next important question is: what about the relationship between a cartesian coordinate system on a plane/space and the homogeneous coordinates which may be defined on their projective extension? Is it possible to define a homogeneous coordinate system so that the original cartesian coordinates are not lost?

Such a relationship exists. Let us have a coordinate system on the Euclidean plane. Let us define the following points on the projective extension in the following way:

O is the origin of the cartesian system;
A1 is the ideal point which belongs to the X axis;
A2 is the ideal point which belongs to the Y axis;
E is the point whose cartesian coordinates are (1,1).

According to Theorem 2.9, there exists a homogeneous system which is generated by these four (projective) points. The relationship between these two coordinate systems is as follows.

- If the point **P** is affine, and its coordinates in the cartesian system are (x, y), the same point may be identified in the homogeneous coordinate system by the (homogeneous) vector $(x, y, 1)$
- If the point **P** is affine, and its coordinates in the cartesian system are (x, y), the ideal point belonging to the line **OP** is represented by the (homogeneous) vector $(x, y, 0)$.

The analogous approach may be taken for a space as well. That means it is extremely easy to make the identification. It is also true that if a point on the projective plane has a homogeneous coordinate (x, y, w) where w is not *zero*, the point is affine, and the corresponding cartesian coordinates may be derived by $(x/w, y/w)$. On the other hand the ideal points are uniquely described by having a homogeneous coordinate $(x, y, 0)$. That means we have an easy way at hand to differentiate between affine and ideal points.

The use of homogeneous coordinates in graphics is a well accepted practice. As we will see a bit later, they provide a good tool to handle different transformations easily and in a compact form [1], [2], [7], [8], [17], [19], [21], [23]. However, in all these cases, the homogeneous coordinates are presented as good "tricks" only, and the coordinates with the last value zero (that is, the ideal points) are never covered properly. Theorem 2.9 shows that the homogeneous coordinates are not just "tricks"; they represent a very deep characterisation of the projective environment.

The homogeneous coordinates have several very practical consequences. For example, some of the well known equations describing geometrical entities in the cartesian system are still valid for describing projective entities as well (a good example of that is the equations for lines). However, care should be taken with these analogies, as we will also see later.

We have also the possibility to have a good visual tool to "illustrate" the project plane (figure 2.2). Such pictorial representations of mathematical models are very important. They give an excellent tool to visualize what is going on, and hence to help our intuition to understand the background to some **mathematical facts.**

Figure 2.2 is just a figurative representation of the identification of cartesian coordinates in a projective environment. In the three dimensional space, the homogeneous coordinates describe those lines which cross the origin (to be very exact, the origin itself is not part of the homogeneous coordinates, but we can now forget about such details). The identification of cartesian coordinates and projective coordinates means that the original Euclidean space corresponds to the plane which crosses the **0,0,1** point and which is parallel to the **XY** plane. In other words, the "usual" Euclidean geometry takes place somehow on this plane. Ideal points are represented by lines which are in the **XY** plane.

The two affine lines **l1** and **l2** are parallel in the Euclidean sense; their intersecting (ideal) point in the projective environment is represented by the line **l3** which is parallel to both of them and which is in the **XY** plane.

Unfortunately, it is not so easy to visualize a projective space. According to the identification scheme, the affine part of the projective space is the three dimensional sub-space

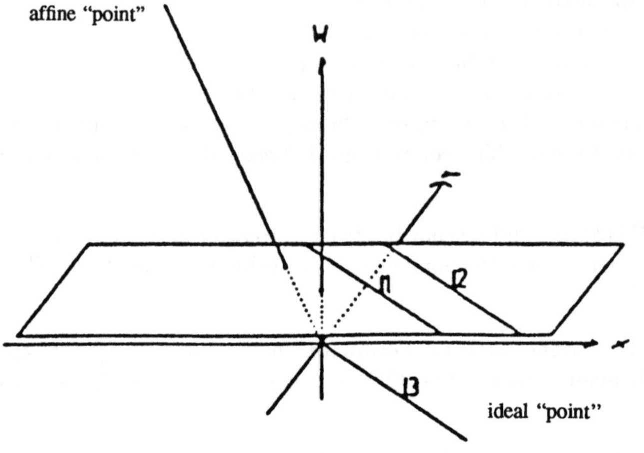

FIGURE 2.2. Cartesian coordinates in a projective environment

W=**1** in the four dimensional space, and we have no good way to represent such a space. This is the reason why Figure 2.2 should be used even for projective spaces; it greatly helps our intuition although it is not exact. As we shall see in the coming chapters, even this tool may be very helpful for us.

2.3.5 Representation of Collinearities

An additional advantage of the homogeneous coordinate system is the fact that the collinear transformations may be described in a very compact manner.

As the projective points are described with three/four element vectors, there is a very well known way to define a number of transformations on these points, namely with the help of matrices. If a 3×3 matrix is given for a projective plane (alternatively, a 4×4 matrix for projective space), the usual matrix-vector multiplication defines a transformation of the projective plane/space as well.

It is relatively easy to see that these transformations are also collinearities, if the determinant of the matrix is non-zero. They transform lines into lines, just as in the classical geometrical environment. Fortunately, a much stronger fact is also true:

Theorem 2.10 *Let us say that a homogeneous coordinate system is defined for the projective plane/space. For each collinear transformation of the projective plane/space there exists a non-singular* 3×3 *(*4×4 *respectively) matrix which describes this transformation. This matrix is unique in the homogeneous sense, that is if two such matrices are given, say,* **M1** *and* **M2**, *there exists a non-zero number* α, *so that* $\mathbf{M1} = \alpha\mathbf{M2}$. *Furthermore, the concatenation of two collinear transformations may be described by the matrix-multiplication of the corresponding matrices.*

This theorem means that all projections which we may encounter in a usual graphic system may be described by matrices. The fact that the representation *exists* (which is a mathematical fact!) makes it possible to use e.g. some standard linear equation solving methods to find the exact form of the matrix; we may know in advance that a solution to the equation does exist.

The analogy of linear transformations and collinearities (that is, projective transformations) gives us the possibility to visualize the effect of a projective transformation

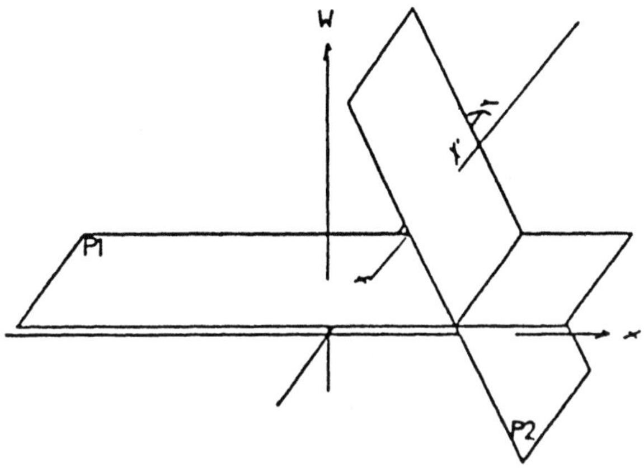

FIGURE 2.3. Projective transformation of the plane

(figure 2.3) of an affine plane. The transformation turns the plane **W=1** into another plane denoted by **P2** of the three dimensional space (this is the effect of the matrix-vector multiplication). The homogeneous vectors of the image points are the intersection points of this plane with the lines crossing the origin of the three dimensional space. If, as a second step, we would like to get the image of the transformation on our original plane, we have to create the intersection points of these lines with the **W=1** plane. This second step corresponds to the division with the last coordinate value (if this is possible); geometrically, it means a central projection of the plane **P2** onto the plane **W=1** with the centre of the projection being the origin (mapping the point $\mathbf{X}'$ to $\mathbf{X}$ in the figure). In the literature, this second step is sometimes called the "projective division".

Another consequence of Theorem 2.10 can also be seen in figure 2.3. The projective transformation of the plane, as we have seen, may be considered as a two-step transformation. In the first step, which is the matrix-vector multiplication (called also the "linear" part of the transformation), a full linear transformation of the three dimensional space is generated. The affine plane (which is, anyhow, in the centre of our interest) is turned into another plane. The *linearity* of this step is of a great importance. On the image plane **P2**, the *Euclidean geometry* is still valid locally, and there is not yet any spectacular effect, such as the appearance of ideal points, or anything like that, which is related to projections. In fact, these effects are generated by the projective division only. We will see the consequence of this later.

2.4 Basic Application for Computer Graphics

2.4.1 Affine Transformations

As far as usual two dimensional graphics are concerned, the graphic packages do not really deal with projective transformations. The relevant transformations are rotations, scales, translations, shears and all those transformations which may be derived by the concatenations of these. In short: the transformations are the affine ones.

The affine transformations, which are special cases for collinearities, have of course their matrix representation. Fortunately, this matrix representation is extremely simple.

Theorem 2.11 *The affine transformations may be uniquely characterized by the fact that the last row of the matrix may be chosen to be* $(0,0,1)$ *(or* $(0,0,0,1)$ *for affine transformations in space).*

The four basic transformation types listed above may be described very easily as follows. (Here and subsequently, the vectors are considered to be column vectors that is, matrix-vector multiplications are used and not vector-matrix multiplications).

Translation If we have a translation vector $T = (T_x, T_y)$ on the plane, the matrix is:

$$\begin{pmatrix} 1 & 0 & T_x \\ 0 & 1 & T_y \\ 0 & 0 & 1 \end{pmatrix}$$

Scale If we have the scaling factors S_x, S_y in the X and Y directions respectively the corresponding matrix is:

$$\begin{pmatrix} S_x & 0 & 0 \\ 0 & S_y & 0 \\ 0 & 0 & 1 \end{pmatrix}$$

Shear A shear in the coordinate X with a factor α may be achieved by the matrix:

$$\begin{pmatrix} 1 & \alpha & 0 \\ 0 & 1 & 0 \\ 0 & 0 & 1 \end{pmatrix}$$

i.e., the value of the Y coordinate is unchanged, and X changes according to:

$$x' = x + \alpha . y$$

Similarly, a shear in the coordinate Y with a factor β may be achieved by the matrix:

$$\begin{pmatrix} 1 & 0 & 0 \\ \beta & 1 & 0 \\ 0 & 0 & 1 \end{pmatrix}$$

Rotation If a rotation around zero is to be described, with a rotation angle α (positive direction is anti-clockwise), the corresponding matrix is:

$$\begin{pmatrix} \cos\alpha & -\sin\alpha & 0 \\ \sin\alpha & \cos\alpha & 0 \\ 0 & 0 & 1 \end{pmatrix}$$

Out of these five kinds of matrix all possible affine transformations may be generated.

In the case of three dimensions, the translation, the scale and the shear matrices are very similar (with one more dimension, of course). However, the rotation matrix becomes much more complicated. In fact, we have to define three different matrices for rotation around the coordinate axis X, Y and Z respectively.

The three matrices are the following (in each case, the angle α is measured anti-clockwise when looking along the rotation towards the origin). Rotation around the Z axis:

$$\begin{pmatrix} \cos\alpha & -\sin\alpha & 0 & 0 \\ \sin\alpha & \cos\alpha & 0 & 0 \\ 0 & 0 & 1 & 0 \\ 0 & 0 & 0 & 1 \end{pmatrix}$$

Rotation around the Y axis:

$$\begin{pmatrix} \cos\alpha & 0 & \sin\alpha & 0 \\ 0 & 1 & 0 & 0 \\ -\sin\alpha & 0 & \cos\alpha & 0 \\ 0 & 0 & 0 & 1 \end{pmatrix}$$

Rotation around the X axis:

$$\begin{pmatrix} 1 & 0 & 0 & 0 \\ 0 & \cos\alpha & -\sin\alpha & 0 \\ 0 & \sin\alpha & \cos\alpha & 0 \\ 0 & 0 & 0 & 1 \end{pmatrix}$$

Out of these matrices a rotation around an arbitrary axis may be composed. For the details see for example [1], [17] or any other standard textbook about computer graphics. Care should be taken over the fact that in the textbooks vector-matrix multiplications are sometimes used instead of matrix-vector ones; also, the rotation angle is sometimes taken as defining a clockwise direction instead of an anti-clockwise one. This may lead to the necessity to transpose matrices and/or to change the sign of some elements.

We have to stress the fact that such a compact form for the transformations cannot be achieved if a cartesian coordinate system is used alone. In fact, this compactness of the formulae was one of the main reasons why homogeneous coordinates have been widely accepted in computer graphics. Usually, the introduction of these coordinates is presented as being just a clever mathematical formulation; from the previous chapters we may see now that these notions reflect much deeper characteristics about projective geometry.

Some of the transformations used by packages like GKS[13] or GKS-3D[14] (e.g., the so called Segment and Insert Transformations) or popular Page Description Languages like PostScript[12] or DDL[4] are affine transformations. In fact, in these cases the transformations are formally specified by 2×3 (3×4 respectively) matrices; this means that full transformation matrices are defined but without determining explicitly the last row (that is, the vectors $(0,0,1)$ and $(0,0,0,1)$ respectively). Mathematically, this is equivalent to the definition of affine transformations. However, these rows are implicitly present when the transformations are effectively in use.

2.4.2 Projections

One of the main functions of 3D graphics systems (e.g., GKS-3D[14], PHIGS[15]) is to perform viewing. This means that a three dimensional object has to be drawn on a two dimensional plane; in other words, some kind of projection has to be performed from the three dimensional object.

Traditionally, two projection types are used within these systems, namely parallel and perspective projections (see figure 2.4 and 2.5). In fact, a much more complicated taxonomy of the projections can also be used in practice, but all the possible projections are sub-classes of these two (see e.g., [11], [24]).

The role of a three dimensional package is not confined to effective viewing. In most of the cases some kind of a Hidden Line/Hidden Surface Removal (HLHSR) method is also required, to create an acceptable three dimensional picture.

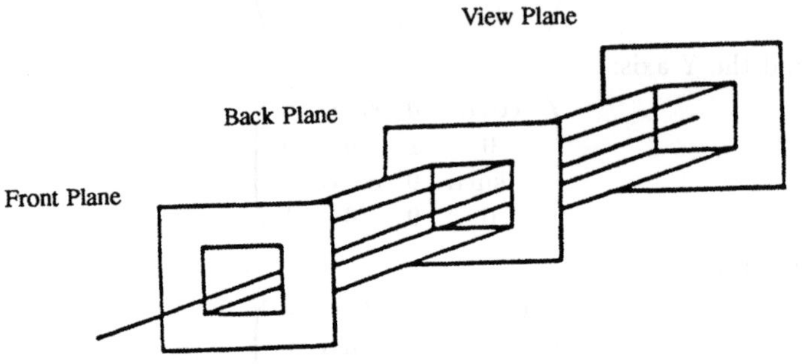

FIGURE 2.4. View Volume: parallel projection

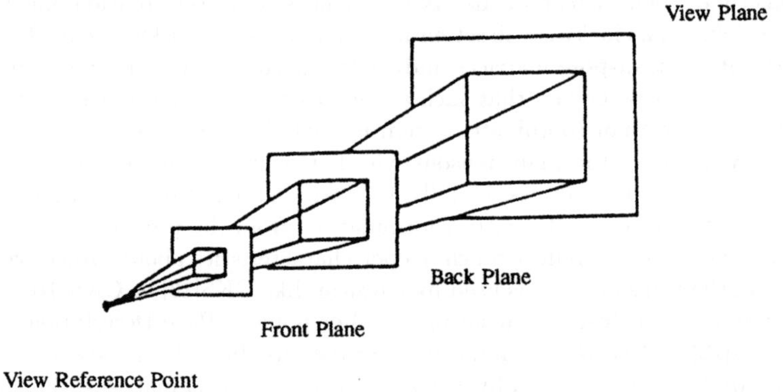

FIGURE 2.5. View Volume: perspective projection

The simplest case of projections is a special parallel projection, in which the view plane is perpendicular to the projection line. If the coordinate system is properly chosen (or, in other words, an appropriate affine transformation is applied on the model which is to be viewed), the direction of the projection may be considered to be the Z axis, with the projection plane being the XY plane and centred around the origin (figure 2.6). Let us for the sake of simplicity, call this kind of projection a "basic" projection. In this case, the projection itself involves simply forgetting about the third coordinate; the HLHSR methods may be applied with visibility in the positive (or negative, depending on traditions) Z direction, which greatly simplifies the necessary calculations.

The situation is much more complicated if a perspective projection is to be applied. For each point to be displayed, the intersection of a plane and a line should be calculated; additionally, the HLHSR methods are also extremely time-consuming. The usual approach therefore is to try to convert the perspective projection into a parallel one. This approach

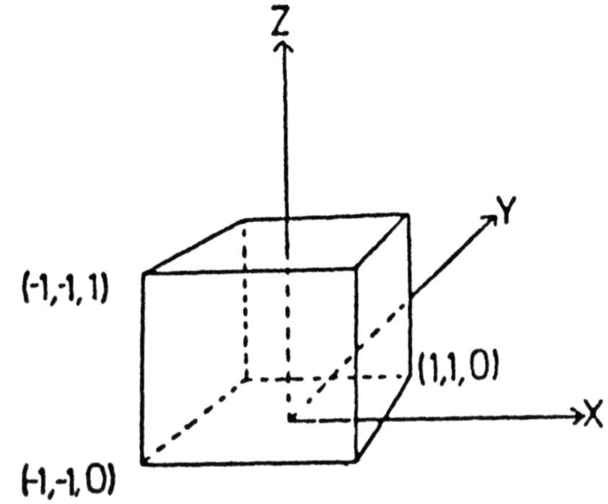

FIGURE 2.6. View Volume: projection reference point

is deeply rooted in projective geometry; there is no possible way to perform this within a usual Euclidean environment. The basic construction is:

A projection is defined by a so called *view volume*. This volume is either a parallelepiped for a parallel projection or a frustrum for a perspective projection (see Figs. 2.4 and 2.5). The edges of the view volume have a common intersection point which is the projection reference point (sometimes called the centre of the projection). In case of a perspective projection this point is an affine one; in case of a parallel projection this is an ideal point (that is, within the framework of projective geometry there is no difference between these two projection types).

In the case of the basic projection, the view volume is a simple cube; the projection reference point is the point $(0,0,1,0)$ (figure 2.6). The possible limits of the view volume may change according to the actual environment; in GKS-3D, for example, the corresponding "basic" view should have a volume within the unit cube, which is slightly different from our agreement. However, these differences are not important; such cubes may be transformed into each other by affine transformations (translations and scalings).

The idea of converting the general projection into the basic one may be reformulated mathematically into the task of finding a transformation which would turn the original view volume into the view volume of the basic projection. It also required that this transformation should keep the linear structure of the original model; the usual models which are drawn by the usual packages are all linear in some sense. Mathematically, this means that the required transformation should be a collinearity. Fortunately, the results of projective geometry give an acceptable solution to this problem. The theorem about the existence of a projective transformation (Theorem 2.8) shows that such collinearity exists.

Let us take in both view volumes four of the vertices so that these vertices would form a point set of general position. For example, three of them would be three vertices at the bottom of the volume, and one on the top ("bottom" and "top" are taken when looking at the volume from the projection reference point). Additionally, let us add to both of these point sets the respective view reference points. According to our theorem cited above, these two point sets define uniquely a projective transformation. Additionally, this transformation may be described by a 4×4 matrix (Theorem 2.10), which gives us a method of computation as well.

As a consequence, the usual three dimensional packages perform the viewing by determining the above matrix first (called the *view matrix* or *view transformation*), and then apply this matrix on the model to be viewed. The objects within the original view volume will be within the view volume of the basic projection; additional calculations, like HLHSR may be performed in this (greatly simplified) environment. In graphic systems like GKS-3D and PHIGS, the viewing is defined by giving the system the appropriate view matrix; theoretically, the determination of the matrix itself is left to the user. Of course, as we will see, the determination of such a matrix is not that easy, and consequently both systems offer a set of utility functions which have the task of calculating the view matrix from the "usual" parameters (view reference points, view plane, etc.).

2.4.3 Determination of the View Matrix

The theorem which asserts the existence of the view transformation (Theorem 2.8) is, of course, of a very high theoretical importance for 3D packages. However, the theorem itself does not say anything about how this matrix is to be generated. From a practical viewpoint, the *unicity* part of the theorem is may be even more important. As a result of unicity, we may try to find different approaches for the matrix generation, approaches which are preferable, in the actual environment because of some special reasons. Even having different methods in hand, we know in advance, that all these generation methods should lead to the *same* matrix (with, eventually, a non-zero multiplicative factor); that means, in very practical terms, that the output on the screen should be the same, independently of the actual approach we use for the determination of the view matrix.

Two methods are presented without the details. There are a number of textbooks and/or articles which present more details for the interested reader, for example [7], [8], [19], [20], [23], or [24]. The first method leads to a series of linear equations. This method has the advantage of being applicable to any kind of projective transformation in general, and not only for determination of view matrices.

According to our theorem, we may have two point sets which generate the transformation. Let us denote the points by **P1, P2, P3, P4, P5** and **P1′, P2′, P3′, P4′, P5′**. The corresponding coordinate values will be denoted by x_{ij} and x_{ij}' respectively (i = 1,...,4 j = 1,...,5). The existence of the matrix is equivalent to the following: There exist non zero $\mu_1, \mu_2, \ldots, \mu_5$ values so that:

$$\sum(c_{ki}.x_{ij}) = \mu_j.x'_{kj} \qquad (k = 1, \ldots, 4\; j = 1, \ldots, 5)$$

The c_{ik} values are the coefficients of the matrix to be generated. This set of equations represents a linear system for 21 unknown variables; the number of equations is 20. However, the exact values of $\mu_1, \ldots, \mu_5$ are not really interesting for us; the homogeneous system allows us to use any kind of multiplicative factor (or, in other words, only the respective *ratios* of the μ_j values are of interest). In other words, the value of, say, μ_1 may be chosen to be 1. As a result of this additional fact, the number of unknown variables is reduced to 20, which is the number of equations.

From the theorem about the existence of projective transformations, we know that this set of equations has a solution. The solution itself may be generated for example by an appropriate utility function solving linear equations (which is available in some environments). If such a utility is not at hand, closed algebraic formulae for the determination of the result may also be generated manually, using for example the classical Cramer rule, or anything similar. Of course, the resulting formulae may be quite complicated and difficult to manage; however, the possibility exists.

In case of a viewing transformation the equations may be simplified by the fact that the x'_{ij} values are relatively simple. If, for example, the basic volume is set to be the cube $(-1,1) \times (-1,1) \times (-1,1)$, the values may be:

$$\begin{aligned}
\mathbf{P1'} &= (1, \; 1, \; 0, \; 1) \\
\mathbf{P2'} &= (-1, \; 1, \; 0, \; 1) \\
\mathbf{P3'} &= (-1, \; -1, \; 0, \; 1) \\
\mathbf{P4'} &= (1, \; 1, \; 1, \; 1) \\
\mathbf{P5'} &= (0, \; 0, \; 1, \; 0)
\end{aligned}$$

which simplify the appropriate formulae (**P5′** is the ideal point belonging to the **Z** axis).

As we have seen, the method of linear equations is not very simple. In case of a viewing transformation another approach is also possible, which has, however, the disadvantage of being applicable exclusively for the determination of a view matrix.

The basic idea is to generate a series of simple transformations which would transform step-by-step the original view volume into the view volume of the basic projection. The aim is to generate a series of transformations, where the concatenation of these transformations results in the view transformation we want to generate. Let us remember the fact that the concatenation of the projective transformations is equivalent to the matrix multiplication of the corresponding matrices; in other words, by multiplying successively the matrices, we arrive to the matrix of the view transformation. At each iteration step the new matrix is generated by applying the matrices of the previous step(s) onto the points of the original view volume and projection reference point. It is very important to represent each elementary step by very simple matrices, which are easy to generate.

There may again be two approaches for the automatic generation of the view matrix. Either an internal matrix multiplication function (or hardware!) is used to multiply the matrices mechanically, or the view matrix may be calculated "on paper", and the resulting formulae may be coded directly into the program. The latter approach may result in faster code, but it is relatively easy to make some disturbing errors in the course of the calculation; the final formulae tend to be very complicated.

In the following we deal with perspective projections only. The analogous parallel case may be derived in a very similar way; in fact, the formulae will be much simpler (the view transformation is affine!).

The view volume of figure 2.5 may be translated, rotated and scaled very easily to arrive to the situation illustrated in figure 2.7. In this case, the base of the (transformed) view volume (the base of the frustrum) coincides with the base of the cube which forms the view volume of the basic projection (that is, the $(-1,1) \times (-1,1)$ square of the **XY** plane). The set of the necessary elementary transformations may be described with the matrices of the previous chapter.

The next transformation which is to be performed is to move the point P of figure 2.7 onto the Z axis, by keeping the basis of the frustrum where it is. This transformation should result in the situation of figure 2.8.

This transformation is a three dimensional shear:

$$\begin{pmatrix} 1 & 0 & -P_x/P_z & 0 \\ 0 & 1 & -P_y/P_z & 0 \\ 0 & 0 & 1 & 0 \\ 0 & 0 & 0 & 1 \end{pmatrix}$$

Let us remark, that up to this point only affine transformations have been used. The next step, which should turn the state of figure 2.8 into a parallel projection is the only non-affine transformation in the chain, and is representated as (see figure 2.8):

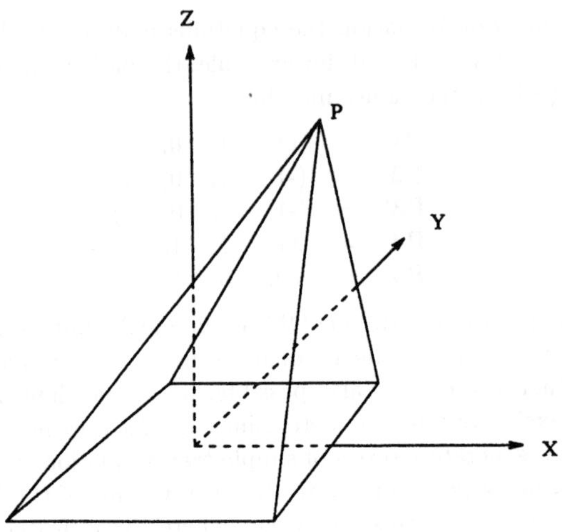

FIGURE 2.7. View volume after translating, rotating and scaling figure 2.4 or 2.5

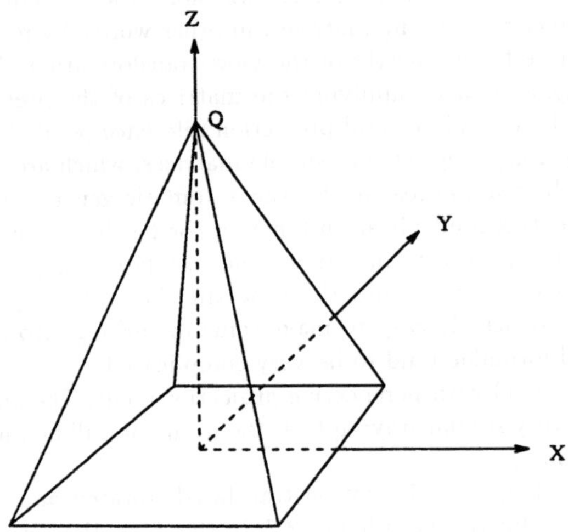

FIGURE 2.8. View volume after shearing figure 2.7

$$\begin{pmatrix} 1 & 0 & 0 & 0 \\ 0 & 1 & 0 & 0 \\ 0 & 0 & 1 & 0 \\ 0 & 0 & -1/Q_z & 1 \end{pmatrix}$$

It can easily be checked that this transformation will move the point **Q** into the ideal point of the **Z** axis (in fact, the image of the plane is parallel to the **XY** plane in the Euclidean sense and which contains the point **Q** will be the ideal plane). The basis of the frustrum remains unchanged. The result is almost the required basic projection; the only possible difference may be that the distance of the top of the cube measured from the plane **XY** is not necessarily 1. If this is the case, an appropriate scaling in the **Z** direction should also be applied to form the last element of the chain.

2.5 Additional Applications

2.5.1 "External" Lines

As we have seen in the previous chapters, projective geometry plays a very essential "intermediate" role in a computer graphics system. The objects which are to be processed are two or three dimensional Euclidean objects; these are the outcome of the application program on the top of the system. The "target" is again Euclidean: usually, a picture has to appear somehow on a two dimensional (very rarely three dimensional) output device, which is essentially Euclidean. However, the graphic system, which acts as a "bridge" between these two endpoints, has to move the Euclidean objects into projective ones, has to process in the projective environment and, finally, has to reconvert the projective data into Euclidean ones. Using projective geometry is somewhat analogous to the use of a subroutine within a program: when a subroutine is invoked, the whole program status (stack content, processor status, register values) are saved, and a new environment is created. Within the subroutine we are not really interested in the exact origin of the data we want to process. Similarly, when we "enter" projective geometry, we forget for a while about the Euclidean origin, to simplify our calculations. "Entering" the projective environment is made by embedding the cartesian coordinates into homogeneous ones; making a "return" means performing a projective division first and then forgetting about the last coordinate value (whose value is 1).

However, it is not that easy to leave the projective environment. The fact is, as we have seen, that only the affine points have their Euclidean counterpart; the ideal points have no meaning there. On the other hand, in the course of a non-affine projective transformation some of the affine points will be transformed into ideal ones; that is, this situation should be taken care of.

The appearance of this problem may be illustrated when trying to keep track of a line segment which is transformed by the system. A line segment is given by two (affine) points **P** and **Q**. It may happen, however, that the image of this line segment will contain an ideal point (this situation is shown in figure 2.9, using our presentation of section 2.3.5). This means that when trying to leave the projective environment, the whole line segment cannot be converted; one point, namely the ideal point, has to be forgotten somehow.

The practical geometrical result of this fact is somewhat surprising at first glance. As we can see in figure 2.9, the (Euclidean) image of the line segment **PQ** will *not* be the line segment $\mathbf{P}''\mathbf{Q}''$, but exactly its "complement", that is the set of two half-lines, one starting at $\mathbf{P}''$ and the other starting at $\mathbf{Q}''$! In fact, the missing point, that is, the ideal point, is just the object which would "link" these two half-lines; the problem is that this linkage has no meaning in the Euclidean sense. In other words, if a (three dimensional) graphic system is realized in a way that the endpoints of the incoming line segments are just processed through the transformation pipeline and, at the end, the line segment generated by the two (transformed) points is drawn, the result may be wrong.

Figure 2.10 shows that this problem is not only theoretical. As we have seen, in the course of the view transformation the image of the plane, which is parallel to the basis of the view volume and which contains the view reference point, will be the ideal plane. That is, all line segments which originally *intersect* this plane will contain an ideal point after the transformation, which means that all three dimensional graphics systems have to be prepared to handle this detail properly, otherwise the picture appearing on the viewing surface will become erroneous. The whole problem is made more complicated because not only (transformed) line segments but (transformed) polygons may also contain ideal points, and, furthermore, in this case the number of such points may be infinite.

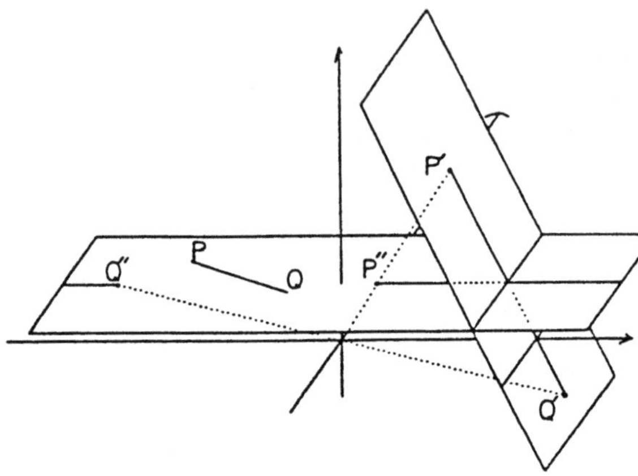

FIGURE 2.9. External lines

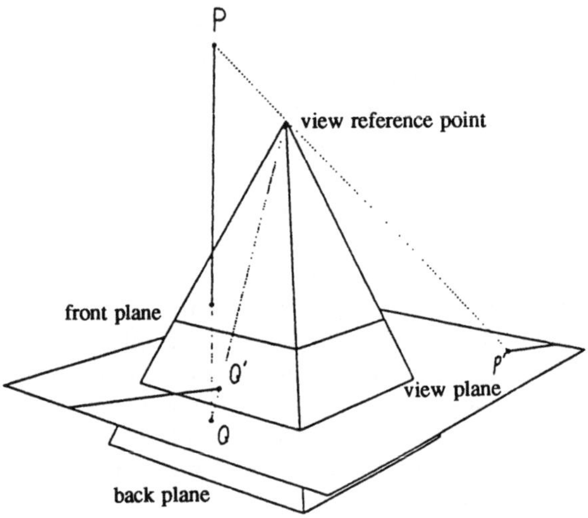

FIGURE 2.10. Viewing pyramid

These kinds of line segments are called "external" line segments in the computer graphics literature. There are several ways to overcome the problem (e.g., [3],[9]); we will present only one of them here, mainly because of the fact that by handling this (very technical) detail we will have some interesting and more general by-products in hand.

The idea is, at first, extremely simple. The origin of our problem is the fact that the line segment (or the polygon) has eventually one (or more) intersection point(s) with a plane, namely the ideal plane. The aim is to get rid of these intersection points, that is to cut the line segment or the polygon in a way that the resulting objects should be free of ideal points. If this has been done, the projective division may be performed without problems.

This way of stating the problem may remind us of a classical task of computer graphics systems, namely *clipping*. Clipping against a line or a plane, to determine which part of a given object is on one or the other side of the line/plane, has a number of nice solutions (see e.g., [8, 18, 19, 25]). The problem of external lines (and polygons) is therefore just a special case of clipping.

However, it is a bit more complicated than that. In fact, all classical algorithms have been developed for Euclidean environments only, and they cannot be adopted so easily to a projective environment; care should be taken when performing calculations with homogeneous coordinates. However, (and here comes the real "trick") we can try to *reenter* a Euclidean environment for the sake of our clipping.

We have seen in figure 2.3 that the detailed effect of a projective transformation in space may be described by considering our affine plane as being embedded in a *four dimensional Euclidean space*. In this case, the image of this sub-space is another sub-space still lying in the Euclidean environment; just its geometrical position is different.

In a computer graphics environment, we are interested in the image of *affine* points only. That is, we can regard our points as being embedded into a four dimensional space; the coordinates of the points after the *linear* part of the projective transformation will all be in an appropriate sub-space of the Euclidean space. In other words, we have applied a new "mathematical subroutine"; we have now temporarily left the projective environment to enter a Euclidean environment, with the significant difference that this new Euclidean environment is of a higher dimension than the original one. (This is the real art of mathematics: you always have to find a good mathematical structure to give an easy description of the problem at hand; sometimes you have to jump from one structure into the other and then back to give an elegant and clear solution.)

All this means that the clipping problem we have is similar to classical clipping, the difference being that it has to be performed in a higher dimension. This seems to be very frightening at first, but fortunately it is not. In fact, a line determined by two points **P** and **Q** of the four dimensional space can be described by the usual and well known equations, like for example:

$$t.\mathbf{P} + (l - t).Q$$

where t is an arbitrary real number. The notion of two vectors, say **v1** and **v2**, being "perpendicular" coincides with the equation:

$$\mathbf{v1.v2 = 0}$$

where "." means of course the scalar product. If a non-zero vector denoted by **n** is given, the equation

$$\mathbf{n.x = 0}$$

describes a three dimensional sub-space of the four dimensional space; the equation is similar to the description of a plane in space. Other possible equations may also be generalized without too many problems.

The intersection of a line with, e.g., the plane $W = 0$ leads to the usual calculations; the only difference is that the formulae should also be used for the fourth coordinate (in fact, the planes $W = \varepsilon$ and $W = -\varepsilon$ are they really interesting ones for our purpose, where ε is an appropriate small number). The polygon clipping algorithms are based on the fact that such an intersection point may be determined (e.g., [25]); in other words, generalizing the polygon clipping is again not particularly difficult.

A number of technical details should also be added to the exact application of this clipping (which we have called "W–clip"); these are not relevant here, and the interested reader should refer to [9]. The important point is that with the application of this W-clip, the problem of external lines can be fully solved.

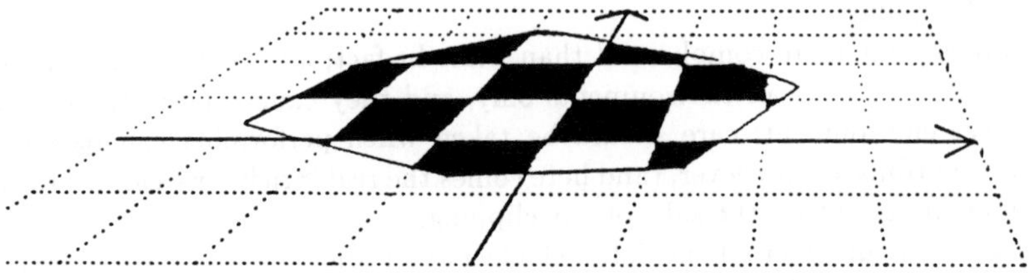

FIGURE 2.11. Patterns under projection

2.5.2 Use of the Four Dimensional Space

In the previous paragraph we had to realize that the use of four dimensional geometry is a natural outcome of projective geometry. However, we might have the feeling that this approach is just advantageous to solve the very particular problem described above, but otherwise it is without interest.

The interesting point is that this is not true, and this is what will be presented in this section. As we will see, the fact that four dimensional geometry has such a close relationship with the projective transformations may lead to faster and better algorithms for computer graphics.

We will elaborate in detail only one problem to clarify the background ideas. The problem which has to be solved is the so called pattern filling of three dimensional polygons. Pattern filling in packages like GKS-3D or PHIGS means that the user may give a color pattern, that is a small parallelogram in space which contains cells of different colours. This pattern may then be used to fill a polygon in space by repeating the pattern along the two sides of the pattern parallelogram in a linear order to achieve pictures like the one in figure 2.11.

Pattern filling is extremely time-consuming, as, eventually, a great number of sub-parallelograms have to be generated, transformed by the projective transformation and displayed. It should therefore be carefully considered where, that is, at which stage of the output pipeline, this generation is to be done.

Unfortunately, the presence of the projective transformation seems to determine the place of this generation: it should be done prior to the projective transformation, that is, in the Euclidean environment of the "top" of the pipeline. The usual argument is that the projective transformation destroys the strict linearity of the generated patterns, and this pecularity of the projection is the one which creates the real three dimensional visual effect of figure 2.11.

Fortunately enough, this argument is not absolutely correct. As we have seen in the previous chapter, the projective transformation may be considered as a transformation moving the original affine space into another affine sub-space of the four dimensional Euclidean space. *This* transformation is linear, that is, if $\mathbf{T}$ is the transformation itself, $\mathbf{v1}$ and $\mathbf{v2}$ are two points of the affine space (that is the $W = 1$ sub-space) and, finally, α and β are arbitrary numbers, the following equation is true:

$$\mathbf{T}(\alpha\mathbf{v1} + \beta.\mathbf{v2}) = \alpha\mathbf{T}(\mathbf{v1}) + \beta.\mathbf{T}(\mathbf{v2})$$

This means that the linear structure of pattern filling remains unchanged. As a consequence of this fact, however, it is absolutely equivalent (mathematically) to perform the pattern filling (that is the generation of the appropriate sub-polygons) prior to or after the transformation *as long as we remain in the four dimensional space.* The pattern generation method itself is linear again: vectors should be expanded linearly, parallelograms should be generated out of vectors, etc. All these calculations are absolutely "portable" among Euclidean spaces of different dimensions, that is, they can be performed in the four dimensional space as well. In other words, instead of generating all sub-polygons prior to the transformation, we may just transform the original polygon and the pattern description; the sub-polygons may be generated afterwards.

The situation is similar to the W-clip. As a start, we have a three dimensional Euclidean environment. For known reasons, we have to move to a projective environment. However, to perform pattern filling, we move into the four dimensional Euclidean space, more exactly into the three dimensional sub–space of the four dimensional space. Once the sub-parallelograms are generated, we return to the projective environment (in fact, this is just a "mental" return, with nothing particular to do) and then we return into the three dimensional Euclidean space (via the projective division).

Why is this tortuous route so advantageous? It is just faster. If the pattern sub-polygons are generated in the four dimensional space, a much smaller number of points are to be transformed, that is, the number of necessary matrix-vector multiplications is greatly reduced. This may result in a significant increase in speed (approximately 15–20%), without making the calculations much more complicated.

The details of pattern filling are without interest here. The important point is the idea behind it: it may be advantageous to perform some of the graphic algorithms in the four dimensional space, to reduce the number of matrix- vector multiplications. In fact, all graphical algorithms which are inherently linear may be done that way. Just to list some of these:

- high precision character generation (STROKE precision in GKS–3D or PHIGS)
- ellipse and elliptical arc generation (described later, see also [16])
- hatching of three dimensional polygons
- a number of general curve and/or surface approximation methods[7].

2.6 Quadratic Curves

2.6.1 Introduction

Besides line segments and polygons, quadratic curves or conics are also frequently used objects in computer graphics. Some of these curves tend to appear as basic output primitives in the functional specification of new graphical standards as well (e.g., CGI, that is [16]). Among the three main classes of quadratics, namely ellipses, parabolae and hyperbolae, the use of ellipses (first of all circles) is the most widespread. Circles and circular arcs are used in business graphics for charts, in mechanical engineering for rounding corners, for holes, etc. Circular arcs may also be used to interpolate curves (see e.g., [22]).

The role of parabolae and hyperbolae is not so important. They may of course occur in some engineering environments, and there are also proposals to use parabolic arcs for curve interpolation (the so called double-quadratic curves, see e.g., [26]) but these are very special cases. However, as we will see, these curves may appear as a result of some projective transformations; that is, we should not forget about their existence either.

One of the main problems in handling these curves is the task of finding a good and compact representation within a graphics system. As a result of the hardware limitations, the curves themselves are usually drawn by approximating them with a set of line segments. However, to achieve an acceptable quality of output, the number of line segments should be relatively high (experience has shown that, for instance in case of a full circle, 60 line segments should be used at least, otherwise the quality will not be acceptable). Consequently, the exact point within the output pipeline where this approximation is effectively performed should be chosen carefully. If the curves are approximated "on the top" of the pipeline, the result will of course be good, but a possibly large number of points will be transformed and, eventually, stored in segments.

Let us have an example. A circle is usually determined by giving its centre and its radius. This is a short and compact form, but it has a severe drawback. If a general affine transformation is applied, the circle may be distorted into an ellipse. However, the information we have in hand (namely the centre and radius) is not enough any more to draw this ellipse. This means that this way of determining a circle is not powerful enough to be really usable with affine transformations.

Projective geometry gives a compact way of describing quadratics and may help us to explain some of their pecularities. Unfortunately, as we will see, it does not give exhaustive tools for the internal representation of such curves. However, some useful approaches may be derived.

2.6.2 Definition and Basic Properties of Quadratics

In the following, we will speak exclusively of quadratic curves in a projective plane. The resulting formulae and properties may be generalized easily for any quadratic curve lying in a sub-plane of a projective space.

In projective geometry, a quadratic curve is defined as follows.

Definition 2.17 *Let us denote the homogeneous coordinates of the (projective) point x by (x_1, x_2, x_3). A quadratic curve is the set of all points, for which the following equation holds:*

$$a_{11}x_1^2 + 2a_{12}x_1x_2 + a_{22}x_2^2 + 2a_{13}x_1x_3 + 2a_{23}x_2x_3 + a_{33}x_3^2 = 0$$

If we accept the equality $\mathbf{a_{ik}} = \mathbf{a_{ki}}(i, k = 1, 2, 3)$, the equation may be written in the following form:

$$\sum\sum a_{ik}x_ix_k = 0$$

Furthermore, if we define the (symmetric) matrix $\mathbf{A}$ as follows:

$$\begin{pmatrix} a_{11} & a_{12} & a_{12} \\ a_{21} & a_{22} & a_{23} \\ a_{31} & a_{32} & a_{33} \end{pmatrix}$$

The equation is equivalent to $\mathbf{x.(Ax)}$, where "." denotes the scalar product of vectors. For the sake of simplicity, the brackets will be omitted and the notation $\mathbf{xAx}$ will be used.

In the following, the matrix $\mathbf{A}$ will be considered to be non-singular, that is, $\mathbf{det(A)} = \mathbf{0}$ will not be allowed.

A number of geometrical properties may be derived for such curves; unfortunately, we cannot go into all details here. The following theorem is, however, of great importance for us as well.

Theorem 2.12 *If $\mathbf{K}$ is a quadratic curve on the projective plane and $\mathbf{l}$ is a line, the number of intersection points of $\mathbf{K}$ and $\mathbf{l}$ may be 0, 1 or 2.*

This theorem is well-known for the usual Euclidean environment. As a result of it, we have also information about the possible number of *ideal* points a quadratic curve may contain: the ideal line is just a normal line in projective geometry, consequently, according to Theorem 2.12, the number is 0, 1 or 2.

The definition of quadratics is a clear generalization of the well known Euclidean case. We will describe the exact relationship between Euclidean and projective curves in what follows.

Each projective quadratic curve has a number of affine points (in fact, as we have just seen, almost all of them are affine). Consequently, a projective quadratic curve determines automatically a curve on the affine plane as well. The following theorem is valid:

Theorem 2.13 *Let* $\mathbf{K}$ *be (projective) quadratic curve,* $\mathbf{K}'$ *be the set of affine points of* $\mathbf{K}$*, i.e., the affine curve determined by* $\mathbf{K}$*, and* n *be the number of ideal points of* $\mathbf{K}$*.*
If n *is 0,* $\mathbf{K}' = \mathbf{K}$*, and* $\mathbf{K}'$ *is an* ellipse.
If n *is 1,* $\mathbf{K}'$ *is a* parabola. *The ideal point of* $\mathbf{K}$ *corresponds to the axis of* $\mathbf{K}'$*.*
If n *is 2,* $\mathbf{K}'$ *is a hyperbola. The ideal points of* $\mathbf{K}$ *are the ones corresponding to the asymptotes of* $\mathbf{K}'$*.*

Consequently, by deriving geometrical properties for projective quadratics, we can automatically deduce a number of common properties for ellipses, parabolae and hyperbolae. This fact is of overall importance for the description of the behavior of these curves. Of course, some of these properties have been known for a very long time; in fact (affine) quadratic curves had already been examined extensively before the birth of modern projective geometry (the names “ellipses”, “parabolae” and “hyperbolae” are, originally, ancient Greek words). However, projective geometry gives an elegant tool to describe and prove these properties very easily, in contrast to previous proofs, which were sometimes very awkward and difficult to follow.

The number of ideal points of a quadratic may also be derived out of its matrix. This fact may have a very important practical consequence, if a graphic system has to deal with such curves. This characterization is as follows.

Theorem 2.14 *Let us consider a quadratic curve, whose matrix is* $\mathbf{A}$*. Let* f *be as follows:*

$$f = \mathbf{a}_{11}\mathbf{a}_{22} - \mathbf{a}_{12}\mathbf{a}_{21}$$

(that is, f *is the determinant of the upper-left* 2×2 *sub-matrix of* $\mathbf{A}$*). Then: if* $f > 0$*, the number of ideal points on the curve is 0, the curve is an ellipse;*
if $f = 0$*, the number of ideal points on the curve is 1, the curve is a parabola;*
if $f < 0$*, the number of ideal points on the curve is 2, the curve is a hyperbola.*

The last general theorem, which is of a great importance for graphic systems as well, describes the relationship of quadratic curves and projective transformations. This theorem is as follows.

Theorem 2.15 *The class of quadratic curves is invariant for collinearities (projective transformations). In other words, the image of a quadratic curve under the effect of a projective transformation will always be a quadratic curve.*

It is not particularly difficult to determine the matrix of the transformed curve. If the original matrix is $\mathbf{A}$, and the matrix of the transformation is $\mathbf{T}$, the new matrix will be $\mathbf{T}'\mathbf{A}\mathbf{T}$, where $\mathbf{T}'$ denotes the transpose of the matrix $\mathbf{T}$ (that is, the element of the matrix are mirrored against the main diagonal).

Let us now see what practical consequences we may derive from these theorems.

According to Theorem 2.14, when applying an affine or a general projective transformation, the image of a quadratic curve will be a quadratic curve. For affine transformations, the situation is even simpler: the affine transformation does not exchange ideal and affine points, and, consequently, the number of ideal points on the image of the quadratic curve will be unchanged. In other words, the image of an ellipse or a circle will be an ellipse, the image of parabola will be a parabola, and, finally, the image of a hyperbola will be a hyperbola. It may be very helpful for an implementor to be sure of these facts in advance (e.g., segment/insert transformations are affine in GKS or CGI!).

Unfortunately, in the case of projective transformations, the situation is much more complicated. The image of, say, an ellipse may be any kind of quadratic curve; if one or two of the points on the original curve is transformed into ideal point(s), the image may be a parabola or a hyperbola. However, if some kind of additional test may be performed regarding the effect of the transformation (for example by using Theorems 2.14 and 2.15), or the system is able to avoid some dangerous situation (e.g., if the ellipse is relatively "far" from the view reference point, none of its points will be transformed into ideal ones) the situation may still be kept under control. Anyway, knowing the mathematical background of these effects may again be very helpful for an implementor.

2.6.3 Conjugate Points and Lines

In this paragraph, we consider a quadratic curve to be defined once for all; as we have already said, its matrix is denoted by **A**. The matrix **A** (or, in other words, the quadratic curve) induces a relationship among the points of the projective plane. This relationship is called *conjugation.* The exact definition is as follows.

Definition 2.18 *Two points, denoted by* **x** *and* **y**, *are said to be* conjugate points *if the following equation holds:*

$$\mathbf{xAy} = 0$$

In other words, we generalize the basic equation for quadratic curves. With this definition in hand we could also say that the points of the curves are those which are *auto-conjugate.*

The definition of conjugation induces an additional relationship among points and lines. In fact, the following theorem is true:

Theorem 2.16 *If* **x** *is a fixed point, the set of all* **y** *points which are conjugate to* **x** *form a line on the (projective) plane. Furthermore, for each line of the plane there exists one and only one point which generates this line in this manner.*

This means that each point generates a line; this line is called the *polar* of this point; additionally, each line generates a point, which is called the *pole* of the line. Generally, the pole of a line does not belong to the line itself. The only case for that situation is when the line is *tangential* to the curve; that means that the number of intersection points of the curve and the line is 1. In this case, and only in this case, the pole of the line is on the line itself; namely, it is the intersection point of the line and the curve.

Finally, to close the set of relationships, another can also be defined for lines:

Definition 2.19 *Two lines, denoted by* **l1** *and* **l2**, *are said to form a* conjugate pair of lines, *if the pole of* **l1** *is on* **l2**, *and, conversely, the pole of* **l2** *is on* **l1**.

All these definitions are, unfortunately, quite abstract. It is not easy to give a more "visual" interpretation of them. Figure 2.13 however, shows, some properties which are valid for these relations, and which may help to give an intuitive feeling about them. In figure 2.12, a line is determined by the two points of the curve, namely **C** and **D**. This line is a chord. Two tangents are generated, one at **C** and one at **D**. The two tangents intersect at the point **M**.

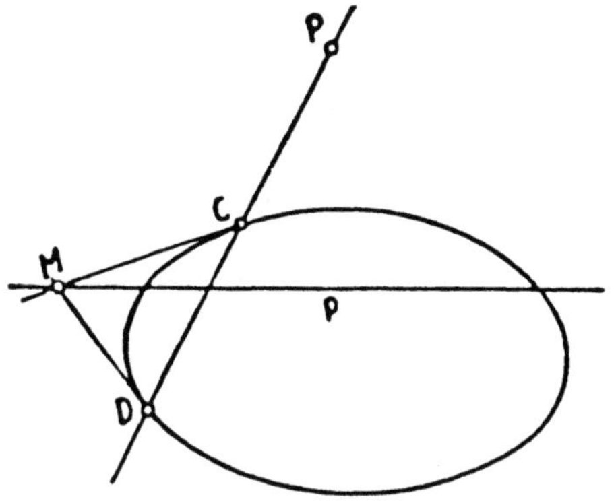

FIGURE 2.12. Conjugate chords of an ellipse

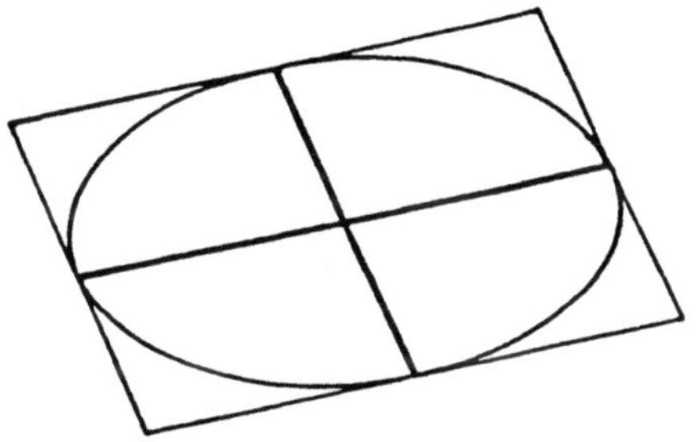

FIGURE 2.13. Conjugate diameters of an ellipse

It can be proved that **M** is the *pole* of the line **DC**. Conversely, if a point **M** is given "outside" the curve (beware: this notion is not absolutely clear for all curves!), by determining two tangents crossing the point **M**, the generated chord (to be more exact, the corresponding line) is the *polar* of **M**. Furthermore, it may also be proved that all lines crossing **M**, like for example the line **p** in figure 2.12, will be *conjugate* to line **DC**. That is for each chord there is a whole set of possible conjugate lines. One additional definition is required, namely,

Definition 2.20 *The pole of the ideal line is called the* centre *of the curve.*

This definition may be surprising, as it defines a well-known notion; it can be proved, of course, that in case of ellipses and hyperbolae, this point is an affine one (this can be deduced easily from Theorems 2.12 and 2.13), and it coincides with the "traditional" notion of centre. In case of parabolae, this point is the (only) ideal point of the curve (the pole of a tangent is the intersection point). We may also use the term *diameter*, denoting all lines which cross this centre.

We have seen in figure 2.12 that all chords (in fact, all lines in general) generate a whole set of conjugate lines. If the original line is a diameter, there exists also one and only *conjugate diameter*, that is, a diameter which forms a conjugate pair with the original one. The only thing we have to do is to connect the point **M** in figure 2.12 to the centre of the curve.

Let us see what a conjugate diameter pair means geometrically in case of an ellipse. If we follow the construction of figure 2.12, as a first step, we have to determine the two tangents. On the affine plane, these two tangents will be parallel (the ellipse is symmetric against its centre). In other words, the conjugate diameter pair will be the one which is parallel to these tangents. Figure 2.13 illustrates this situation; in fact, the conjugate diameter pair will determine a parallelogram, which contains the ellipse, with its edges tangents to the ellipse. In case of a circle, for example, the conjugate diameter pairs are the perpendicular diameters. The conjugate diameters for ellipses have great importance in computer graphics. In fact, the following theorem is true.

Theorem 2.17 *If* $\mathbf{C}$ *is the centre of the ellipse,* $\mathbf{CP}$ *and* $\mathbf{CQ}$ *are two vectors pointing from the centre to the endpoints of a conjugate diameter pair (*conjugate radii*), the points of the ellipse may be described with the following equation:*

$$\mathbf{x}(t) = \mathbf{C} + \mathbf{CP}.\cos(t) + \mathbf{CQ}.\sin(t) \qquad (\mathbf{0} \le t \le \mathbf{2}\pi)$$

The approximation of an ellipse with line segments may be performed by using this equation with appropriate t values. This way of defining an ellipse has been adopted for example by the ISO CGI functional description [16] and, eventually, it may also appear in later revisions of GKS as well. It is therefore important to know exactly what this definition really means.

A similar equation may also be derived for hyperbolae; the main difference is that instead of trigonometric functions so called hyperbolic functions are used. The equation is of the following form:

$$\mathbf{x}(t) = \mathbf{C} \pm \mathbf{CP}.\mathbf{ch}(t) + \mathbf{CQ}.\mathbf{sh}(t) \qquad (-\infty \le t \le \infty)$$

The exact meaning of $\mathbf{CP}$ and $\mathbf{CQ}$ should of course be defined; however, this equation (and hyperbolae in general) is not really important in the field of computer graphics, and consequently, we do not enter into the details here.

Why is this equation so important? This becomes understandable in relation with projective transformation. In fact, the notion of a conjugate point is invariant under the effect of a projective transformation. That is, if the points $\mathbf{x}$ and $\mathbf{y}$ are conjugate points in respect to a given curve $\mathbf{k}$, then $\mathbf{T}(\mathbf{x})$ and $\mathbf{T}(\mathbf{y})$ are also conjugate points in respect to the curve $\mathbf{T}(\mathbf{k})$. That means that the relationship of poles, polars and conjugate pairs of lines are all invariant. A pair of conjugate chords will become a pair of conjugate chords.

What about conjugate diameters? One has to be very careful here. The question is: what happens with the centre? Of course, the image of conjugate diameter pair is a conjugate chord pair, and the image of the centre will be the intersection point of these chords. Consequently, the question is whether the image of the centre remains a centre or not.

In the case of an affine transformation, the image of the ideal line is still the ideal line. Consequently, the image of the pole of the ideal line will be the pole of the ideal line; that is, the image of the centre is the centre. In other words, the image of the conjugate diameter pair is a conjugate diameter pair! This is very important: this means that the equation above is affine invariant. In practice, it is enough to apply the affine transformation on the centre and the conjugate radii endpoints, and to apply the equation afterwards, it is not necessary to perform an approximation before the transformation. In other words, the determination of ellipses with conjugate radii seems to be a good answer to the problem we raised in section 2.6.1.

While the situation is very advantageous for affine transformations, non- affine transformations destroy the conjugate diameter pair. If the transformation is non-affine, the

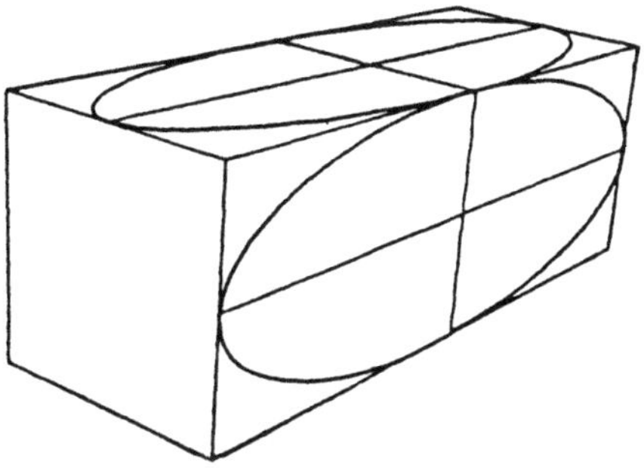

FIGURE 2.14. Transformation of conjugate diameters

image of the ideal line will be affine; the image of the pole of the ideal line will be the pole of an affine line; that is the image of the centre is not the centre.

The situation is well illustrated in figure 2.14. The ellipses, when seen as ellipses in space, are drawn around parallelograms in space, that is, the chords are conjugate diameter pairs. On the other hand, if we regard these figures as projected figures, as *planar* figures, the chords are clearly *not* diameters (although the curves are still ellipses).

In other words, care should be taken when applying projective transformations against ellipses (and quadratic curves in general). Hopefully, by gaining a deeper understanding of the mathematical backgrounds of the possible problems, new and more powerful approaches may also be found (as an example, the method described in section 2.5.2 may be applied for the generation of ellipses as well, based on the above equation). It is, however, very important for this purpose to have a real and precise understanding of the mathematical background of graphical systems; if the present tutorial may help at least some of the readers to make this more understandable, the work which was necessary to write it has not be superfluous.

Acknowledgements:

I would like to thank all my colleagues and friends in Insotec Consult GmbH, Germany, who have helped me to form the final version of this tutorial. I am also grateful to my professor of geometry, Dr Mátyás Bognár, who, in the years 1974–75, introduced me to geometry in the University of Budapest. In fact, my own personal notes of his courses were the most important reference when writing this tutorial.

2.7 References

[1] R D Bergeron. Introduction to Computer Graphics (Part III). In P J W ten Hagen, editor, *Eurographics Tutorial '83*, Eurographic Seminar Series. Springer-Verlag, 1983.

[2] H Bez. Homogeneous Coordinates for Computer Graphics. *Computer Aided Design*, 15, 1983.

[3] J F Blinn and M E Newell. Clipping Using Homogeneous Coordinates. In *ACM SIGGRAPH Proceedings*, pages 245–251, 1978.

[4] IMAGEN Corporation. *DDL Tutorial.* IMAGEN Corporation, 1986.

[5] A Dürer. *Underweysung der Messung mit dem Zirckel und richtscheyt in Linien Ebnen und Ganzen Corporen.* Nürnberg, 1525. Facsimile reprint Josef Stocker/Schmind, 1966.

[6] Euklid. *Die Elemente, Buch I-XIII.* Akademische Verlagsgesellschaaft, Leipzig, 1975.

[7] I D Faux and M J Pratt. *Computational Geometry for Design and Manufacture.* Ellis Horwood, 1979.

[8] J D Foley and A Van Dam. *Fundamentals of Interactive Computer Graphics.* Addison-Wesley, Reading, Massachusetts, USA, 1982.

[9] I Herman and J Reviczky. A Means to Improve the GKS-3D/PHIGS Output Pipeline Implementation. In *Proceesings of Eurographics '87.* North-Holland, 1987. Also in Computers and Graphics **12** 1988.

[10] D Hofstadter. *Gödel, Escher, Bach: An Eternal Golden Braid.* Penguin Books, 1981.

[11] R J Hubbold. Introduction to Computer Graphics (Part I). In ten Hagen [1].

[12] Adobe Systems Inc. *Postscript Language Tutorial and Cookbook.* Addison-Wesley, 1985.

[13] International Standards Organisation (ISO). *ISO: Information Processing Systems — Computer Graphics — Graphical Kernel System (GKS) — functional description*, 1985. IS 7942.

[14] International Standards Organisation (ISO). *ISO: Information Processing Systems — Computer Graphics — Graphical Kernel System for Three Dimensions (GKS-3D) — functional description*, 1986. IS DIS 8805.

[15] International Standards Organisation (ISO). *ISO: Information Processing Systems — Computer Graphics — Programmers Hierarchical Interactive Graphics System — functional description*, 1987. IS DIS 9592/1.

[16] International Standards Organisation (ISO). *ISO: Information Processing Systems — Computer Graphics — Interfacing techniques for dialogues with graphical devices functional description*, 1987. IS DP 9636/1-6.

[17] N Magnenat-Thalmann and D Thalmann. Introduction à l'informatique Graphique. In Enderle et al. [18].

[18] S P Mudur. Mathematical Elements for Computer Graphics. In G Enderle, M Grave, and F Lillehagen, editors, *Advances in Computer Graphics I*, Eurographic Seminar Series. Springer-Verlag, 1986.

[19] W M Newman and R F Sproull. *Principles of Interactive Computer Graphics.* McGraw–Hill, London, second edition, 1979.

[20] M A Penna and R R Patterson. *Projective Geometry and Its Applications to Computer Graphics.* Prentice-Hall, 1986.

[21] R F Reisenfeld. Homogeneous Coordinates and Projective Planes in Computer Graphics. *IEEE Computer Graphics and Applications*, 1, 1981.

[22] M Sabin. The Use of Piecewise Forms for the Numerical Representation of Shape. Dissertation 60/1977, Computer and Automation Institute of the Hungarian Academy of Sciences, 1977.

[23] R Salmon and M Slater. *Computer Graphics—Systems and concepts.* Addison-Wesley, 1987.

[24] K M Singleton. An implementation of the GKS-3D/PHIGS Viewing Pipeline. In A A G Requicha, editor, *Proceedings Eurographics '86*, pages 325–355, Amsterdam, 1986. Eurographics, North-Holland. *Winner of award for best submitted paper at the Eurographics 1986 Conference.*

[25] I E Sutherland and G W Hodgman. Reentrant Polygon Clipping. *Communications of the ACM*, 17:32–42, 1974.

[26] T Várady. Basic Equations and Simple Geometric Properties of Double-Quadratic Curves and Surfaces. CAD Group Document 117, Cambridge University Engineering Department, 1984.

3 GKS-3D and PHIGS — Theory and Practice

Roger Hubbold and Terry Hewitt

ABSTRACT
Since this tutorial was presented both GKS–3D and PHIGS have become international standards. PHIGS PLUS extends PHIGS to include lighting and shading capabilities. This chapter examines these systems and looks at some simple applications. In practice, PHIGS can be quite difficult to use – a fact which is not apparent until one actually tries to do something with it. For many applications, GKS–3D would be a better choice, unless lighting and shading are necessary. It seems, however, that many manufacturers are ignoring GKS–3D and are only supporting PHIGS for 3D applications. The chapter concludes with advice and information about implementations.

3.1 Introduction

Two standards for three-dimensional interactive graphics are nearing completion:

- GKS-3D — an extension of the Graphical Kernel System, GKS
- PHIGS — the Programmer's Hierarchical Interactive Graphics System.

Both systems contain functions for defining and viewing three-dimensional primitives, and for controlling input devices. In GKS-3D, facilities for structuring graphical data are quite limited but can be implemented fairly easily. For many applications GKS-3D is relatively straightforward to use, but its limitations become apparent when more complex displays are required — a robot arm, for example.

In contrast, PHIGS permits graphical data to be hierarchically structured. It is targeted at high-performance displays and workstations and caters for dynamic picture editing and display. This, however, makes it more difficult to use.

Many graphics system suppliers have indicated their intention to support these emerging standards and some preliminary implementations are already available. As yet, however, there is little experience of using them for applications development.

This tutorial looks at the facilities offered by these systems and examines how they might be used for some typical applications. PHIGS implementations can be very complex and correspondingly expensive! It is necessary to know what to look for when choosing a system and how to compare different implementations; this is also considered.

The tutorial is divided into two main parts:

1. Review of facilities offered by GKS-3D and PHIGS:

    - Standards for computer graphics
    - Primitives and attributes for defining three-dimensional pictures
    - Picture structuring capabilities of GKS-3D and PHIGS
    - Controlling views and projections
    - Input
    - Editing pictures
    - PHIGS PLUS.

2. GKS-3D and PHIGS in practice:

   - Case studies of application programs to compare and contrast GKS-3D and PHIGS
   - Reference model and relationship with other systems
   - Survey of GKS-3D and PHIGS hardware and software systems
   - Assessing GKS-3D and PHIGS implementations.

We have not included information on current implementations in these notes because this information dates rapidly. However, it will be presented during the tutorial.

3.2 Standards for Computer Graphics

3.2.1 GKS

GKS (the Graphical Kernel System) [12, 8] became the world's first international standard for computer graphics in 1985. Over a decade of effort went into its development. GKS is a standard for two-dimensional graphics — it does not attempt to address 3D graphics, although application programs could choose to perform their own 3D picture generation and utilise GKS to output 2D projections.

GKS is becoming well-known, a reasonable number of implementations are available, and there is some experience of writing application programs with this system.

3.2.2 GKS-3D

GKS-3D — the Graphical Kernel System for Three Dimensions [13] — is an extension of GKS, designed to provide similar functions but for three dimensions. It has been developed by the ISO committees and became an International Standard in 1989. [1]

A number of implementations exist, but as yet there is little experience of writing application software with GKS-3D. Several manufacturers have chosen to by-pass GKS-3D and go straight to PHIGS for 3D graphics.

3.2.3 PHIGS

PHIGS is the Programmer's Hierarchical Interactive Graphics System [16, 4]. PHIGS has been adopted in the USA as an ANSI standard and became an ISO standard during 1989. Many manufacturers and software suppliers have committed to supporting PHIGS, and several prototype implementations exist. Experience of using these early implementations is also increasing [19].

3.2.4 PHIGS PLUS

Although GKS-3D and PHIGS do define a mechanism for removal of hidden surfaces, they do not provide any facilities for generating complex shaded images. Having recognised this, a group in the USA set about defining a set of extensions to PHIGS called *PHIGS PLUS*. This now the subject of a standards proposal, but is expected to be widely adopted by manufacturers. PHIGS PLUS defines additional primitives for modelling curved surfaces, and for simulating lighting conditions in order to generate shaded images.

---

[1] Some places in these notes refer to GKS. This can be regarded as synonymous with GKS-3D.

3.2.5 Other Standards

The *Computer Graphics Metafile (GGM)* is a definition of a file structure for medium and long term storage of pictures, and for transporting them between different computers and devices. The CGM for 2D graphics was the second proposal to receive full International Standard status. The standard does not support all features of GKS — segments are not included, for example. Work is currently in hand to produce future standards to support GKS fully, and subsequently GKS-3D.

The *Computer Graphics Interface (CGI)* is an attempt to specify a standard interface between a device independent graphics system, such as GKS, and a particular device. This is clearly a difficult task, given the huge variations between different devices, and a number of seasoned observers believe the CGI specification is in something of a mess. It currently has the status of a Draft Proposal within ISO. For graphical output, the *PostScript* language [11] for raster devices seems destined to become a de-facto standard before CGI becomes accepted. Originally developed as a page description language for laser printers and other high-resolution devices, PostScript is now being used for other purposes (see below).

3.2.6 Window Managers

The standardisation process is painfully slow. Not surprisingly, a number of other developments have overtaken it. This raises the question of how these other things relate to GKS and PHIGS. One area not addressed by the standards, but of fundamental importance in a market increasingly populated by high-performance workstations, is that of window management. None of the standards mentioned has any facility for window management, which is leading to questions about how GKS-3D and PHIGS can be run in a workstation/window environment.

The *X11 Window System* [17], developed at MIT with support from IBM and DEC, has recently been adopted by a large number of manufacturers. It provides fairly minimal support for graphics, but is designed to run on low-cost displays.

NeWS is the Network extensible Window System, designed by Sun Microsystems [1]. This is based on an extended version of the PostScript language. It is expected that other companies (e.g. AT & T) will adopt NeWS. The use of PostScript gives NeWS greater descriptive power for graphics than X11, and also makes it programmable.

Neither of these window systems is based on the kind of reference model used by GKS-3D or PHIGS, nor do they support 3D graphics. However, DEC have been developing a set of 3D extensions to X11, called *PEX* which provides PHIGS functionality [14].

3.3 Primitives and Attributes for 3D

Both GKS-3D and PHIGS have primitives which are very similar to those provided by GKS, except that they are extended to three dimensions. The following tables list all primitives and attributes. Those which are new in GKS-3D and PHIGS are marked with an asterisk and are explained in the subsequent notes. The majority of the above are very similar to their GKS counterparts. Co-ordinate specifications are given in 3D by the addition of a z value. Thus, a polyline connects a sequence of points in three dimensions. Similarly, polymarker positions, text positions and vertices for filled areas are given with (x, y, z) values. GKS is a subset of GKS-3D in which the z value is zero.

The *fill area set* primitive allows a number of logically related filled areas to be specified as a single primitive. This permits areas containing holes to be correctly rendered. It is

Polyline	Polyline index Linetype Linewidth scale factor Polyline colour index View index (*) HLHSR identifier (*) Pick identifier Name set (*) *(PHIGS only)*	 Linetype ASF Linewidth scale factor ASF Polyline colour index ASF
Polymarker	Polymarker index Marker type Marker size scale factor Polymarker colour index View index (*) HLHSR identifier (*) Pick identifier Name set (*) *(PHIGS only)*	 Marker type ASF Marker size factor ASF Polymarker colour index ASF

TABLE 3.1. Attributes of polyline and polymarker

Text	Text index Text font Text precision Character expansion factor Character spacing Text colour index Character height Character up vector Text path Text alignment View index (*) HLHSR identifier (*) Pick identifier Name set (*) *(PHIGS only)*	 Text font and precision ASF Character expansion factor ASF Character spacing ASF Text colour index ASF
Annotation text (*) *(PHIGS only)*	*The above attributes, plus:* A. text character height A. text character up vector A. text path A. text alignment	

TABLE 3.2. Attributes of text and annotation text

Fill area	Interior index	
	Interior style	Interior style ASF
	Interior style index	Interior style index ASF
	Interior colour index	Interior colour index ASF
	Pattern size	
	Pattern ref. point	
	Pattern ref. parallelogram (*)	
	View index (*)	
	HLHSR identifier (*)	
	Pick identifier	
	Name set (*) *(PHIGS only)*	
Fill area set (*)	*Same as* fill area, *plus:*	
	Edge index	
	Edge flag	Edge flag ASF
	Edgetype	Edgetype ASF
	Edgewidth scale factor	Edgewidth scale factor ASF
	Edge colour index	Edge colour index ASF

TABLE 3.3. Attributes of fill area nad fill area set

Cell array	View index (*) HLHSR identifier (*) Pick identifier Name set (*) *(PHIGS only)*
Generalised drawing primitive (GDP)	*Any of the preceding attributes, plus* View index (*) HLHSR identifier (*) Pick identifier Name set (*) *(PHIGS only)*

TABLE 3.4. Attributes of cell array and GDP

also possible to specify various styles of edge for this primitive; edges of fill area sets may be drawn or not, in a different colour from the interior, and in different widths.

The *text, fill area, fill area set* and *cell array* primitives are all planar, and may be viewed from different angles. When seen from behind these primitives will appear back to front (assuming that they are visible at all). The *HLHSR identifier* attribute permits *hidden line/hidden surface removal* to be requested. The attribute can be set to indicate which implementation dependent algorithm is to be employed for this.

Text, already a complex primitive in GKS, has some new parameters in order to allow its orientation in space to be specified. As well as the text position, two text direction vectors are given, which together define the plane in which the text is to be written. The first vector forms a "baseline" and is used to create a third vector, in the text plane, which is at 90° (anti-clockwise) to the first one. The *character up vector* is measured relative to this third vector.

PHIGS has introduced the notion of *annotation text.* Unlike normal text, this does not get transformed by the process of viewing a scene from different angles. The starting position of the text does get transformed, but a string written with annotation text will still appear the correct way round wherever it is viewed from. This is valuable for labelling parts of a picture, such as names of atoms on a molecular model.

A pattern to be used within a filled area must be mapped on to the plane of the *fill area*, or *fill area set*, primitive. This is achieved by specifying three *pattern reference points;* the first of these is the origin of the pattern, and the other two define two sides of a pattern box. This pattern box is obtained by projecting the 3 reference points along a normal on to the plane of the area primitive. The resulting box may be a parallelogram. A series of cells are now constructed whose sides are parallel to the sides of the pattern box and whose dimensions are the *pattern size.* These cells are then filled with colors determined from the *pattern array.* The pattern can be transformed subsequently by viewing it from different angles. Thus, this provides a means to place simple textures upon a surface. (It might be used to simulate bricks on a wall, for example.)

The plane of a *cell array* is specified by three points, which may form a parallelogram. The filling of the cells within this parallelogram is similar to the use of a pattern in a filled area.

The *view index* and *name set* attributes are discussed in section 3.5.3.

3.4 Structured Pictures

The facility to introduce structure into picture descriptions is of particular importance for interaction. Basically, by *structure* we mean the ability to group output primitives into meaningful named entities, which can be manipulated independently of one another. (Note, however, that PHIGS uses the term *structure* in a very specific way; see below.)

The structure of a picture may also include relationships between different parts. GKS-3D does not permit these relationships to be expressed, but PHIGS does, to some extent.

3.4.1 Structuring in GKS-3D

Structuring in GKS-3D is achieved by grouping primitives into *segments.* This will not be described here in any detail, since it is the same as in GKS. In summary, the operations permitted on segments are: create (open/close), delete and rename. Segments may have a number of *segment attributes*: transformation, visibility, highlighting, detectability, and priority.

In GKS-3D the segment transformation is a three dimensional one. It is described in the section of these notes dealing with transformations.

3.4.2 Structuring in PHIGS

PHIGS differs significantly from GKS in the facilities it provides for structuring pictures:

- PHIGS allows primitive elements to be grouped into *structures*, in much the same way that primitives in GKS can be grouped into segments. However, whereas segments may not call (invoke) each other, structures can. This means that hierarchical picture definitions can be composed in PHIGS

- A structure is composed of *structure elements.* These include (among others, see later for full list):

  - Primitives and aspects, as in GKS-3D
  - Structure invocations. These are calls to other structures

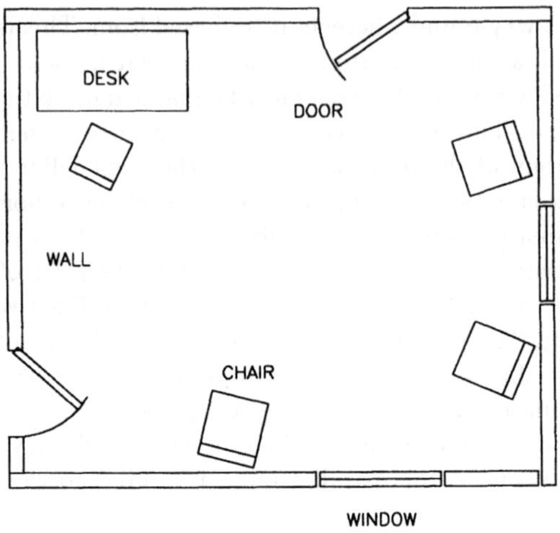

FIGURE 3.1. Plan of a room layout

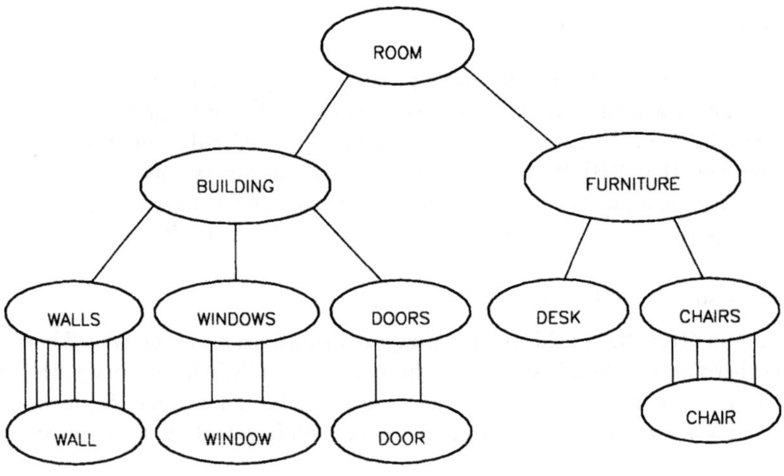

FIGURE 3.2. Structure network for room layout

- Modelling transformations. With the hierarchical picture defintions, PHIGS allows nested, hierarchical transformations. These will be discussed in a later section
- View selections for different orientations and projections. These too are described later.

• Hierarchical descriptions are represented by *structure networks.* A structure network is an acyclic directed graph in which the nodes represent individual structures and the arcs represent calls from one structure to another. A structure may have *parents*; these are other structures which reference this structure. It may also have *children*; these are the structures referenced by this structure. At the highest level in each structure network there is a *root structure.*

An example of such a hierarchy is shown in figure 3.1 and figure 3.2 which illustrate a simple room layout problem. The room comprises things which belong to

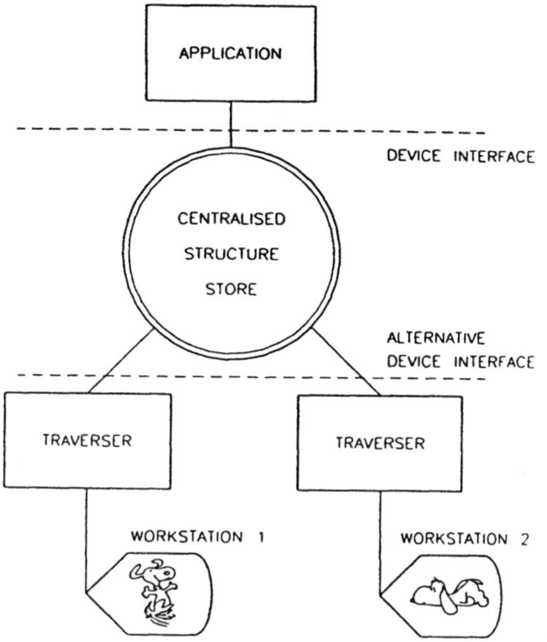

FIGURE 3.3. PHIGS Centralised Structure Store

the building's structure, such as walls, windows and doors, and things which are classified as furniture, such as chairs. The picture contains a number of *instances* (invocations) of each type of item. These can be positioned, rotated and scaled to their correct locations by applying geometric (modelling) transformations (see next section)

- Conceptually, a structure network is stored centrally and is shared by all workstations. This differs from GKS, in which segments may be stored at each individual workstation in a *segment store.* (GKS does have a centralised *workstation independent segment store* (WISS), but this fulfils a different role. For example, input devices cannot interact directly with a WISS)

- Each structure within a network *inherits* attributes and transformations from its parent. It may then modify these, the modified values being inherited, in turn, by its child structures. Note that this inheritance is performed during structure traversal (see below), not at the time structures are defined

- In GKS, segments are displayed as they are defined. In PHIGS, the definition of structure networks and their display are separated. Once a network has been defined it can be displayed by *posting* its root to the active workstations. Once a network has been posted, it is displayed until it is explicitly *unposted.* Posting a root structure causes its network to be *traversed.* This traversal process is conceptually continuous, so that if any changes are made to the structure definitions the displayed picture is immediately updated to show the changes. In figure 3.3 a centralised structure store (CSS) is shown. A description of a design and implementation of a CSS can be found in [9]

- Where in GKS the segment is the lowest level at which pictures can be edited, in PHIGS the contents of individual structures can be modified (see section 3.7.2). Thus, a modelling transformation can be altered within a structure and this will affect any primitives subsequently encountered, either in that same structure, or in its children.

3.5 Transformations and Viewing

Coordinate transformations have two main rôles in computer graphics:

- For positioning, scaling and orientating different components of the picture. This can be used, for example, to draw copies of a library picture at different locations — such as symbols on a map or circuit diagram. We refer to these as *picture construction transformations*

- For *viewing* — that is, for looking at 3D scenes from different viewpoints. We refer to these as *viewing transformations.*

Typically matrices are used to implement coordinate transformations in GKS-3D and PHIGS. Theory on homogeneous coordinate transformations can be found in any of the main text books on computer graphics.

GKS-3D and PHIGS have a common *Viewing Pipeline* which implements transformations for viewing objects from different positions in space, and for obtaining different projections (such as perspective).

3.5.1 Segment Transformations in GKS-3D

GKS-3D provides two transformation mechanisms for picture construction:

- *Normalization transformations* which perform mappings between the user's *World Coordinates (WC3)* and *Normalised Device Coordinates (NDC3)*, which represent an idealised, device-independent set of axes for describing screen layouts.

  These transformations map a cuboid in WC3 to another cuboid in NDC3; thus they can perform translation and scaling, but not shear or rotation. They are useful for simple conversions between a local coordinate system and NDC3

- *Segment transformations* which permit stored picture segments to be manipulated. The segment transformation is specified as a 4×3 matrix, permitting scaling, shearing, rotation and translation. Utility functions are provided to assist in calculating these matrices, or the programmer is free to compute his/her own:

  - `Evaluate Transformation Matrix 3` will generate a matrix, given a *fixed point, shift vector, rotation angles* for x, y and z, and *scale factors* for x, y and z. The *fixed point* and *shift vector* can be given in WC3 or NDC3
  - `Accumulate Transformation Matrix 3` accepts the same parameters, but modifies an existing matrix.

3.5.2 Modelling Transformations in PHIGS

PHIGS uses a much more flexible transformation scheme than GKS-3D. It permits quite complex hierarchical transformations to be constructed. PHIGS uses *modelling transformations.* There are two kinds:

- *Global modelling transformations.* During structure traversal, when a structure is invoked it *inherits* a current transformation matrix from its parent. This matrix becomes the global modelling transformation for the structure. It can be modified by calling the appropriate PHIGS function (`Set Global Transformation`), but, as with other attributes, when the traversal eventually returns to the parent structure its value will be reset to its original value

- *Local modelling transformations.* These do not affect the global value but are concatenated with it in the order $G \cdot L$ to produce a composite *current transformation matrix (CTM).* Local transformation matrices can be *pre-concatenated* ($CTM = CTM \cdot L$), *post-concatenated* ($CTM = L \cdot CTM$), or *replaced* ($CTM = L$). These transformations permit quite convenient local sets of axes to be used to define pictures. Typical applications of this include all kinds of instancing schemes, and molecular modelling. Examples are given later in these notes. Note that the *Local modelling transformation* is applied before the *Global modelling transformation.*

3.5.3 The GKS-3D and PHIGS Viewing Pipelines

A major addition in GKS-3D, shared by PHIGS, is the *viewing pipeline*, which permits different views and projections of 3D scenes to be generated. A full description of the GKS-3D and PHIGS viewing pipelines can be found in the standard, but there have been two interesting papers published in Eurographics Conference Proceedings about the pipeline [18, 7].

The viewing mechanism corresponds to what is normally termed the *camera model*, where the position of the camera is the viewpoint. A view is specified by setting the position and direction of the camera in space, together with a *view plane* on to which the picture will be projected. Both parallel and perspective projections are permitted. This is effected by computing two transformations — a *view orientation* and a *view mapping.* The former rotates the NDC3, or world, coordinate system to orientate the picture to correspond to the direction in which the camera is pointing. The latter projects this rotated view on to the view plane. The resulting view can be clipped to a view volume.

Two new coordinate systems have been introduced in order to simplify the specification of the view orientation and view mapping. These are *View Reference Coordinates* (VRC) and *Normalised Projection Coordinates* (NPC). The VRC system is effectively a rotated, translated of version of NDC3 (in GKS-3D) or WC3 (in PHIGS) and is established by the view orientation transformation. The xy plane of VRC is parallel to the view plane. The rotated view of the scene is then projected on to the view plane by the view mapping, which converts VRC into NPC.

The complete GKS-3D and PHIGS *output pipelines* are shown in figure 3.4.

To summarise, the GKS-3D output pipeline is as follows:

- Primitives are defined in 3D world coordinates (WC3)

- These are transformed (by a normalization transformation) into normalised device coordinates (NDC3)

- Primitives stored in segments may be transformed (scaled, rotated, sheared, translated) by a segment transformation. This takes place in NDC3

- An optional (user-specified) clip to a viewport in NDC3 is performed

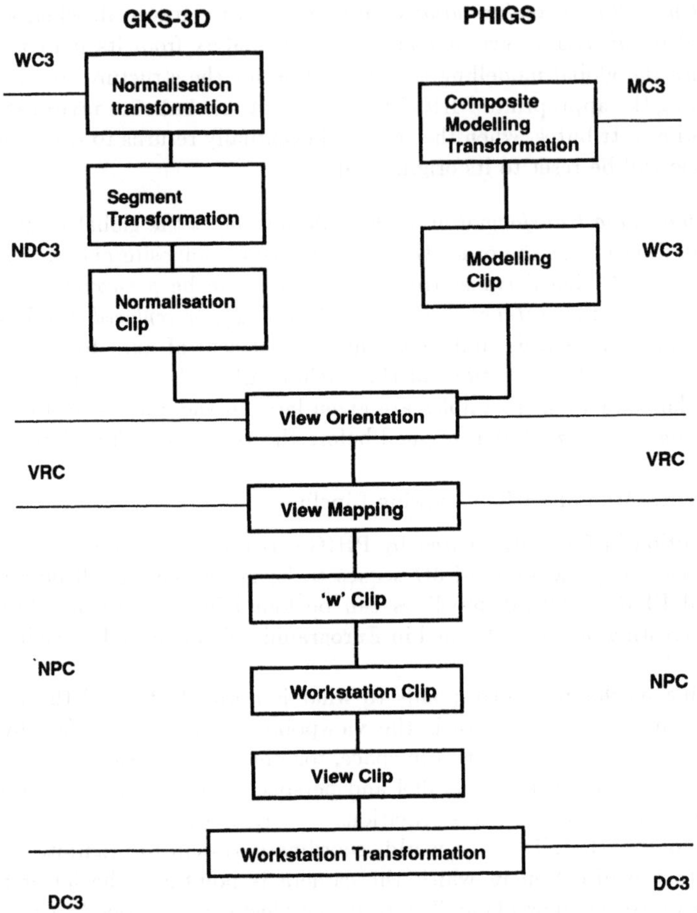

FIGURE 3.4. The GKS-3D and PHIGS Output Pipelines

- A view orientation transformation rotates and translates the NDC3 coordinates to obtain a desired orientation in which the picture is to be viewed. This transformation converts NDC3 into VRC
- A view mapping transformation applies a parallel or perspective mapping, and converts VRC into NPC
- Primitives are then clipped to a view volume in NPC. The clipping limits, and whether they are active, can be controlled by the application program
- Optional hidden line or hidden surface removal (HLHSR) is performed in NPC using an implementation defined method. The method of removal is bound to individual primitives, so that, even within a segment, some primitives may be subjected to hidden surface removal and others not
- A workstation transformation maps the view volume to the output device, converting NPC into device coordinates (DC3)

- A (mandatory) clip to the workstation viewport.

To summarise, the PHIGS output pipeline is as follows:

- During traversal (certain) structure elements create primitives in 3D modelling coordinates (MC3)
- These are transformed (by the current composite transformation) into World coordinates (WC3)
- An optional (user-specified) modelling clip to a volume in WC3 is performed
- A view orientation transformation rotates and translates the WC3 coordinates to obtain a desired orientation in which the picture is to be viewed. This transformation converts WC3 into VRC
- A view mapping transformation applies a parallel or perspective mapping, and converts VRC into NPC
- Primitives are then clipped to a view volume in NPC. The clipping limits, and whether they are active, can be controlled by the application program
- Optional hidden line or hidden surface removal (HLHSR) is performed in NPC using an implementation defined method. The method of removal is bound to individual primitives, so that, even within a segment, some primitives may be subjected to hidden surface removal and others not
- A workstation transformation maps the view volume to the output device, converting NPC into device coordinates (DC3).
- A (mandatory) clip to the workstation viewport.

The viewing parameters — the view orientation and view mapping matrices, the clipping limits in NPC, and clipping indicators — are stored in a *view table* at each workstation. The index to this table is bound to the individual primitives. This mechanism provides considerable flexibility:

- Different primitives, even within a single segment, can be bound to different viewing transformations
- Since there are separate view tables for each workstation, the same pictures can be viewed differently on different workstations.

3.5.4 Clipping

Logically there are several places in the output pipeline where primitives are clipped. It is possible, however, for an implementation to combine all the transformations together and, by transforming and combining the clipping limits, to perform a single transformation a single clip (to multiple planes). The clipping volume is a convex polyhedron.

- **Normalisation Clip (GKS-3D).** The user may request that all information outside the *normalisation viewport* is excluded

- **Modelling Clip (PHIGS).** The modelling clip in PHIGS is quite powerful. The user may specify an arbitrary number of half-planes and intersect them, creating a convex space, and use this as a clipping region. These half-planes are structure elements, specified in modelling coordinates, and so are subjected to the composite modelling transformation and the other aspects of traversal, implying that the clipping volume can change during traversal. This facility is useful for selecting a portion of some complex object that is available for viewing (though it may subsequently be clipped by the view clip). Furthermore, because the clipping planes are transformed, the same portion of, for example, a robot arm is always visible, even when the whole object is transformed
- **View Clip (GKS-3D and PHIGS).** The view clip specifies the region in NPC outside which no picture may appear. The limits, specified in NPC, define a cuboid, aligned with the principal axes. This contrasts with many other graphics systems, where the view clipping volume is a frustum
- **Workstation Clip (GKS-3D and PHIGS).** This is the only clip which is mandatory, and is to ensure nothing is drawn outside the device limits. The days of "wrap around" are gone.

The workstation clipping volume and the view clipping volume are aligned, and can be combined quickly and easily. The modelling clip (PHIGS) and the normalisation clip (GKS-3D) are not usually aligned with the view and workstation clip, so, if these are active, then clipping to a convex object rather than a cuboid must take place. Combining all the clipping and transformations into one transformation and one clipping volume needs special care as described in [2, 18, 7].

3.6 Input

3.6.1 Logical Input Classes

Both GKS-3D and PHIGS use a very similar input model to GKS, but extended for three dimensions. Input devices are divided into logical classes:

LOCATOR, which returns an (x, y, z) position in WC3, a view index, and a normalization transformation number

STROKE, which returns a sequence of points in WC3, a view index and normalization transformation number

VALUATOR, which return a real number

CHOICE, which returns a choice status and an integer indicating a selection

PICK, which returns a pick status and:

- For GKS-3D: a *segment name* and *pick identifier*
- For PHIGS: a *pick path depth* and a *pick path.* The pick path gives the name of the structure picked and all its ancestors.

STRING, which return a character string.

3D coordinates must be converted from DC3 to WC3 by passing them through the inverse of the workstation, viewing, and (for GKS-3D) normalization transformations, using the highest priority projection viewport and normalisation transformations. Note that there is a possibility that the viewing/projection transformation matrix is singular and has no inverse. This situation can only occur if the projection reference point is placed in the view plane), and will not normally happen if the appropriate utility function is employed to compute the matrix (`Evaluate View Matrix`).

3.6.2 Input Modes

These are the same as GKS:

REQUEST: synchronous input. Program execution is suspended until the request is satisfied by the operator

SAMPLE: returns the current, or last known, status of the input device, without waiting for any operator action

EVENT: all input is placed in a time-ordered queue. The application may inspect the queue, remove input from it, and perform cleaning up operations such as flushing events from the queue.

Different devices may be switched between the various modes supported by the implementation under application program control.

3.7 Editing and Manipulation of Pictures

Pictures may be edited in four principal ways:

- Primitives may be edited in order to alter the content of a picture
- Segment or structure attributes can be changed, affecting their visibility and how they are displayed
- Attributes of individual primitives can be altered, affecting their appearance
- Viewing parameters can be altered to change views and projections of 3D scenes.

3.7.1 Editing Picture Content in GKS-3D

In GKS-3D the picture segment is the lowest level at which a picture can be edited. It is *not* possible to edit the contents of a segment. There is thus a difficult decision to be made when designing highly interactive programs, such as schematic layout editors, or drawing programs.

One possibility is to make almost every primitive within the picture a separate segment, so that each can be created, transformed, deleted, independently of all the others. However, segments incur large penalties in most implementations because so much status information has to be stored for each one. For complex pictures it is therefore usual to chunk information together, but to use *pick id* attributes to label the individual primitives. Editing at the level of individual primitives then requires that the appropriate segments are regenerated by the application. This is not as inefficient as it might sound provided that attention is paid to the initial chunking strategy.

- Additional primitives for display of planar and curved surfaces
- Lighting models for simulating different lighting conditions
- Shading and interpolation techniques for rendering shaded surfaces
- Facilities for depth-cueing — a popular technique in some applications, such as molecular modelling
- Additional control of colour specification and intepretation.

The fact that PHIGS PLUS has such extensions explains why a number of manufacturers are emphasising their support for it rather than for GKS-3D. These notes deal only very briefly with PHIGS PLUS. Our aim is to give something of the flavour rather than a detailed recipe. Before examining the extensions we review, briefly, some of the techniques for generating shaded images.

3.8.1 Shaded Image Generation

Generally, the display of shaded images involves a number of steps which form an output pipeline:

- A model of the scene to be portrayed is constructed. The visible components of the picture are represented by primitives and associated attributes which control their appearance, such as colour. Shaded images usually depict scenes containing solid objects and surfaces, so primitives suitable for defining these are required. The system may also permit a variety of modelling effects, such as the transformations found in PHIGS and PHIGS PLUS. The latter defines some additional primitives for surface display
- Next some means is required for specifying a view. In PHIGS PLUS this is achieved in the same way as in PHIGS
- The viewing parameters are applied to generate the desired view, and then hidden surface computations are performed. On raster-scan displays, of the type used to display images, scan-conversion algorithms are necessary in order to render an image. Hidden surface removal is also usually performed during scan conversion, or with a z-buffer
- Once the surface which is visible at a particular pixel has been determined, it is necessary to compute a colour and intensity for it. A *lighting* model is used to simulate different lighting conditions. This allows the amount of light incident upon the various primitives in the scene to be found and used to compute how much light is reflected from an individual pixel towards the viewer — the so-called *reflectance, or shading, calculation.* A variety of lighting models have been reported in the graphics literature, including spot lights, strip lights and lights with a large area. The most usual reflectance models are the Lambertian one for diffuse reflection, and the empirical Phong method [5] which simulates specular highlights.

  Curved surfaces are often represented by a collection of approximating polygonal facets. This is because the equations for planar primitives are easily implemented in hardware. If faceted models are employed, it is common to interpolate intensities or

process by permitting application data to be stored inside structures. A typical use of this would be to store a pointer to some application data structure, or data base, within a structure, such that when the structure is modified in some way (or perhaps picked by an operator) the program can immediately access the relevant related information. Other uses would include storing things such as material properties, masses etc

- *Generalised structure element (GSE's).* These are similar to *escape* functions in GKS. They offer a "standard way of being non-standard".

3.7.3 Altering Segment Attributes in GKS-3D

Although the contents of segments cannot be changed, the attributes of complete segments may be altered. These include visibility, highlighting, detectability, priority, and transformation.

The priorities of segments determine the order in which they are drawn by a workstation. On raster devices this can be exploited to utilise overpainting — one application for this being hidden surface removal by the painter's algorithm. A priority can also be associated with a structure in PHIGS when it is posted.

3.7.4 Changing Primitive Attributes in GKS-3D and PHIGS

In GKS-3D it is possible to alter the appearance of primitives by changing their attributes. However, since the contents of a segment cannot be edited it is only possible to do this retrospectively by using *bundled attributes,* and altering the definitions in the bundles.

In PHIGS, however, it is possible to edit attribute elements within a structure, as well as by changing values for bundled attributes.

3.7.5 Altering Viewing Parameters in GKS-3D and PHIGS

In both GKS-3D and PHIGS, *viewing* is separated from picture definition. One a picture has been defined and stored in segments or structures, different views can be created by altering the various view orientation and view mapping transformations. These are stored in the *view table* and can be accessed by a *view index.*

3.8 PHIGS PLUS

For nearly two decades a major area of research in graphics has been the generation of "realistic" shaded images. Such images help us to perceive depth in what are, in reality, flat displays. For example, shading gives clues about the curvature of surfaces, whilst cast shadows show the relative depths of objects in a scene. Largely inspired by work on computer animation for television and film, various techniques have been developed for depicting texture, tranparency and reflection. Increasingly, these techniques find application in other areas, such as mechanical and architectural CAD, molecular modelling, and visualisation of multi-dimensional data — that is, in precisely those areas addressed by PHIGS.

Whilst GKS-3D and PHIGS have been progressing, manufacturers responding to end-user requirements have designed, and now market, increasingly powerful systems capable of displaying quite sophisticated images. Neither GKS-3D nor PHIGS provides facilities for accessing these advanced capabilities. Having recognised this, a group in the USA defined a set of extensions to PHIGS, called PHIGS PLUS [15]:

- Additional primitives for display of planar and curved surfaces
- Lighting models for simulating different lighting conditions
- Shading and interpolation techniques for rendering shaded surfaces
- Facilities for depth-cueing — a popular technique in some applications, such as molecular modelling
- Additional control of colour specification and intepretation.

The fact that PHIGS PLUS has such extensions explains why a number of manufacturers are emphasising their support for it rather than for GKS-3D. These notes deal only very briefly with PHIGS PLUS. Our aim is to give something of the flavour rather than a detailed recipe. Before examining the extensions we review, briefly, some of the techniques for generating shaded images.

3.8.1 Shaded Image Generation

Generally, the display of shaded images involves a number of steps which form an output pipeline:

- A model of the scene to be portrayed is constructed. The visible components of the picture are represented by primitives and associated attributes which control their appearance, such as colour. Shaded images usually depict scenes containing solid objects and surfaces, so primitives suitable for defining these are required. The system may also permit a variety of modelling effects, such as the transformations found in PHIGS and PHIGS PLUS. The latter defines some additional primitives for surface display
- Next some means is required for specifying a view. In PHIGS PLUS this is achieved in the same way as in PHIGS
- The viewing parameters are applied to generate the desired view, and then hidden surface computations are performed. On raster-scan displays, of the type used to display images, scan-conversion algorithms are necessary in order to render an image. Hidden surface removal is also usually performed during scan conversion, or with a z-buffer
- Once the surface which is visible at a particular pixel has been determined, it is necessary to compute a colour and intensity for it. A *lighting* model is used to simulate different lighting conditions. This allows the amount of light incident upon the various primitives in the scene to be found and used to compute how much light is reflected from an individual pixel towards the viewer — the so-called *reflectance, or shading, calculation.* A variety of lighting models have been reported in the graphics literature, including spot lights, strip lights and lights with a large area. The most usual reflectance models are the Lambertian one for diffuse reflection, and the empirical Phong method [5] which simulates specular highlights.

  Curved surfaces are often represented by a collection of approximating polygonal facets. This is because the equations for planar primitives are easily implemented in hardware. If faceted models are employed, it is common to interpolate intensities or surface normals across the facets to create the illusion of a smooth surface. Bi-linear interpolation in image space is employed, staring with the intensities at the facet

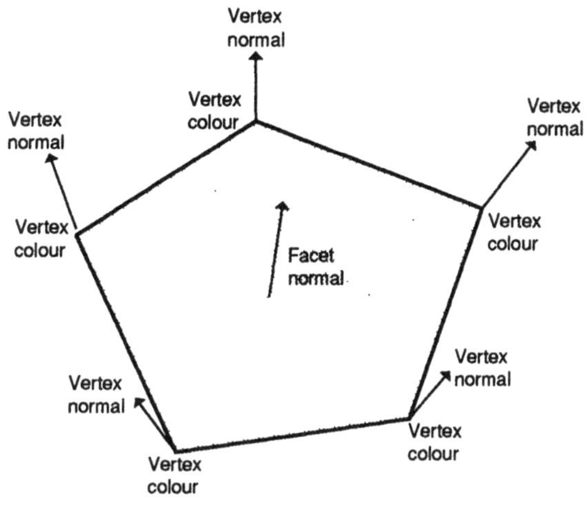

FIGURE 3.5. Fill Area Set with Data Primitive

vertices (known as Gouraud interpolation [6]), or with the normal vectors at vertices (using the Phong method).

Other effects, such as depth cueing can also be incorporated at this stage. Depth cueing is a technique which modulates the intensity, and sometimes the colour, of primitives according to their distance from the viewer in order to create an impression of depth

- A *colour mapping* phase determines how the resulting colours are displayed.

3.8.2 New Primitives

PHIGS PLUS is defined so that *colours* as well as intensities can be interpolated across a surface. This could be used, for example, to colour code a stress value, with colours at interior points on a polygonal surface interpolated from known values at key points, such as the vertices of the polygon (e.g. finite elements). This requires that colour values be specified at vertices — this is provided by having *with data* variants of the standard PHIGS functions:

POLYLINE SET 3 WITH DATA. Colour information is supplied at the points in the polyline sequence. Colours are linearly interpolated along segments of the polyline

FILL AREA 3 WITH DATA. The additional data are colours at vertices, vertex normals, and a facet normal, as shown in figure 3.5. If the *colours* are specified then they are used during scan conversion to compute the colour of each pixel. If not given, then the fill area colour is used (as in PHIGS). If the *vertex normals* are supplied, they can be used to perform intensity interpolation. (Whether they are *actually* used depends on the shading method selected during traversal.) If vertex normals are not specified, the *facet normal* can be used for shading. In the event that none of the normals are given then a geometric normal is computed from three non-collinear points. (As PHIGS PLUS is currently defined, this may give problems if the three points correspond to a concave part of the area's boundary. In this case the normal will point in the wrong direction!)

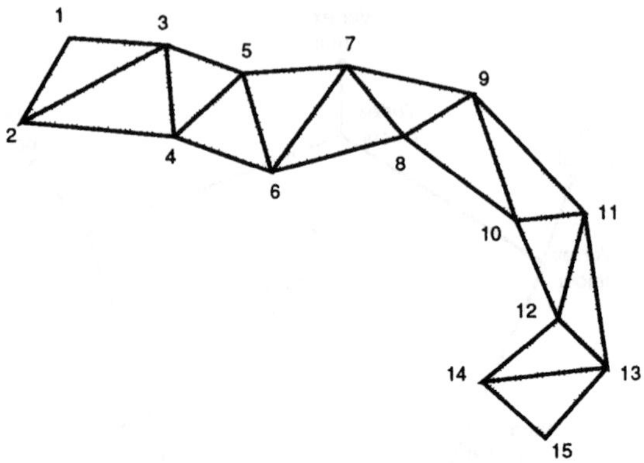

FIGURE 3.6. Triangle Strip Primitive

The facet normal, or geometric normal, can be employed to perform *back face culling*. This is the removal (i.e. non-display) of faces which point *away* from the viewer. This can substantially reduce (typically by 50%) the amount of work required to render a shaded image. Alternatively, back-facing surfaces can be rendered in a different colour.

FILL AREA SET 3 WITH DATA. The additional information is similar to that of **FILL AREA 3 WITH DATA**

EXTENDED CELL ARRAY 3. Permits a general colour to be specified for each cell. (PHIGS allows only an *indexed* colour for each cell.)

PHIGS PLUS also defines a number of *new* primitives:

TRIANGLE STRIP 3 WITH DATA. This comprises a strip of triangular facets as seen in figure 3.6. The data is more compact than with **FILL AREA** because each vertex is only specified once, as shown in the sequence numbering in the figure

QUADRILATERAL MESH 3 WITH DATA. This primitive, like triangle strip, offers a reduction of data needed to represent a surface, defined as a grid of quadrilateral elements. An example is given in figure 3.7. Note that a quadrilateral facet cannot be guaranteed to be planar. PHIGS PLUS will render non-planar facets as two triangles. A mesh of $(M-1) \times (N-1)$ quadrilaterals is passed as a two-dimensional array of $M \times N$ vertices, as shown in the numbering in the figure

POLYHEDRON 3 WITH DATA. This is a short-hand method for generating a series of facets. For a polyhedron, vertices may be shared by several adjacent facets. These vertices are only specified once and are accessed via sets of indices associated with facets. Such a set of indices is assumed to define a closed boundary. Non-planar facets are dealt with in an implementation-dependent manner

NON-UNIFORM B-SPLINE CURVE. PHIGS PLUS allows definition and display of B-spline curves, which can be controlled by the following parameters: the *spline order* (e.g. 3 for a cubic), a set of *knot numbers*, a set of *control points*, a *parameter range* which specifies which part of the defined curve is to be drawn, and a *type* which may take the values `RATIONAL` (control points are given as 4-D homogeneous

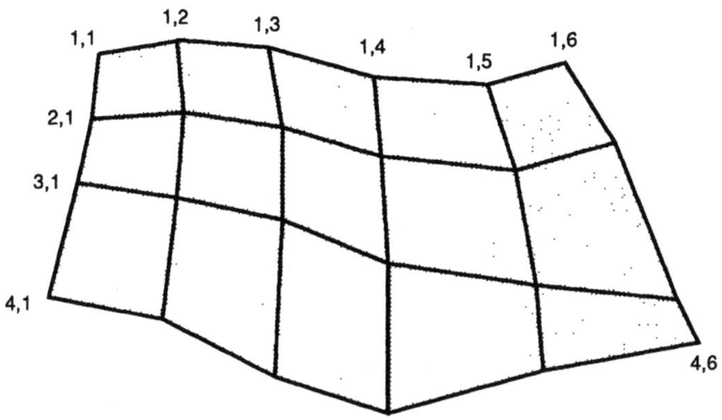

FIGURE 3.7. Quadrilateral Mesh Primitive

PHIGS modelling coordinates), or NON-RATIONAL (control points are given in ordinary PHIGS modelling coordinates). The same attributes which apply to polylines also apply to these curves (e.g. width, colour)

PARAMETRIC POLYNOMIAL CURVE. This is a curve defined by the follwing parameters: a *basis*, for which there are currently two possible values (1 = draw a Uniform B-spline, 2 = draw a piecewise Bézier curve), a *curve order* (e.g. 3 for a cubic), a *type* (RATIONAL or NON-RATIONAL), and a set of *control points*

NON-UNIFORM B-SPLINE SURFACE. This primitive is used to define curved surfaces. It is possible to control the accuracy with which the surface is rendered by adjusting a tolerance value. The parameters which control the primitive are: a *spline order* for each of the u and v directions, a set of *knots* for each direction, a set of *control points*, a *range* for each direction, a *trimming definition*, and a *surface type* (either RATIONAL or NON-RATIONAL). The trimming definition provides a method for displaying a part (or parts) of the defined surface. It comprises a list of *trimming curves* which are themselves non-uniform, rational B-spline curves defined in the uv parameter space of the surface. These curves form a closed loop which, in effect, defines a curved clipping boundary on the surface

PARAMETRIC POLYNOMIAL SURFACE. This primitive defines a surface with the following parameters: a *surface form* (uniform B-spline, or piecewise Bézier), an *order* for the u and v directions, a rectangular array of 3D *control points*, and a type (rational or non-rational).

3.8.3 Lighting Models

PHIGS PLUS supports the following types of lighting:

AMBIENT. The shading of surfaces is not dependent on the relative positions of light sources, although the light sources have a *colour* which does affect the result

DIRECTIONAL. Light sources have both *colour* and *direction*, but are located conceptually at infinity. (That is, all rays from a particular direction will be parallel, so each light can be represented by a single direction vector)

POSITIONAL. These are located at finite positions. Rays from such lights to different parts of a scene will *not* be parallel. *Attenuation coefficients* may be used to simulate the inverse square law governing the reduction in energy of incident light according to the distance from the source

SPOT. This is a positional light source which has some additional controls: a *direction*, a *concentration* and a *spread angle*. The spread angle defines a cone of illumination centred around the specified direction. Within this cone, the intensity of light varies as a cosine function raised to a power (the *concentration* parameter). This yields an illumination profile similar to the highlights determined by the Phong reflectance method. (Setting the concentration to a very small value yields almost constant illumination emitted from the cone, setting it to unity gives a cosine distribution, and setting it to a high value gives a concentrated, small spot light.)

3.8.4 Shading and Interpolation Techniques

Effectively, PHIGS PLUS employs the Phong reflectance method described in numerous text books. The implementation must cater for positional and spot sources by calculating the appropriate direction and energy of incident rays.

During scan conversion, both intensities and colours can be interpolated across the surface of filled areas, the following options being supported:

NONE. A single intensity is computed for each area or facet — this is sometimes termed *constant shading*. It is equivalent to the *ambient* term in the Phong reflectance model

COLOUR. The reflectance calculation is performed at each vertex of a facet. From this a colour is derived, and this colour is interpolated across the facet — this corresponds to *Gouraud interpolation*

DOT. The vector dot product of a vertex (or facet, or geometric) normal and incident light ray is computed at each vertex. The dot products and vertex colours (if given) are interpolated across the facet and these interpolated values are used to compute a colour at each pixel. This is sometimes called *cheap Phong interpolation*

NORMAL. Here, both the vertex colours and normals are interpolated. This is equivalent to the full *Phong interpolation* method. An interesting problem arises because PHIGS allows arbitrary 4 by 4 transformation matrices to be applied to primitives. Such transformations permit perspective distortions to be applied, as well as shear and asymmetric scalings. The simple way to interpolate normals is in device coordinates during scan conversion. However, if a perspective transformation is applied this cannot be done because the interpolation *after* transformation is no longer linear in the z direction. It is therefore necessary to perform an inverse mapping into world coordinates in order to correctly perform the vector computations on the interpolated normals in the linear world coordinate system. Many present day graphics displays cannot cater for this inverse mapping because it does not conform to the usual transformation pipeline. A consequence is that some implementations of PHIGS PLUS may not be able to correctly support shading for pictures which include perspective distortions.

3.8.5 Depth Cueing

This technique alters the colour of a primitive according to its z coordinate in NPC space. It is commonly employed to reduce the intensity of primitives which are further

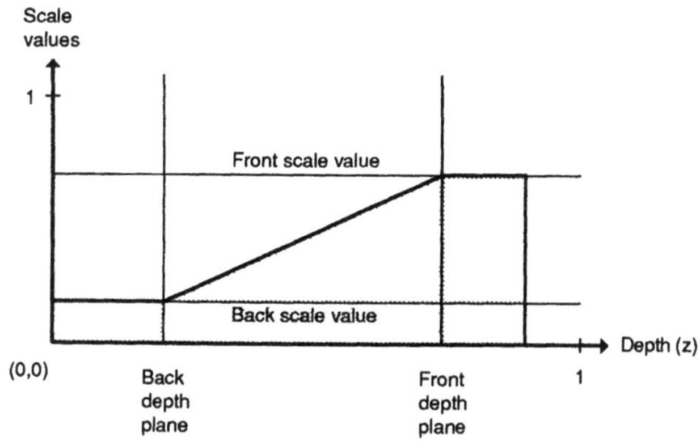

FIGURE 3.8. Depth Cue Profile

from the viewer, thereby creating an illusion of depth. However, in PHIGS PLUS a more general formulation is adopted in which the depth controls the mixing of the primitive's colour and a depth-cue colour. The parameters which affect this are: *front depth cue* and *back depth cue reference planes* which determine the range of z values over which a linear interpolation is performed, and *front* and *back scale factors* which determine the ratio in which the primitive's colour and depth cue colour are mixed for points in front of and behind, respectively, the front and back planes. This is illustrated in figure 3.8. One application of this general approach is to allow colours to de-saturate with depth (distance) — as happens in the real world.

3.8.6 Colour Specification and Mapping

PHIGS uses the notion of *indirect* colour specification. Colours are accessed by means of an *index* which points at a *colour table.* In addition to this, PHIGS PLUS allows colours to be specified *directly.* Direct colour specification means that colours are defined by a triple of values within some colour model. The models supported are RGB, CIE, HLS, and HSV.

Once colours have been defined, directly or indirectly, they are used to render primitives. Whatever method is used to render primitives, the result is a direct colour for a pixel — obtained as a result of complex lighting, shading and depth cueing, or by simple look-up in a colour table. This colour must be mapped in some way to the colours available on a workstation. To make this as flexible as possible, PHIGS PLUS has a *colour mapping* stage which supports true colour mapping, and two kinds of pseudo colour mapping.

With *true colour mapping* the desired colours are passed to the workstation and must be reproduced as faithfully as possible using an implementation-dependent method. Possible solutions include use of full colour devices (e.g. 24 bits per pixel), and use of dithering. Interestingly, dithering is common on hardcopy devices, but unusual on displays, although it can be used to very good effect on the latter, provided techniques which rely on altering a look-up table dynamically (such as colour table animation) are not required.

With *pseudo colour* a weighting function is used to combine the given colours to form an index. This index is then employed to access a *pseudo colour table* from which the final colours are found. These must then be represented as faithfully as possible by the workstation. In effect, pseudo colour offers a mechanism for controlling the final mapping of colours. As an example, consider a picture specified with HSV parameters. The weighting

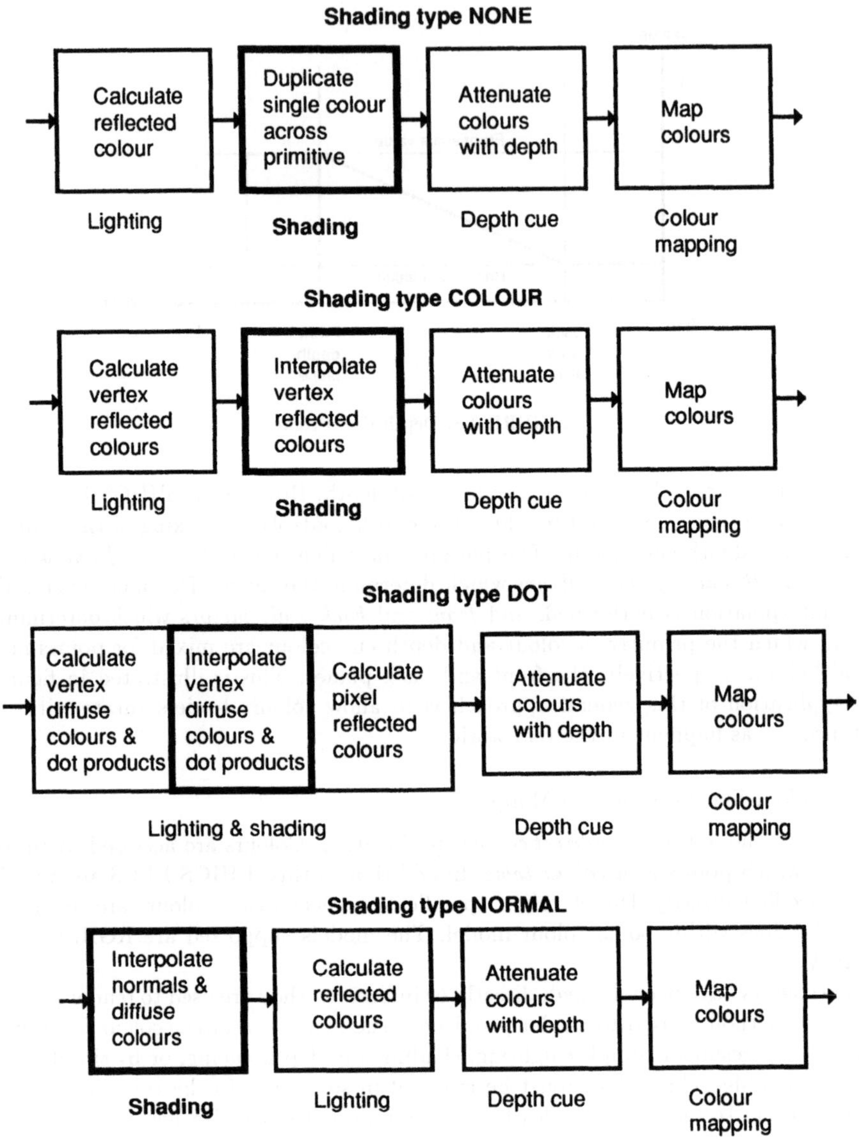

FIGURE 3.9. PHIGS PLUS Rendering Pipline

function parameters could be defined such that only the V values are of any consequence, providing a simple way to map a colour picture to a greyscale display.

A variation of pseudo colour is *pseudo-3 colour*. Here, three look up tables are employed, one for each colour component, and separate indices are computed for each. Clearly, this gives greater flexibility.

3.8.7 Summary of the PHIGS PLUS Rendering Pipline

How the PHIGS PLUS rendering pipeline is organised is shown in figure 3.9 for the various shading options (NONE, COLOUR, DOT and NORMAL).

3.8.8 Things PHIGS PLUS Does Not Do

PHIGS PLUS is aimed at *interactive* graphics. With current systems this rules out techniques which are especially time consuming or complicated to implement. Thus, PHIGS PLUS does *not* define ways to handle translucency, tranparency, texture, reflections, and cast shadows, nor is it specifically geared to methods such as ray tracing, or radiosity lighting/relectance models.

However, it does not preclude the inclusion of such capabilities in implementations. It would be feasible to have a range of renderers each working from a common description of the structures. For example, one might have a wire-frame renderer for very fast interaction, a shading renderer which capitalised on hardware capabilities to perform shading in near real-time, and a ray tracer for final production of high-quality hard copy. This approach is already being adopted in systems such as Doré from Ardent Computer [3].

3.9 Case Studies

The main emphasis in this section will be on some examples which illustrate how GKS-3D and PHIGS can be used for different tasks. We examine how the different aspects of these systems described previously can be used in practice by looking at some small-scale case studies.

3.9.1 Modelling

Writing real application programs, as distinct from simple demonstrations, requires that various aspects of the problem are modelled. Ideally, we wish to use graphics as a "window" on to our model — that is, as a tool for viewing the model and understanding its behaviour. A graphics system helps us to do this by allowing the display of representations of the model and by providing various input techniques, such as pointing, entering coordinates, monitoring dials and buttons, and so on.

A fundamental feature of GKS and GKS-3D is that *modelling* is separated from display and interaction. The decision to make this separation emanated from the historic meeting on standards held at Seillac in France, in 1976. It was thought that this would make it easier to obtain international agreement, in that graphics and interaction were felt to be understood better than modelling.

Modelling has to do with representation of application-dependent aspects of a problem. The term modelling has very widespread usage in computing. Examples include financial modelling, weather forecasting, molecular modelling, simulation (e.g., flight simulators).

The ease with which graphical data, needed to display a picture of our model, and other application-dependent data can be tied together is really quite important for designing good interactive interfaces. By separating graphics and modelling we may achieve the goal of device-independence but make it harder to have a well integrated interface. The facilities for structuring the graphical data, for editing it, and for relating the graphical and non-graphical data to one another are therefore key issues when implementing graphics systems.

When we examine a variety of applications we do find similarities between them:

- There is often a need to represent *topology* — the fact that different parts of a picture are connected or related in some way. Examples include drawing programs, where moving a reference point may cause all related information (such as lines which meet at the point) to be automatically updated, piping design programs, printed circuit layout programs, public utility networks, etc. Neither GKS-3D nor PHIGS tackle

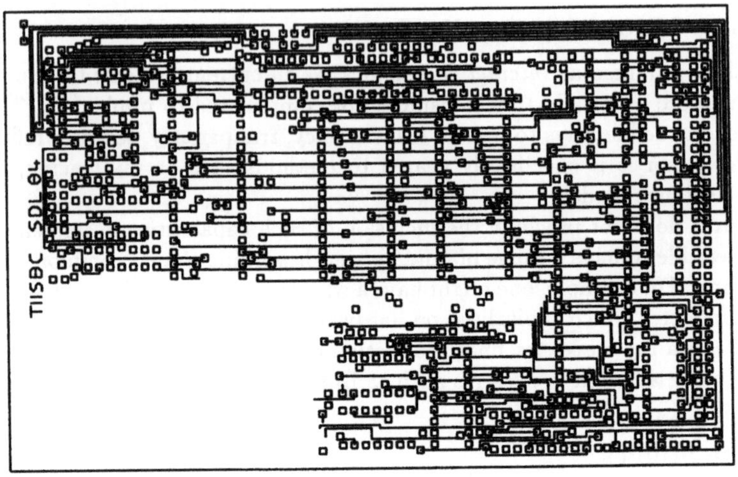

FIGURE 3.10. PCB Layout

this problem, which they regard as the responsibility of the application. The first case study illustrates the difficulties which can arise from this

- Many problems have a natural hierarchical structure and result in pictures which contain sections, or parts, which are copies of each other, or are smaller/larger or rotated versions of master parts. Examples include symbols on maps and circuit diagrams, and drawings from a library in a drafting package. It is therefore useful to be able to define a master copy of a picture and then create different *instances* of this. The first case study also illustrates this

- Geometric transformations are frequently needed to scale, rotate and position elements of a picture. GKS-3D provides segment transformations and viewing utilities. PHIGS has extensive modelling transformations, and a viewing model similar to that in GKS-3D. The second case study illustrates viewing, whilst the third and fourth illustrate modelling transformations.

3.9.2 Case Study I — Printed Circuit Board Editor

This example uses only 2D graphics[2], but illustrates rather well the difficulties of representing relationships between parts of a picture. The problem is essentially similar to many drafting programs, schematic diagram editors, mechnism design programs, and (in 3D) piping design programs. The problem is a real application at the authors' institution, and figure 3.10 shows a typical example.

The application program to be implemented has the following features:

- A printed circuit board (PCB) is to be displayed and edited interactively. The editor is part of a larger suite of programs allowing a design to be carried through from schematic capture to production of control files for manufacture (photoplotting and drilling, for example). The PCB editor interfaces to these other programs, and allows minor corrections to be made, or — for small, special designs — the whole layout can be worked out by the designer and input with the editor

[2]Throughout this section the name GKS-3D has been used, but in practice GKS would suffice.

- The editor has a library mechanism which allows commonly used components to be accessed via a menu. The designer can add new components to the library, such as re-useable sub-layouts, and additional pin configurations for new IC's
- Components accessed in this way are then placed on the board, interactively, with a mouse or tablet
- Components are connected by drawing constrained (to 45° angles) tracks of appropriate width. During drawing, other constraints must also be applied, such as prohibiting lines from entering "no-go" areas of the board
- The board may have multiple layers — typically 4, but up to 10 — which the designer may display singly or together.

Primitives

Both GKS-3D and PHIGS provide similar primitives and attributes, so there is little to choose between them on this basis. For this application, the standard polyline and fill area primitives are used to display components and tracks. It is also convenient to define a window and viewport to delineate the area of the board we wish to view. Zooming in or out is easily accomplished by altering the window.

There is, however, one problem with zooming. Tracks on the PCB are displayed as polylines of appropriate width. Zooming requires that line widths are scaled to show their true dimensions. Neither GKS-3D nor PHIGS will scale the line widths automatically. Fortunately there is a work-around for this problem. It is usual for PCB designs to use only a limited number of different track widths. Therefore we define a number of different width lines using the bundled attribute method and plot each track using the appropriate line style index. The line width for each entry in the line style bundle table is then recomputed and set each time we zoom in or out.

An alternative solution would be to draw tracks as filled areas. These will get scaled automatically during zooming. Unfortunately, most systems take very much longer to render filled areas than polylines — not only does the interior have to be filled, but clipping is also more complicated. This is a disincentive for this approach.

Picture Structure

GKS-3D provides segments and pick identifiers for structuring and naming. Therefore, it seems logical to divide each part of the layout into segments, allowing them to be manipulated independently. The segment is the lowest level in GKS-3D at which editing operations can be performed so if we wish to be able to edit individual parts of a track or component we must place each one in a separate segment. Unfortunately, in many implementations segments carry a high overhead, so if we use a separate segment for each part of the picture we will soon have thousands of segments. This overhead may pose severe problems, and may also be apparent by a reduced interactive performance.

This problem is quite a nuisance. We wish to use segments because they allow us to edit the picture and to identify parts of it with a *pick* device. A compromise solution may be necessary. One approach is to group parts of the layout into segments, and to use *pick_id* to differentiate between components inside a segment. Now, when we wish to edit a single component which is stored in a segment along with other components, we must break the segment down into a number of new segments each containing a single component. These new segments can then be manipulated separately.

Clearly, this is quite inconvenient for the application programmer who must keep track of all the segment names in use, but works tolerably well. It is typical of the kind of compromise between efficiency and convenience which characterise real applications.

The representation of the hierarchical nature of the data must be performed entirely by the application program. GKS-3D provides no assistance with this at all. In effect, the task is once again to keep track of segment numbers, so that these can be mapped back to the application data. The most straightforward way to manage repeated parts of a picture with GKS-3D is to use workstation-independent segment store (WISS). At program start-up the definitions of pre-defined components are read from a metafile and stored as segments in the WISS. These can then be inserted from the WISS into new segments on the active workstation as and when required.

The separation of the PCB design into layers can again be achieved with segments. Once more, the application program must remember which segments belong to which layers. Layers can then be made visible/invisible by changing the segment visibilities, and brought to the front by changing segment priorities.

PHIGS provides more flexibility than GKS for representing hierarchically structured data, and for detailed, low-level editing of pictures. Now we are able to represent our small components with structures, and to build these up into larger groupings to represent sub-circuits. A logical hierarchy is to have a structure for the whole board, which invokes a structure for each layer. The content of each layer comprises structures whose hierarchy reflects the way circuits and sub-circuits are assembled. At the lowest levels will be IC's, with one structure for each different type.

Altering the visibility and depth order of layers is easy — we simply post the different layers with a suitable priority, and can control visibility by means of name sets. However, we do have to select suitable priority values — which implies that we know in advance how many levels we are likely to need. With this application this is not a problem. Note that priorities can *only* be associated with *posted root structures.* Thus, if we wish to use structure priorities to alter the overlaying of our layers, we must post the layer structures — that is, we cannot call them from a parent structure which represents the whole PCB, as suggested previously. An alternative would be to assign different z values to each layer and to apply transformations to alter the depth order.

In GKS-3D we have seen that there may be some pressure to minimise the number of segments; similar arguments hold with structures in PHIGS. Generally, we will be able to use many less structures in PHIGS than segments in GKS-3D, because of PHIGS' more powerful editing capabilities. We can edit the contents of structures, so there is no longer any need to artifically subdivide the picture purely for the purposes of editing, as happened with GKS-3D. Each part to be manipulated can be created with its own *pick_id*, and *labels* can be used to access individual parts quickly and conveniently. However, inside the PHIGS CSS these editing functions are complicated to implement and eat up space at an alarming rate for storing pointers and administrative data, leading to very large data structures. Generally speaking, the space occupied by a PHIGS CSS will be many times larger than the equivalent GKS-3D segment store.

Both GKS-3D and PHIGS use a single name space for segments or structures. This gives us a further problem — choosing segment or structure names. Our GKS-3D program is designed to work with various utilities, such as a menu management package. This package is also implemented with GKS-3D and uses segments. We must exercise care that names used by our application do not clash with those used by our utilities. We can do this with an inquiry function to find out what names are already in use. We encounter a similar problem with structure names in PHIGS. We have seen that the management of segment names in GKS-3D imposes quite a burden on an application programmer. Segment and *pick_id* names provide the only direct mechanism in GKS for tying the graphical data to other application data. With PHIGS a similar situation holds, but PHIGS does have one additional facility which can be useful in this regard: the storage of *application data* inside structures. One use for this is the storage of pointers to application data structures.

Transformations

With GKS-3D, to place components, or sub-circuits, on the board we can use segment transformations. These allow us to scale and rotate pictures and to place them at positions entered with a locator. However, we cannot have hierarchical structures, so as sub-circuits are built up we can only apply a single transformation to each one. This makes segment transformations of limited use where complex hierarchies are designed. In effect, an application program will have to keep track of the hierarchy of transformations, accumulate the component transformations and apply them to each segment — quite a lot for the application programmer to think about!

In contrast, PHIGS' modelling transformations appear to provide exactly what is required. As new levels are created we can associate modelling transformations with them in order to scale rotate and translate them to the desired locations. These transformations can be concatenated with those at other levels in the hierarchy, making the display of hierarchical designs relatively straightforward. This is hardly surprising because it is precisely what PHIGS is designed to do.

Interaction

The main tasks to be accomplished here are adding new components from a library, placing them on the board using a locator, *creating* new components by drawing pads (connectors) and tracks, and editing a design by moving, deleting or otherwise modifying component positions and tracks. We have already seen how the structuring facilities of GKS-3D and PHIGS play a part in this process. Here, we concentrate on the problem of adding new components and of drawing tracks with a locator.

Drawing tracks entails:

- Starting to draw the track in the vicinity of a connector on the component from which the track starts. A smart system would be able to find the closest connector and to know that this is where the track should start

- Drawing a series of constrained lines, starting with a line from this point and finishing with a line which connects to another component.

Problem — where *are* the starting and finishing points? Because we have relied on GKS-3D or PHIGS to apply transformations to position the components we don't know the current coordinates of the points on the components, but we *need* to know them in order to decide where to start drawing our track. How do we find out?

There is no way an application program can find out from GKS-3D where these points are! To solve this, the application must perform its own simulation of the segment transformations. Using a pick device we can find out which component has been picked, apply our transformation to the original data points belonging to this component, find the one closest to the locator, and use this as the starting position for the line. Not all implementations support pick; with those that do not the spatial search must be made with *all* the primitives in the picture. With problems of this type it is usual to round coordinates to a grid. Whilst this may make finding the appropriate points easier, the onus is still on the application program to know the actual coordinates of the connectors.

With PHIGS, the situation is even worse because we have several levels of modelling transformation to worry about. At first sight the problem looks horrible — the application, having capitalised on PHIGS' ability to display complex hierarchical models, must now do all its own transformations, including all the concatenation of matrices, in order to find out the coordinates of points in world coordinates, so that they can be compared with points entered with the locator.

Fortunately, PHIGS comes to the rescue here with its *incremental spatial search* facility. Using this, it is possible to find out which structures are close to a position entered with a locator. However, this facility requires that a 3D point is given as the *search reference point*. Usually, with a locator driven by a graphics tablet or mouse we only know the 2D coordinates of a point. For 2D applications, such as our PCB editor, the search facility offers what we require — the z coordinate is simply assumed to be zero, unless we have used different z values for different layers. However, for 3D applications, such as a piping design program, we would have to use a genuine 3D locator in order to apply the PHIGS spatial search. In other words, for 2D the spatial search does what we want by accident rather than design.

Whilst drawing tracks we would like to draw rubber-band lines which are constrained, both in angle and to stay outside no-go areas. Most GKS-3D implementations have no adequate way of doing this. At best, we may be able to get rubber-band lines, as an echo type for locator, but not ones with arbitrary (application-dependent) constraint checks applied. Since a segment is the lowest level in GKS which we can edit, we must hope that we can place our line in a segment and update the segment definition sufficiently rapidly to get dynamic feedback. This will only be feasible if we have a fast display which can update the picture at least 10 times a second. The following pseudo-code illustrates the technique for *replacing* the track:

```
Open_Segment(track);                {Make a dummy segment}
Close_Segment;
repeat
    Sample_Locator(pnts[2].x,pnts[2].y);
                               {Define the end of the line
                               assumes start already defined}
    Constrain(pnts);           {Apply constraints to angle etc}
    if Inside_NoGo(pnts) then
        Warning_Message('Inside No Go area!!!')
    else
    begin
        Create_Segment(temp); {Create a temporary segment}
        Set_Polyline_Index(tracktype);
                               {Select correct width etc}
        Polyline(pnts);        {Uses point returned by locator}
        Close_Segment;
        Delete_Segment(track);{Delete old line segment}
        Rename_segment(temp,track)
    end
until not Mouse_Button;
pnts[1] := pnts[2];            {Ready to draw next track}
```

One can think of numerous extra things the program would have to do. It is merely meant to illustrate the technique of replacing segments, with an application-dependent constraint check as part of the interactive loop. (The procedure `Constrain` adjusts the entered point so that the track is at a 45° angle, and rounds to a grid if required, whilst the logical function Inside_NoGo checks for violation of no-go areas.)

Because PHIGS supports editing of primitives *within* structures, by means of element pointers, we can edit our pictures by *replacing* existing primitives by new versions, thus solving the problem of constrained rubber-banding. Provided that the amount of space required in the structure store by the new data is the same as the old, it should be possible for this to be done quite efficiently. However, PHIGS has overheads of a different kind — namely complex data structures needed to permit this kind of low-level editing. More complicated edits may require a fair amount of pointer manipulation, garbage collection and the like, making them correspondingly slow for changes of this kind. The following code illustrates the technique:

```
Open_Structure(component);        {Open structure for editing at
                                  end of current data}
Set_Polyline_Index(tracktype)     {Select correct width etc}
Label(track);                     {Make a label for subsequent use}
 repeat
    Sample_Locator(pnts[2].x,pnts[2].y);
                                  {Define the end of the track
                                   assumes start already defined}
    Constrain(pnts);
    if Inside_NoGo(pnts) then
        Warning_Message('Inside No Go area!!!')
    else
    begin
        Set_Elementptr_To_Label(track);
                                  {Set pointer to track}
        Polyline(pnts)            {Uses point returned by locator}
                                  {Assumes editing mode is REPLACE}
    end
until not Mouse_Button;
pnts[1] := pnts[2];
```

Some Conclusions

As we have seen, for this type of application PHIGS may have some advantages over GKS-3D. It has more appropriate structuring capabilities which require less work of the application program in managing the mappings between application data and graphics data.

However, the problems of using segment transformations or modelling transformations to position different parts of the picture, which must subsequently be connected together, raise real doubts about their suitability for this. PHIGS is marginally better, but the real problem is that neither system was designed to address the issue of *connectedness* of different parts of the picture. It is clear that if the application program must know where the parts of the picture are positioned then it must perform its own transformations. The only value of having PHIGS or GKS perform them is the ability to re-display the picture rapidly when changes are made.

In order to have a responsive user interface, both GKS and PHIGS will require a fairly fast (i.e. expensive) display. In general, both systems will regenerate the entire picture when some small part of it is altered, such as when rubber-banding a constrained line. Both systems will only work really well on displays which support some kind of display

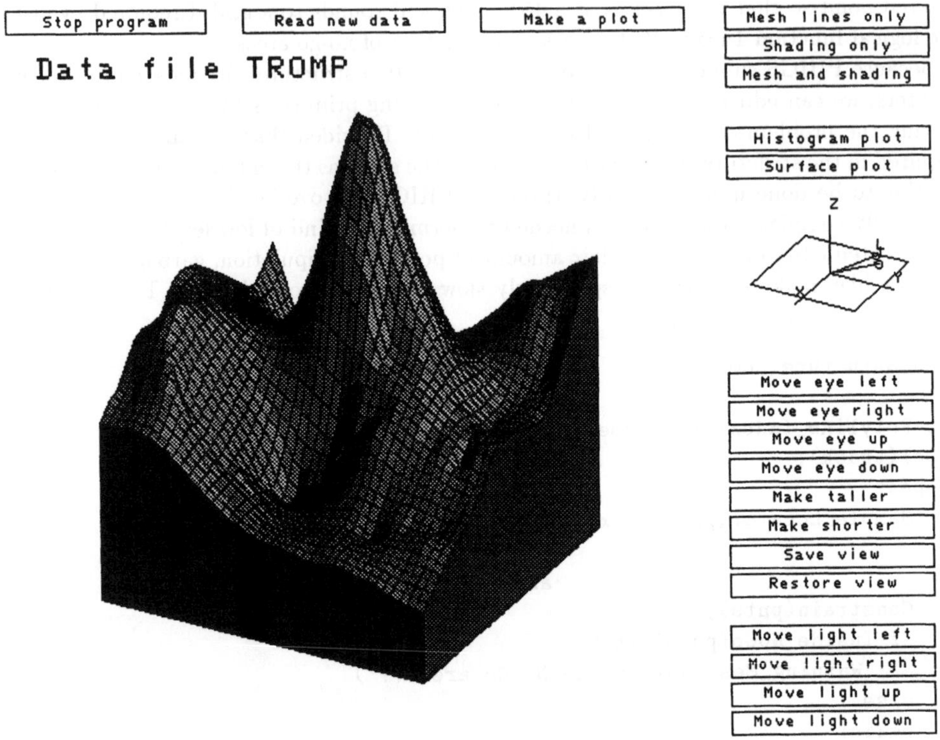

FIGURE 3.11. Screen Dump from Surface Display Program

list and can regenerate the picture from this very rapidly. For reasonable interaction a redraw rate of about 10 frames per second is necessary.

One must conclude that for this type of application a better approach might be to use a bit-mapped colour display, to store data for different layers of the PCB in different video memory planes, and to use raster-op instructions to manipulate the picture. This solution, of course, would very likely be device-dependent. Also, such techniques do not work for 3D graphics, so it is perhaps unfair to use this as an argument against GKS-3D and PHIGS, although it could justifiably be levelled against GKS.

3.9.3 Case Study II - Interactive Surface Display with GKS-3D

This example illustrates the use of GKS-3D for viewing a 3D surface described as an array of spot heights. The data could represent some function, or might be gathered experimentally.

Figure 3.11 shows a dump of a typical display. (The original image is in colour.) Here, the surface shows variations in the earth's gravitational field at Trompsberg in South Africa [3]. The original program was designed to operate with a straightforward raster display with no segment store or transformation hardware (a Genisco GCT3000). A subsequent version of the program was implemented for a high-performance display with a segmented display list and very fast transformation hardware (a Spectragraphics 1500). Both implementations

[3] Thanks to Dr W.T.C. Sowerbutts for data.

were constructed with a graphics system called GINO-M [10]. It was felt that it would be instructive to see if the same application could be implemented easily with GKS-3D.

First, a description of the original Genisco version. The user interface features the following objects and actions:

- The main object is a surface. This is the large display. The purpose of the program is to manipulate this display by adjusting the viewing direction and the position of a single light source
- In the middle of the right hand side of the picture is an icon providing feedback to the user. This is employed in adjusting the view and light source position by mimicing the orientation of the main display. The control object is simple enough that it can be redrawn quite rapidly using software transformations. This allows dynamic manipulation, so that the user can adjust the orientation until it looks correct
- The other objects in the scene are menus which specify actions to be carried out. There are 5 principal menus:

    1. **Stop program, Read new data, Make a plot**. These should be self-explanatory. The user selects a menu by "clicking" on the display with a mouse or tablet
    2. **Mesh lines only, Shading only, Mesh and shading, Base Block, No base block**. These control how the surface is displayed — as a wire frame diagram, as a shaded surface, as a combination (as in the figure), and with or without a base.
    3. **Histogram plot, Surface plot**. Selects whether the surface should be shown as a carpet plot (as in the figure) or as a 3D histogram
    4. **Move eye left** etc., **Make taller, Make shorter**. These commands alter the viewer's position relative to the surface, and allow the vertical scale to be adjusted. (By default the picture is automatically scaled to fit the available space.) Here, the initial menu choice is performed by clicking, but the program is designed to keep selecting this option as long as the cursor remains inside the menu box and the mouse button is depressed. Each time the command is executed, the icon is updated to show the correct view. The main display is only redrawn when the **Surface** or **Histogram** options are selected
    5. **Move light left** etc. These manoeuvre a light source relative to the surface, producing different shading effects. These menus also operate in a "continuous" mode. Again, the icon is updated.

This version of the program does not use segments. Each area of the screen is repainted as and when required. The icon uses two image memory work planes and a double-buffering technique to animate the orientation of the surface and the position of the light source. This is a device-dependent feature which only works under certain circumstances and is not available with GKS-3D.

The Spectragraphics version of the program was simplified by removing the icon, since the new orientation could be displayed by changing a hardware transformation. This version did not support shading.

Initially, we examine how to view the surface without shading. The implementation of this program with GKS-3D is as follows:

- Each of the different parts of the picture is stored in one or more segments. This allows the definitions of the parts to be edited independently

- The parts of the picture are built using the standard GKS-3D primitives. The surface plot can be drawn with polylines, or with quadrilateral facets (fill areas). The surface is displayed with a particular *view transformation*

- The main plot and the icon can each be rotated to show different viewpoints by altering appropriate viewing transformations. This will work quite well in *wire frame* mode but will only work for a *solid surface* if the GKS-3D implementation supports hidden surface removal. With this application a simple hidden-surface method suffices, such as those based on depth sorting, or a z-buffer

- Although segments have been used, menu selection is performed by comparing cursor coordinates rather than by using pick. (This was because the original system did not support picking on the Genisco.) The cursor (locator) is initialised to *request* mode to make an initial selection. In the event that one of the "continuous" mode menus is selected the locator is switched to *sample* mode to test when the cursor moves outside the menu area. The mouse buttons are also sampled to test whether the button has been released. When either condition is met the locator is switched back to *request* mode.

The original program also performed shading of the surface facets, so next we examine the impact of this on our GKS-3D solution.

Neither GKS-3D nor PHIGS includes any lighting or reflectance models for surface shading. Therefore, the shading must be done by the application program. This is not too complicated, since the orientation of the surface and the position of the light are both known. It entails computing a surface normal vector for each facet and deriving an intensity/colour for filling the facet. The quadrilateral facets of the surface may not be exactly planar. The program assumes that they are small enough that this does not matter.

However, there is a problem. If the light source is moved relative to the surface, then the computed intensities will change. We have no option in GKS-3D other than to re-generate the picture, computing new fill colours for each facet. There is no method for editing the fill colours of previously generated facets.

If the viewpoint is changed we can differentiate two cases:

- The lights remain fixed relative to the *surface*. There is no need to change the fill colours, so we can alter the viewing transformation and rely on GKS-3D to regenerate the picture

- The lights remain fixed relative to the *viewer*, so they move relative to the surface. The facet intensities will alter, so we must regenerate the picture with new fill colours.

Clearly, each time the light source position is altered we will have to regenerate the picture.

Conclusions

In some respects GKS-3D is quite well suited to this problem. Its window/viewport facilities can be employed to map different parts of the picture, such as menus and main display area, to different parts of the display.

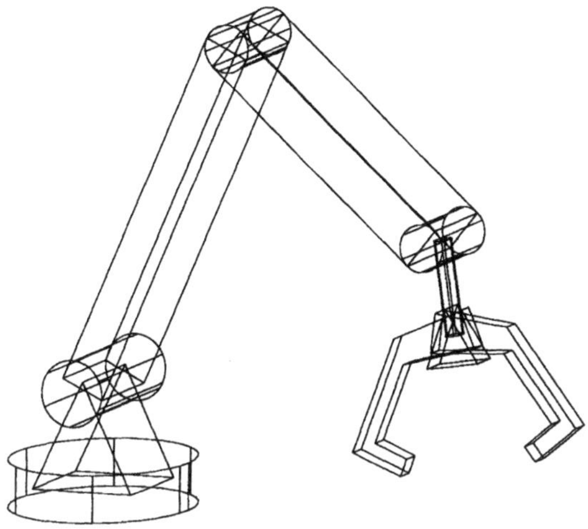

FIGURE 3.12. PHIGS Robot

The surface can be viewed from different angles by changing a single viewing transformation, and GKS-3D will update the display accordingly. This should work in all cases for a wire frame view. A (constant) shaded surface can also be depicted provided that the implementation supports hidden surface processing.

This program nicely illustrates the need for a display with a segment store to get the most out of GKS-3D. Even quite simple surfaces have a few hundred facets, and more complicated ones have several thousand. It is very tedious for an application program to have to regenerate the data for a picture of this complexity. Most GKS-3D implementations are not intelligent enough to refresh only those parts of a picture which have actually changed (which is not easy to do for 3D problems) so making minor alterations usually results in the entire display being redrawn. Without special 3D hardware it can take a long time to display pictures like these. On the more advanced Spectragraphics the program works well. Changes of viewpoint can be accomplished simply by altering the viewing transformation.

However, if a variably shaded surface is required then GKS-3D cannot calculate the required facet shades, so the application program must do this and regenerate the picture each time the viewpoint or light source position is altered.

In fact, similar conclusions can be drawn about PHIGS. The structuring capabilities of PHIGS are not really needed for this problem, and PHIGS cannot do the shading calculations. However, PHIGS PLUS is specifically designed for this and would be the obvious choice if shading were deemed necessary.

3.9.4 Case Study III — A Simple Robot

This example is frequently employed to illustrate the value of hierarchical models and transformations in PHIGS. The task is to display and manipulate a simple robot arm, shown in figure 3.12.

The robot is constructed from a number of parts which are mostly instances of basic shapes, such as a cube and a cylinder:

- A cylindrical base (structure `Rob`)
- An arm pivot, which is a triangular prism (`ArmPivot`)

- A cylindrical shoulder joint (`ShoulderJoint`)
- A square section upper arm (`UpperArm`)
- A cylindrical elbow joint (`ElbowJoint`)
- A square section lower arm (`LowerArm`)
- A cylindrical wrist joint (`WristJoint`)
- A circular section wrist (`Wrist`)
- A triangular section gripper pivot (`GripperPivot`)
- A pair of grippers (`Gripper1, Gripper2`).

The basic parts are defined to be of unit dimensions and are scaled to the required size by local *modelling transformations.* Transformations are also employed to position the parts relative to each other. For example, the lower arm is positioned using a local modelling co-ordinate system whose origin is located at the shoulder joint and whose x axis points along the upper arm. Then when the upper arm is rotated about the shoulder, the co-ordinate system used to define the lower arm also moves, so that the lower arm itself moves. Similarly, the wrist is positioned relative to the lower arm, and the grippers are positioned relative to the wrist.

In order to animate the robot, we define rotation transformations at each joint. As each transformation is changed the robot moves. Because each part is defined in a relative fashion, moving one part causes all parts further along the structure (lower down the PHIGS hierarchy) to move appropriately.

With most applications such as this there are several ways in which the PHIGS structures could be defined; figure 3.13 shows one possibility. There are four basic building blocks which can be instanced, each defined as a structure: `Cylinder, Pivot, Cube` and `Gripper`. Associated with each structure invocation is a modelling transformation, labelled `M` in the figure, and for each joint which can be manipulated there is a transformation labelled `X`.

The rotation angles can be read from a suitable input tool, such as a set of *valuators* (e.g., dials). The application program *samples* the dials, recomputes any matrices which have changed since the last sample, and edits the structures in order to *replace* the old transformations by the new ones. Checking for valid angles (for example to ensure that the robot does not collide with itself) is the responsibility of the application program.

Now consider how we could get the robot to pick up a glass of beer without spilling it. When a single angle is altered, for example to move the upper or lower arm, the remainder of the machine moves in a rigid body fashion, which would cause the glass to get tilted. This can be avoided by keeping the orientation of the grippers constant (except when the beer is to be poured!). To do this, we must keep track of any accumulated angles of rotation at the shoulder and elbow, and apply an *equal and opposite* rotation at the wrist. This necessitates changing *at least two* transformation matrices at once. In order that updates to the displayed image remain synchronised, we must use the `Set Display Update Status` function of PHIGS so that refreshing of the screen is only performed after all the relevant matrices have been redefined.

Much more seriously, if we wished to simulate a real robot, which has to obey certain constraints, such as avoiding objects in its way, not moving through the ground etc, then PHIGS is of almost no value whatsoever for anything other than actually displaying the

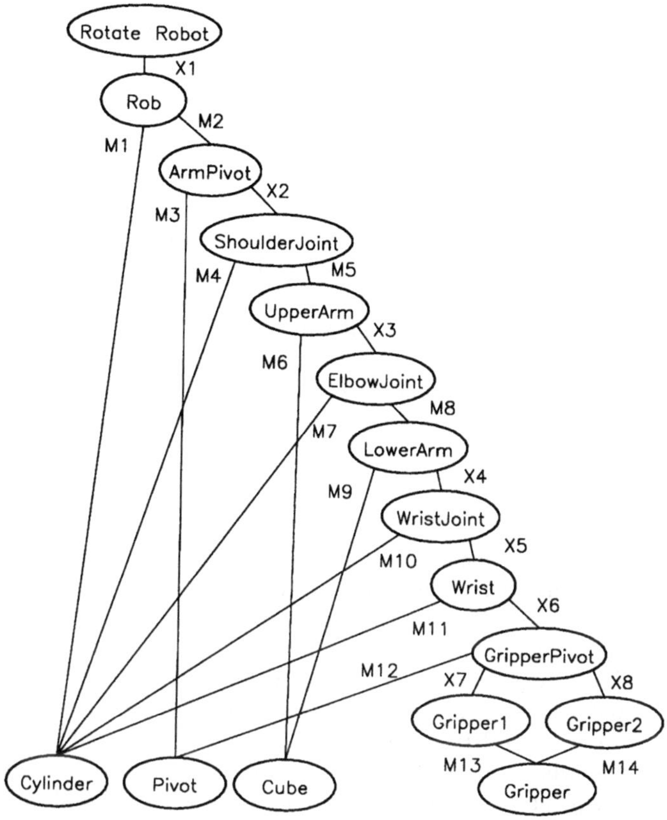

FIGURE 3.13. PHIGS Structures and Transformations for Robot

picture. As with the PCB problem, there is no way of finding out from PHIGS where, in world coordinates, the end of the robot is; the application program will have to compute this for itself. It will have to perform the equivalent of a traversal, applying transformations in order to track the movements of the parts. If the application must do its own transformations in order to apply constraint checks, and if the constraints must be applied during manipulation, then some of the speed with which pictures can be manipulated will be lost — PHIGS will have to wait while the application carries out its own computations.

Some Conclusions

This is a nice example of the use of PHIGS to display dynamically articulated structures, and is often to be seen in demonstrations by manufacturers. However, it must be said that the degree of interaction and editing required to change the robot's position is trivial, requiring only that a smallish number of transformations be *replaced*. Thus, although a nice demonstration, a program like this cannot be used to judge how well PHIGS will perform for more complex editing. An indication can be gained by seeing how long a particular implementation takes to *build* the structure store for the robot — it is often quite a long time, and the space required can be staggering, often tens or even hundreds of kilobytes.

Although the robot is ideally suited as an example of PHIGS in action, the design of the structure hierarchy is non-trivial and requires time and effort to get right. PHIGS does not know *how* the robot parts are connected together. In fact, in the PHIGS representation they are *not* connected. The fact that they appear to be relies entirely on getting the transformations correct, so that the different parts appear to move as one. The way in which one would describe the robot to PHIGS and the way one would describe it to another person are very different. For example, verbally, one would probably explain that the robot has certain parts with particular shapes which are *connected* at certain points. This notion of things being connected occurs again and again in real applications where we are trying to explain how things are constructed. It is the lack of such a concept in PHIGS which makes it difficult to use directly for this sort of task. It is probable that a layer of software on top of PHIGS could be used to overcome this problem, but at present the effort needed to design correct structures for 3D problems should not be underestimated.

The claim that PHIGS performs *modelling* can be seen to somewhat overstating its value. It *is* good at *displaying* dynamically changing pictures and editing them, but not at applying constraints or supplying any useful geometric information back to the application program.

3.9.5 Case Study IV — Molecular Modelling Example

The sort of rigid body movement exemplified by the robot is quite common in molecular graphics applications. Here, the task is typically to carry out a "docking" operation between two molecules, or to modify the structure of a molecule by rotating a part of it about some particular bond. These operations differ from the robot example because the specific manipulations required may not be known in advance. It is here that the editing capability of PHIGS comes to the fore by allowing new local sets of axes to be introduced.

We assume here that the PHIGS structures are initially created with an appropriate hierarchy, such that when a molecule is to be "broken" at a particular bond (to perform the rotation) subsequent elements encountered during traversal will be affected by the transformations introduced at this point.

The transformations are introduced as follows:

- When the molecules are originally displayed, each bond is assigned a different *pick identifier*
- A bond about which a rotation is to be performed can then be indicated with a *pick* device. The *pick_id* is used to access the relevant application data
- The two end points of the bond define the axis about which the rotation is to be made and an origin for a new *local* co-ordinate system. We might assume, for example, that the bond is the x axis in the new system
- Assume that the co-ordinates of the atoms to be rotated are specified with respect to some local co-ordinate system and that this system is subject to a transformation T_1. We compute a matrix which relates the *new* local system to the original system. Call this matrix T_2; it will comprise translation and rotation terms
- To perform the desired rotations, we must introduce at the position of the bond within the PHIGS structure a sequence of three transformations:
  1. A transformation to make the new co-ordinate system the current one (T_2)
  2. A transformation to *rotate* about the x axis of the new local system (T_3)
  3. A transformation to relate the new local system back to the original axes (the inverse of T_2).

Note that this sequence is the *logical order* in which the transformations must be applied to the world co-ordinates, which means that the sequence which must be introduced into the structure, assuming column vector notation for the coordinates of points, is

$$T_2^{-1} \cdot T_3 \cdot T_2$$

These transformations can be specified in whichever order is most convenient by using either pre- or post-concatenation. By keeping the three transformations separate in the structure, an application can effect the desired rotation by replacing just T_3.

Note that having edited the structure to include these transformations, they must be left there. At some stage the application program may need to modify its own copy of the co-ordinate data to take account of the new atom positions resulting from interactive manipulation. It is also worth noting that care must be taken when computing such matrices to keep a *master transformation* computed from a fixed base position. Otherwise, if transformations are applied incrementally, rounding errors can creep in, so that after some minutes of manipulation the transformed objects begin to distort.

In trying to update the application data structure's copy of the coordinates we encounter the same problem as before — there is no way we can find out from PHIGS where the transformed points are. In molecular modelling we also wish to be able to check distances between structures and once again the application will have to perform all the transformations itself.

3.9.6 General Conclusions

There is little to choose between GKS-3D and PHIGS in terms of the display primitives they offer. However, PHIGS does have the ability to represent more closely than GKS-3D the hierarchical nature of some types of picture. PHIGS PLUS offers the advantages of more powerful primitives and illumination and shading, which can helpful for some problems. If shaded images are deemed necessary then PHIGS PLUS is the obvious choice (other than alternative rendering systems), since neither GKS-3D nor PHIGS will perform shading.

The editing functions of PHIGS are superior to the crude segment replacement offered by GKS-3D. The division of a picture into a hierarchy of structures can be made on the basis of the problem being modelled, rather than for secondary reasons like editing strategies which may be needed when using segments. However, it must be recognised that segment storage in GKS is simpler to implement and maps readily on to a number of display architectures. In contrast, PHIGS is complex and will almost invariably require special hardware to operate efficiently. The space required for the CSS in PHIGS will often be several megabytes, and will generally be much larger than the equivalent GKS-3D segment store.

The design of PHIGS structures is not easy. It requires practice to design optimal networks for articulated objects, due to the need to think of the structure of objects in PHIGS terms, rather than as one might describe them to another person.

The claim that PHIGS is concerned with *modelling* should be treated with some scepticism. PHIGS is good at *displaying* complex models, but is useless for answering geometric queries or applying application-dependent constraints. The latter must be performed in parallel by the application program and, in practice, this may mean that the power of PHIGS cannot be fully exploited. Because of this, GKS-3D may be a better choice for

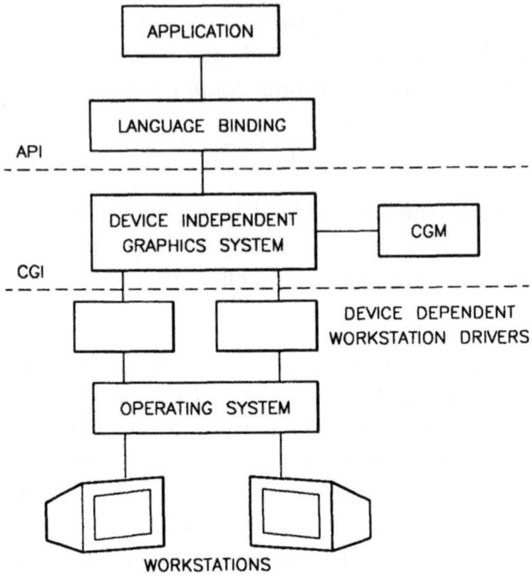

FIGURE 3.14. Reference Model for Computer Graphics

some applications, bearing in mind that it will generally run on cheaper hardware and be cheaper to buy and maintain than PHIGS.

3.10 A Reference Model for Computer Graphics

In figure 3.14 reference model for the structure of a device independent graphics system such as GKS-3D or PHIGS is illustrated. The following interfaces (dotted lines) of the structure are worthy of mention:

- **The Application Programmer Interface.** The application program includes modules from a graphics software library (such as GKS-3D). These modules are accessible from different programming languages — such as FORTRAN, C, Pascal, Ada — via a *graphics language binding.* For each language, the binding represents the *application program interface (API).*

  The application program contains specifications of the objects within the application data structure. Generally, data other than graphical entities must also be stored, such as physical properties, component numbers, stock control data, and cross-references. The rules for processing the objects (design rules) must also be encapsulated. The application program describes to the graphics system in geometric terms (lines, arcs etc.) the portion of the users's world (i.e., the application data strucutre) to be viewed. Non-graphical properties of the (application) objects must be presented to the graphics system, e.g., colour to represent tensile strength, or displaying a transistor as a circle and some lines. The application program must analyse and respond to input provided by the operator

- **The Graphics Systems Interface.** This is sometimes termed the *Computer Graphics Interface*, or CGI. The device independent graphics system communicates with idealised devices, termed *workstations.* In practice, each workstation maps on to one or more physical devices via a device dependent *workstation driver.* A number of logical workstations might map to a single physical device, such as a physical workstation running a window manager.

Information for display is presented to the graphics system as graphic output primitives. (Application) objects can only be drawn using these primitives. The graphics system checks the validity of the data and converts the device independent description to instructions or order codes that are communicated to the display device. All operating system dependencies are hidden from the application programmer; it is the task of the GKS or PHIGS implementor (the graphics programmer) to deal with these. The graphics system may also have to simulate capablities not found on the particular device to which the output is being sent. This includes generating text for devices that do not have hardware text, generating line segments to simulate dashed lines, generating the interior points of polygons to fill them. The application program must supply information about how the graphics primitives are to be viewed, such as the position of a 'camera' in the world of the application data structure.

For the application to create another 'view' of the same objects is easy: it invokes the `SET VIEW REPRESENTATION` function. If the hardware has stored the segments or structures then the graphics software merely passes the relevant parameters to the hardware. If however the segment store or CSS is mainatined in the host CPU, then the graphics system must perform a traversal each time a change is made and re-transmit display instructions to the workstation. Systems which operate like this tend to be slow!

The device independent graphics system may provide a method for storing pictures — usually a *metafile* (GKS-3D), or a *structure archive* (PHIGS). This is labelled CGM (for Computer Graphics Metafile) in the figure

- **The Operating System Interface.** Once the graphics system has converted from the device independent picture specification to the form suitable for the device, it must be transmitted to the device. In general, various operating system services are called upon to achieve this. Many different ways of connecting are possible, but the most common ones are a simple terminal line (RS232), a parallel interface (e.g. VME bus), or an Ethernet connection. RS232 interfaces are popular for cheap devices but are generally inappropriate for high-perfomance systems. Nonetheless a number of 'GKS terminals' employ this kind of connection, with I/O routed through the system's terminal driver. In this case, the graphics programmer accesses devices through *WRITELN* (Pascal) or *WRITE* (Fortran) functions. Where more complex interfaces are used the operating system device driver must perform any necessary processing, such as handling interrupts, and protocol checking for networks.

  To absolve the graphics programmer from having to worry about such details, many graphics equipment manufacturers provide a utility package which hides all the operating system's dependencies. Obviously the number of versions of this depends on how many different computers the manufacturer is asked to support. This extra layer brings about portability and ease of implementation but the cost may be slower execution of the transmission of data.

3.11 Assessing GKS-3D and PHIGS Implementations

A GKS-3D or PHIGS implementation cannot be assessed on its own. GKS-3D and PHIGS are only tools used by the application. The true measure of success is how well the total combination of hardware, graphics software and application program does the job intended. Thus one may consider radically different solutions to a problem. It may be necessary to consider application X/graphics system Y/hardware Z, against application A/graphics system B/hardware system C, rather than application X on Y/Z versus application X on B/C.

It is therefore necessary to be clear at the outset about both the criteria used to select the GKS-3D or PHIGS implementation and their relative importance:

- **Audience.** Why are you buying this system? For whom? Do they have the same selection criteria as you? Is it just the software you are purchasing or the complete system?
- **Cost.** Capital and Recurrent. Any other costs: delivery, installation, building works, training?
- **Functionality.** How close is the proposed system to a Standard? Is this important? Do you need GKS, GKS-3D, PHIGS, or another graphics package?
- **Installation issues.** How easy is it to install, and upgrade? What software do you need to develop? How easy/hard is it to add additional device drivers? What is the availability of these?
- **Portability.** Do you want the software itself to be portable, and/or the device drivers and/or the application program?
- **Lifetime.** How long is the manufacturer going to support this software? Will you get upgrades as the software improves? Will it be able to support new hardware that becomes available in say 1 year, or 5 year's time?
- **Hardware.** Is the underlying hardware important for cost, performance? Are there any constraints: e.g., it's got to run on this hardware because the company has already bought thousands of them?
- **Expandability.** Once a system has been bought it is almost inevitable that at some time the demand will be such that the system will need upgrading. Unless you are certain that this will not be the case (or you are prepared to ensure that this will not be the case) then the 'upgradability' of a system should be examined.

There can be no substitute for seeing the system carrying out the application you are interested in. It's obviously useful to see prepared demonstrations, but don't forget you will *not* be shown things that the system is slow at, or can not do. During the demonstration ask questions like "What if...?" Again, find out exactly what is being shown: is it the model/system you need or the bigger and better one you can't afford?

3.11.1 The Manufacturer

- **Beware The Glossy Brochure!** This will present the product in the best light! It will give performance figures for the *top of the range* product in the *ideal* situation, and then give you then entry level price. **You** must remember that you are not buying the model in the brochure. These brochures are designed to be ambiguous. You should question everything; things which look as though they are standard could well be optional extras. The only consolation is that most manufacturers are as bad (or as good) at this as each other. Brochures are a useful starting point, and do provide some pretty pictures

- **The Salesman.** An extension of the brochure. This person believes that the product will solve all of your problems all of the time
- **The Technical Manuals.** The best source of information about the functionality of the system, though not the performance. However, there's usually a lot to read
- **The Technical Support Person.** Now you are getting close to the answers you really want. This person believes that their product is the one you should buy but will answer your technical questions effectively. Has a habit of changing your problem to one he/she knows their company can solve
- **The Designer.** If you get this person, then you are winning! The manufacturer is worried that they have not satisfied you that their product is the best, so they are bringing in the big guns. The problem is the person will not show you how to solve your problem with the product (by bypassing most of the system); he or she will promise that the next model (in the research lab) will do everything you require.

3.11.2 Benchmarks

To help us understand these complex and expensive systems, manufacturers come up with one number to prove their machine is the fastest etc. From the manufacturer's viewpoint the benchmark should be vague and open to all sorts of interpretation. Don't be fooled by performance figures: the system will only be as good in practice as it's weakest part. It is no use having 500,000 3D transformations per second if the device has an abysmal polygon fill rate which you are dependent upon. Big numbers look impressive until you realise that 10 to 20 frames per second are needed for adequate animation/manipulation.

The only good benchmarks are the ones you write and run yourself. Benchmarks tend to be aimed more at the hardware performance, rather than the software performance. In PHIGS and GKS-3D systems its usually the number of 3D transformations per second. The brochures never make it clear exactly what that entails. Does it include clipping, perspective projection, transmission times, time for the application program to create these vectors? If clipping is included how many vectors were actually clipped; recall most clipping algorithms are much quicker if all the vectors tested are completely inside the clipping volume! Typically it is for vectors that produce 10 pixels on the screen, but one manufacturer has been known to quote the time for one pixel vectors!

The only way to judge several systems is to devise your own simple benchmarks. Rather than one big one it is better to use several simpler ones. It can be difficult to interpret the results if it's a complex benchmark. Remember, you are effectively measuring the slowest part in a complex chain of operations, involving the application software, graphics software, operating system, and the underlying (graphics) hardware. Most manufacturers are very helpful when it comes to letting you run tests, but make sure you run them on the configuration you are thinking of purchasing!

3.11.3 Conformance Certificates

A significant problem with graphics standards is ensuring that they match the functional specification. From the day a standard becomes a project within ISO to the day it becomes an International Standard, many versions (paper specifications) of it are produced. These specifications are readily available. Thus, many manufacturers, trying to get an edge on the competition, produce an implementation of the standard before it is finalised. It's always worth asking which version of the PHIGS or GKS-3D document they used. (At the time of writing neither the PHIGS or GKS-3D IS text has been published). One way to verify what is on offer is to ask to see a conformance certificate. To date (June 1988) these are only available for GKS systems, but there are projects under way to develop test suites for GKS-3D and PHIGS.

3.12 Survey of Available Implementations

This information is *not* included here because it rapidly becomes out of date. However, it will be presented during the tutorial.

Acknowledgements:

We should like to thank our colleagues in the Computer Graphics Unit for their highly valued support and comments: Tony Arnold, David Finnegan, Toby Howard, Manjula Patel and Karen Wyrwas.

3.13 References

[1] *NeWS Manual, Version 1.1.* Mountain View, CA 94043, 1987.

[2] J F Blinn and M E Newell. Clipping Using Homogeneous Coordinates. *Computer Graphics (Proc Siggraph 78)*, 12(3):245–251, 1978.

[3] B Borden. *Doré (Dynamic Object Rendering Environment) Description.* Ardent Computer, 880 W. Maude Ave., Sunnyvale, CA 94086, 1988.

[4] M D Brown. *Understanding PHIGS.* Template, San Diego, 1985.

[5] Phong Bui-Tuong. Illumination for Computer Generated Pictures. *Communications of the ACM*, 18(6), 1975.

[6] H Gouraud. Computer Display of Curved Surfaces. Technical Report UTEC-CSc-71-113, University of Utah, 1971.

[7] I Herman and J Reviczky. A Means to Improve the GKS-3D/PHIGS Output Pipeline Implementation. In *Proceedings of Eurographics '87*, Amsterdam, 1987. North-Holland.

[8] F R A Hopgood, J R Gallop, D A Duce, and D C Sutcliffe. *An Introduction to the Graphical Kernel System (GKS).* Academic Press, London, second edition, 1986.

[9] T Howard. A Shareable Centralised Database for KRT^3 – a hierarchical graphics system based on PHIGS. In *Proceedings Eurographics 1987*, Amsterdam, 1987. North-Holland.

[10] R J Hubbold and P J Bramhall. A Flexible, High-Performance, Interactive Graphics System. *Computer Graphics (Proc. Siggraph 78)*, 12(3), 1978.

[11] Adobe Systems Inc. *PostScript Language Reference Manual.* Addison-Wesley, Reading, Massachusetts, 1985.

[12] International Organisation for Standardisation (ISO). *ISO–7942 Information Processing Systems — Computer Graphics Graphical Kernel System (GKS) functional description*, 1985.

[13] International Organisation for Standardisation (ISO). *ISO 8805 Information Processing Systems — Computer Graphics, Graphical Kernel System for Three Dimensions (GKS–3D) functional description*, 1988.

[14] International Organisation for Standardisation (ISO). *ISO IEC JTC 1, N2, PEX Protocol Specification*, 1988.

[15] International Organisation for Standardisation (ISO). *ISO IEC JTC 1, N3, PHIGS+ Functional Description Rev 3.0*, 1988.

[16] International Organisation for Standardisation (ISO). *ISO 9592 Information Processing Systems — Computer Graphics, Programmers' Hierarchical Interactive Graphics System (PHIGS)*, 1989.

[17] R Schiffler and J Gettys. The X Window System. *ACM Transactions on Graphics*, 5, 1986.

[18] K M Singleton. An implementation of the GKS-3D/PHIGS Viewing Pipeline. In *Proceedings Eurographics 1986*, Amsterdam, 1986. North-Holland.

[19] B R Thatch and A Mycklebust. A PHIGS-based Graphics Input Interface for Spatial Mechanism Design. *IEEE Computer Graphics and Applications*, 8, 1988.

4 Special Modelling

André Gagalowicz

ABSTRACT
Texture modelling and synthesis are first studied in a general framework. Models for planar black and white textures are extensively studied. This work is then generalized to the colour case and to textures lying on 3D surfaces.
Graftals may simulate various plants and trees. They are based upon the use of parallel rewriting grammars. Various *fractal* synthesis techniques are described next. A particular interest is given to Barnsley's IFS (integrated function system) model. *Particle systems* of W Reeves simulate beautifully fires, plants and trees. *Solid texturing* is a new possibility to produce textured objects. A solid texture block is sculptured in order to obtain the desired object contour. We present the most striking applications of this technique by Perlin to the design of marble, glass objects, fires and bubbles, and for clouds by Gardner. *Botanistic models* of the French botanist De Reffye who discovered a model available for most types of trees and for their growth, are implemented by various French researchers from the AMAP group. *Planar graphs* of Lienhardt and Françon are used to model leaves and flowers grown more geometrically.
This tutorial is dedicated mainly to advanced users and developers, and presents some non-standard techniques tractable to present particular objects (trees, plants etc...) and the texture of their surface.

4.1 Texture Modelling and Synthesis

This part is dedicated to a texture approach guided by a psychovisual texture model. We first describe texture modelling considerations and then apply them to computer graphics. The models obtained are such that artificial textures constructed from these models are hardly visually discriminable from real samples on which the models were computed. The models obtained from a texture sample can be used to cover a 3D object of any shape and size.

The texturing method has the ability to reproduce a vast class of textures and presents the following advantages:

- a small amount of data characterizes each texture (40 to 2000 coefficients)
- there is no repetition in the synthesized texture
- as synthesis presents a feedback loop it is rather insensitive to aliasing effects
- mapping difficulties are not so crucial as synthesis is done in a pixelwise way and controlled by a texture model
- synthesis may be achieved theoretically on any surface described either analytically, by patches or by a set of points.

As a result of being model driven, the rendering of these textures may be slightly degraded.

We briefly review previous work citing some of the existing problems in order to highlight the features of the method proposed here.

Catmull [16] presented the first technique to generate texture information on parametric surfaces. As no model of texture existed, he decided to directly map images of planar natural textures onto 3D surfaces. Catmull's mapping technique establishes a one-to-one continuously differentiable mapping from a rectangular patch of the natural texture plane onto its curved version lying on the 3D surface. It remains, then, to partition the texture plane into rectangles and to do the equivalent partitioning on the 3D surface. This latter partitioning may be difficult to perform and has to be done manually in some cases. Catmull does not mention any surface partitioning algorithm, which is a limitation for his technique.

Blinn and Newell [10] generalized this work while incorporating reflections on objects. Then Blinn [8] made it feasible to create undulations on the surface while perturbing locally the normal to the surface. This simple heuristic method is limited to the rendering of a small class of textures.

On the contrary, mapping techniques may be used for any type of texture, but they mainly work well for simple surfaces and for certain types of textures due to aliasing effects and other spatial distortions (when a surface patch is small, the mapped texture is small; when a patch is larger, the texture is also larger, so that it does not appear as being homogeneous).

Blinn [8] and Feibush et al. [31] implemented good anti-aliasing algorithms. Catmull and Smith [17], Norton et al. [68], Williams [88] and finally Crow [20] obtained better computing costs using various filtered versions of the planar texture.

Based on the previous advances on mapping techniques, it is possible to achieve a very good rendering of natural textures on 3D surfaces. Nevertheless there still remain a number of drawbacks and shortcomings:

- No solution seems to be available in the case of general 3D surfaces where there is no partitioning procedure for the construction of the various curved patches on the surface
- It is necessary to use large image data bases since planar textures have to be mapped directly onto a surface which may be of any extent; otherwise a planar texture must be repeated on the surface. This repetition is then perceived and may be very displeasing
- In the case of the use of planar texture of sufficient extent, additional computation is needed still to connect and smooth the borders of the curved patches so as to avoid repetitions
- It is necessary to use anti-aliasing techniques which may be computationally intensive
- The mapping itself may be also computationally intensive
- As mentioned above, spatial distortions appear due to the different sizes of the surface patches.

All these reasons motivated the computer graphics community to investigate other techniques. Following Mandelbrot's research [66], numerous researchers were attracted by *fractal* models in order to represent textures. The interest of such a method is obvious as one then has a model. This approach will be studied in section 4.2.1. Fractal models generally only use a few parameters (2 or 3) to reproduce a fractal texture of any extent and at any resolution. Thus, the texture data base is small; aliasing effects are not crucial.

Their great simplicity also leads to their limitations: textures are not all fractal, and there is no analytical methodology with which one can determine the type of fractal model and the parameter values for a desired fractal which approximates a given texture. The use of fractals is thus ad hoc. It happens that such models reproduce rather well scenes consisting of mountains, valleys, fires and some plants. To our knowledge, this is the extent of the usefulness of fractal models for texture reproduction (see Fournier, Fussell and Carpenter [33] for an adaptive implementation of such algorithms).

A syntactic approach, very different from the former methods, has been used to represent various plants and trees. This approach, studied in chapter 4.2.2 is based upon rewriting parallel grammars developed by Lindenmayer [61]. The elements of the language, called *graftals* possess a certain similarity with fractals in that they are also defined at various resolutions. The realism of Smith's applications of this technique is striking in Smith [80].

Reeves [76] used a very different model: *a particle system* flowing through surfaces, to reproduce plants and trees. Results are splendid but at a high cost in computing time (see 4.2.3).

Solid texturing (section 4.2.4) was a very different approach which gave birth to new types of pictures too. Texture was no longer considered as a surface property but as a volumetric entity. Peachey [72] produced wood and marble effects. Ken Perlin[74] obtained wonderful images of marble, glass and waves and Gardner [44], simulations of clouds.

Recently, research has moved to an analysis of the physical phenomena creating specific textural effects. Weil [87] Terzopoulos et al. [82] studied the appearance of cloth suspended at certain constraint points to design very realistic pieces of cloth. Peachey [73] and Fournier, Reeves [34] were themselves interested by ocean waves.

Natural trees have been synthesized from botanical models by the AMAP group. This will be the object of section 4.2.5.

The extension of the previous work to leaves and flowers models using free surface modelling and evolution was performed by P. Lienhardt. This work is discussed in section 4.2.6.

Texture modelling has been extensively studied in Computer Vision. A bibliography of the literature concerning this topic may be found in [38]. Most publications have provided a choice of ad hoc parameters to solve classification and segmentation problems. As an example, we provide samples of this literature concerning year 1987 [89]-[23]. Approaches concerning models aimed at faithfully reproducing a given texture are not so common. We want to mention work performed by Julesz [15, 50, 51, 52] which is noteworthy in this direction.

In this first part, we study textures and start with texture modelling considerations.

4.1.1 General Considerations

We may consider that two different types of information exist in images:

edges which are due to the contours of the various regions of an image. This information is 1-dimensional information (lines)

texture which is the spatial information of each region. This information is 2D.

It is easy to model numerically edge information but the reader may be easily convinced that to propose a numerical model usable for texture detection and/or recognition is not so obvious. It is this task that we want to solve.

Qualitatively, texture may be considered as *hierarchical* information with two levels in general (it is possible to conceive of more than two). The *lowest* level consists of the various primitives (or basic patterns) which may be found in the texture. Each primitive has a homogeneous non structured aspect. The *highest* level is the information necessary to describe the arrangement of the primitives. Some textures consist of only one pattern extended over the whole texture image which is then not structured; such textures are called *microscopic*. All other textures are called *macroscopic*.

As an example consider the texture of a brick wall. The lowest level of this macroscopic texture consists of brick and cement texture. The highest level is the organization of the various bricks and cement layers forming the wall.

The study of macroscopic textures is a priori difficult as it is related to shape modelling and recognition. Study of microscopic textures seems to be much simpler as it is related to low level perception and no shape recognizer is necessary. For these reasons we decided to *study first microscopic textures.*

Fundamental Properties of Microscopic Textures

We restrict ourselves to the case of visual textural information (we do not consider tactile textures, or other sensorial ones). If we observe an object surface, and eliminate perspective effects (we observe it in the direction of the normal to the surface), we perceive the same visual information. Thus microscopic texture information is *HOMOGENEOUS* all over its surface: we have the same visual impression whatever the origin we choose for our foveal axis on the texture surface. We could believe that texture perceived by the visual system only depends on the object state of surface but, texture information also depends on the object lighting and on the relative eye/object position, so that it becomes clear that texture information also depends on our visual system. The limits of a texture zone are given by our visual system. They do not compulsorily correspond to the object limits. Many attempts to model texture do not take the visual system into consideration. We do believe that it is a limitation for the methods studied.

When the visual system observes texture, it makes some measurements which are invariant under translation (homogeneity). An interesting problem is: what are the parameters computed by the visual system to characterize texture? (They are space invariant!)

Guidelines for the Construction of a Microscopic Model

We shall assume that texture is a realization of a homogeneous stochastic process (and shall verify a posteriori that it is a correct assumption). The interest of such an hypothesis is that stochastic processes produce images which are likely to fit microscopic textures, and stochastic processes are entirely defined by well known statistics where it is easy to "search" eventual models.

Our strategy is to search, among all possible statistics, the *set of discriminating parameters*, i.e., the minimal set of parameters such that, for two textures presented simultaneously on a screen, if this set is practically the same, the two textures will look the same, but if, at least one parameter varies significantly from texture to texture, then both textures look different. The *model proposed* will be simply *the set of discriminating parameters* (to find).

Analysis and Synthesis Duality

Analysis of the validity of a set of discriminating parameters requires the development of texture *synthesis* methods allowing control of this set of parameters: we need to be able to change any parameter value of a given texture at our convenience in order to perform visual comparisons between textures having different parameter values. Such synthesis methods have also the advantage of making the model tractable.

General Definitions

Texture is a realization of a stochastic process $X.X$ is defined on N^2 where N is the set of integers, and takes its values on the finite set L of grey levels $L = \{0, 1, \ldots, L-1\}$.

m is a loose expression for the location (i,j) of a pixel in the discrete image plane ($m = (i,j)$, where i is the row index and j the column index).

X_m is the random variable at location m, or the pixel value at this location (realization of the random variable) depending on the context.

4.1.2 Planar Textures

We first summarize previous results on planar textures.

Models Proposed for Planar Black and White Textures

Psychovisual experiments [35] have proven that texture discrimination is *local* so that the models proposed must be considered only *locally*. We have verified experimentally that a window of size 20 × 20 pixels may be a reasonable size of the local domain where our models are computed visually (it is the case when a human observer is at a position where he begins to discriminate pixels).

Model 1: Second Order Spatial Averages

We have shown [35, 36] that *all* microscopic textures may be modelled by a set of second order spatial averages $p_\Delta(L_1, L_2)$ where:

$$p_\Delta(L_1, L_2) = \frac{1}{I} \sum_{i=1}^{I} \delta(X_i - L_1)\delta(X_{i+\Delta} - L_2) \tag{4.1}$$

$\Delta = (\Delta x, \Delta y)$ is a translation of the plane, where Δx and Δy are integers defining the x-y coordinates of the translation

$L = \{0, 1, \ldots, L-1\}$ is the set of grey levels

δ is the Kronecker delta

$\mid \Delta \mid$ stands for the norm of Δ

$i + \Delta$ is the location of i translated by Δ

I is the number of pairs of points $(i, i+\Delta)$ in the texture plane.

The second order spatial average $p_\Delta(L_1, L_2)$ counts the number of co-occurrences of pairs of grey levels (L_1, L_2) when both grey levels are translated by Δ from one another. We have shown that this texture model must include $p\Delta(L_1, L_2)$ values corresponding to all possible pairs (L_1, L_2) and to all translations Δ such that their length $\mid \Delta \mid$ does not exceed 9′ (solid angle). It is a view-centred distance (see figure 4.2). Such a model may be easily computed on a texture sample (see figure 4.1). For any Δ considered, we scan the texture image as shown in figure 4.1 and add one to the counter $p\Delta(L_1, L_2)$ as soon as a pair (L_1, L_2) is met at the origin and extremity of the various vectors. This model is too rich. It is an upper bound of our set of discriminating parameters.

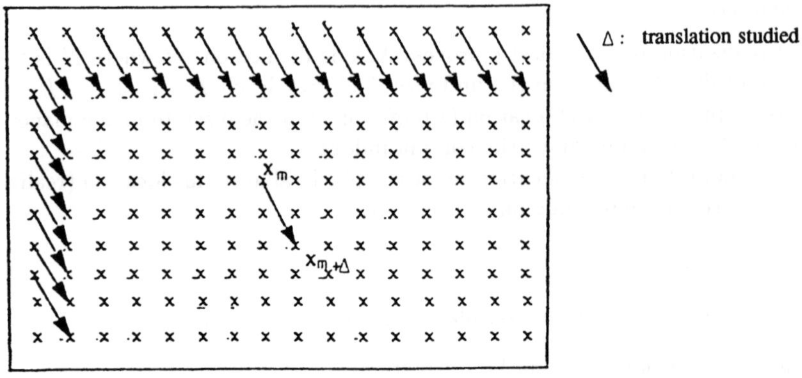

FIGURE 4.1. Computation of the second order spatial averages

Model 2: Autocovariance + Histogram

We have also proposed a second model [38, 36]. This model consists of autocovariance parameters $M_2(\Delta)$ where:

$$M_2(\Delta) = \frac{1}{I}\sum_{i=1}^{I}(X_i - \eta)(X_{i+\Delta} - \eta)/\sigma^2 \tag{4.2}$$

η is the texture mean and σ^2 its variance, for all translations Δ such that $\mid \Delta \mid \leq 9'$.

It is necessary to add the texture histogram $H(L_1)$ to the model, where:

$$H(L_1) = \frac{1}{I}\sum_{i=1}^{I}\delta(X_i - L_1) \tag{4.3}$$

Thus, model 2 is a *hybrid model*: $\{M_2(\Delta), H(L_1)\}$ defined by all the translations Δ such that $\mid \Delta \mid \leq 9'$ and all grey levels $L_1, (L_1 \in L)$.

For a given Δ translation, this second model considers only one parameter, instead of L^2 (a co-occurrence matrix) in the former model.

It is the most compact model but unfortunately it is not a set of discriminating parameters [38, 36]: it is possible to construct microscopic texture pairs which have the same autocovariance and the same histogram and which are nevertheless visually discriminable (even if this situation seldom occurs). Model 2 is thus a lower bound to the set of discriminating parameters.

Model 3: $S_{p,q}$ Moments + Histogram

Statistical considerations brought us to the study of $S_{p,q}$ statistics [40]. These statistics define the various moments available at the level of a pair of random variables X_i and $X_{i+\Delta}$:

$$S_{p,q}(\Delta) = \frac{1}{I}\sum_{i=1}^{I}(X_i - \eta)^p(X_{i+\Delta} - \eta)^q/\sigma^{p+q} \tag{4.4}$$

$S_{p,q}(\Delta)$ is simply the average cross product of lowest *powers* of the centred pixel values X_i and $X_{i+\Delta}$.

If we consider all integer pairs p and q, it is possible to show [63] that this set is equivalent to the set of second order spatial averages $p\Delta(L_1, L_2)$. On the other hand, we verify immediately that $S_{1,1}(\Delta)$ is nothing else but the autocovariance parameter $M_2(\Delta)$,

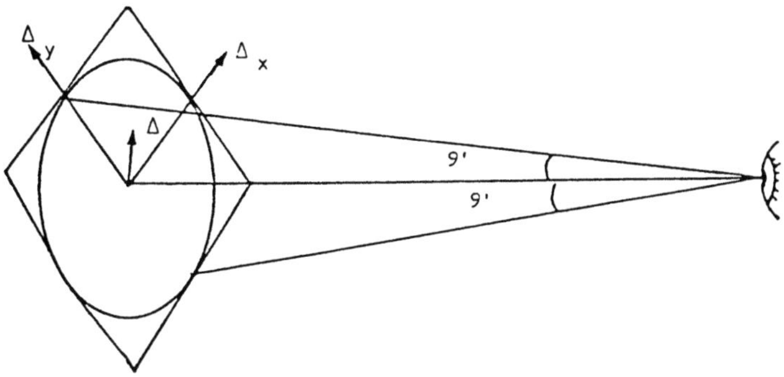

FIGURE 4.2. Set TC of translations Δ sensitive to the visual system

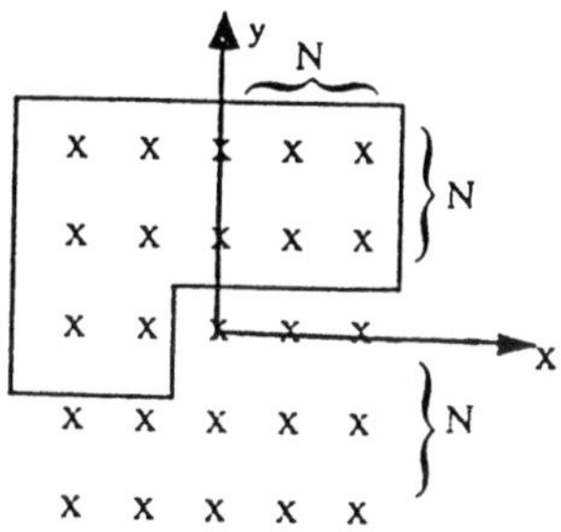

FIGURE 4.3. Set TC of translations used

so that a subset of $S_{p,q}(\Delta)$ moments for $2 \leq p+q \leq m$, is a model lying between the two former limits (model 1 and model 2).

Model 3 is also a hybrid model $\{S_{p,q}, H\}$ which consists of $S_{p,q}(\Delta)$ for $\mid \Delta \mid \leq 9'$, and $2 \leq p+q \leq 4$, and of the histogram $H(L_1)$, for $L_1 \in L$. Model 3 contains 6 parameters per translation Δ : $S_{1,1}, S_{1,2}, S_{1,3}, S_{2,1}, S_{2,2}, S_{3,1}$ instead of L^2 for model 1 and only one for model 2.

Psychovisual experiments have shown [40] that $S_{p,q}$ *moments for* $m = 4$ *and the histogram* $H(L_1)$ *define the desired set of discriminating parameters.*

This model corresponds to the best solution that we propose for microscopic textures with regard to compactness and psychovisual requirements.

We have to recall that the performances of the previous models may not be satisfactory for highly structured textures, i.e., macroscopic textures.

Practical Considerations

Use model 2, which is the simplest model. In the case where the results are not satisfactory, model 1 or 3 should be utilized equivalently. The three possible models are indexed by a set of translations Δ. This set is a circle centred on the intersection of the optical axis with the texture image and has a radius of 9′ as shown in figure 4.2.

For simplicity we replace this set by a square containing this circle. In normal conditions of view, where we watch a 256 × 256 image on a TV screen at a distance of 4m (distance where we begin to discriminate pixels), this square is 20 pixels wide (see figure 4.3).

Models	$N = 4$	$N = 8$	$N = 12$
1	2560	9216	19968
2	248	152	320
3	248	872	1880

TABLE 4.1. Size of the models proposed

As:

$$\begin{cases} p_\Delta(L_1, L_2) & = & p_{-\Delta}(L_2, L_1) \\ M_2(\Delta) & = & M_2(-\Delta) \\ S_{p,q}(\Delta) & = & S_{p,q}(-\Delta) \end{cases}$$

we restrict ourselves to half of the domain (see the surrounded region of figure 4.3). If we consider a square of thickness N pixels, as shown in figure 4.3, the total number of translations is $2N\ (N{+}1)$.

Size of the Different Models

$$\begin{cases} \text{MODEL } 1 & = & L^2\, 2N(N+1) \\ \text{MODEL } 2 & = & 2N(N+1) + L \\ \text{MODEL } 3 & = & 12N(N+1) + L \end{cases}$$

Psychovisual experiments [36] have shown that it is sufficient to quantize grey tone microscopic textures with from $L = 4$ to $L = 8$ values (we usually use 8).

Table 4.1 gives the sizes of the various models for 3 commonly used values of N. The reader has to be aware of the fact that this number of parameters which may seem too large, is sufficient for modelling *all* microscopic textures sensed at a given distance (to which N is related). Our theory does not say that such an amount is *necessary* for a given texture; a given texture may be usually modelled with many less parameters according to our own experience.

Synthesis Method

The problem to solve is: given the set of parameters of a given model (one of the three), synthesize a texture field such that if we compute the model parameters on the texture field as described in figure 4.1, we shall obtain parameter values which are *equal* (or almost equal) to the a priori given values. We describe a sequential procedure (a parallel one is also available [64]) for model 1. We leave it to the reader to generalize it to the case of the two others as almost all considerations still hold.

Input Checks

The input is a set of $2L^2N(N+1)$ real numbers stored in a vector denoted by B. We suppose (and check) that this set is organized in the following way: all second order spatial averages are stored translation after translation in the normal line by line scanning order of figure 4.3, so that each subset of L^2 successive values is positive and has a sum equal to one.

From

$$H(L) = \sum_{L_2=0}^{L-1} p_\Delta(L_1, L_2) \qquad \forall L_1, \Delta, \tag{4.5}$$

we have also to check that any consecutive sum of L parameter values is periodic (a period of L^2 in B). It corresponds to a histogram parameter; all L histogram parameters are obtained by the L successive sums of L consecutive parameters in B.

Initialization

Using the L histogram parameters $H(L_1)$ obtained from (4.5), we first generate an image of TX (a matrix of size $M \times M$) of white noise having the same histogram as the parameters previously computed. This texture image is not correlated but it is constructed very easily. We use a random number generator, choose randomly among the L possible values according to the repartition function (integral of the histogram) and fill in the successive pixels of TX.

As TX consists of independant random variables we may deduce, without computing it on TX, that the various second order spatial averages of TX are:

$$p_\Delta^{TX}(L_1, L_2) = H(L_1)H(L_2) \qquad \forall L_1, L_2, \Delta \tag{4.6}$$

As these values usually do not correspond to the input values, we have to proceed to the second phase of the synthesis algorithm where we only *update* the image TX previously synthesized to achieve our goals.

Second Phase of the Synthesis

The purpose of the texture updating is to have all second order spatial averages $p_\Delta^{TX}(L_1, L_2)$ converge to the desired ones $p_\Delta(L_1, L_2)$ from B.

TX is modified point by point sequentially while minimizing the mean square error $\| B - B^{TX} \|^2$, where B^{TX} is the parameter vector associated to the actual synthesized texture TX.

Suppose that we want to update the pixel situated at location i whose actual grey level is L_1. The purpose of the algorithm is to replace L_1 by L_1* such that $\| B - B^{TX} \|^2$ will be minimum. How to choose L_1*?

As the various second order statistics are related to translations defined by TC (see figure 4.3), the solution is to compute all possible $B^{TX}(L_1')$ vectors corresponding to the possible updatings of L_1 by $L_1'(L_1' \in L)$ and to choose L_1* such that:

$$\| B - B^{TX}(L_1*) \|^2 \leq \| B - B^{TX}(L_1') \|^2 \qquad \forall L_1' \in L \tag{4.7}$$

The problem to solve becomes: how to compute the various $B^{TX}(L_1')$?

We know B^{TX} using (4.6) after the initialization phase. We shall now show how to update B^{TX} when we replace L_1 by L_1' for any location i, supposing that we know $B^{TX}(L_1)$ (which solves the problem).

For a given translation Δ, the modifications in the computing of the second order spatial averages only come from the pairs $(X_{i-\Delta}, X_i)$ and $(X_i, X_{i+\Delta})$. If we replace L_1 by L_1', $p_\Delta(.,.)$ is then updated in the following way (4.8):

$$\begin{cases} p_\Delta(X_{i-\Delta}, L_1) & \Rightarrow & p_\Delta(X_{i-\Delta}, L_1) - \frac{1}{I} \\ p_\Delta(X_{i-\Delta}, L_1') & \Rightarrow & p_\Delta(X_{i-\Delta}, L_1') + \frac{1}{I} \\ p_\Delta(X_{i-\Delta}, L_1) & \Rightarrow & p_\Delta(X_{i-\Delta}, L_1) - \frac{1}{I} \\ p_\Delta(X_{i-\Delta}, L_1') & \Rightarrow & p_\Delta(X_{i-\Delta}, L_1') + \frac{1}{I} \end{cases} \tag{4.8}$$

$B^{TX}(L_1')$ is simply obtained from $B^{TX}(L_1)$ using (4.8) for all Δ translations.

Practical Considerations

The convergence of the procedure is insured by the fact that the mean square error is monotonically decreasing, but there is no evidence that the limit error is *zero*. In order to spread out the error uniformly on the image (which is the condition to obtain homogeneous texture) we use a random scanning procedure: each pixel to update is chosen successively, randomly among the set of remaining, not yet updated pixels.

Visual experiments [36, 38] have shown that the texture models which are designed for microscopic textures also hold for some macroscopic textures made of only one pattern repeated rather regularly (wool, material, rattan...).

The models hold if the translation set TC (figure 4.3) used to control the correlation distance has a size adapted to the pattern size (i.e. at least equal).

Models for Macroscopic Textures

We divide macroscopic textures in two different classes:

weakly structured corresponding to textures having very irregular structure (bark, ground,...) or having only one pattern repeated rather regularily

hierarchical textures corresponding to textures with several patterns.

Model 4: Weakly Structured Macroscopic Textures

We have shown [42] that weakly structured textures may be modelled by a *statistical* model similarly to the microscopic case. This model, called model 4, consists of model 2, plus third order moments: $M_3(\Delta_1, \Delta_2)$, (see (4.9)):

$$\begin{cases} H(L_1) \\ M_2(\Delta) \\ M_3((\Delta_1, \Delta_2) = \frac{1}{I}\sum_{I=1}^{I}(X_i - \eta)(X_{i+\Delta_1} - \eta)(X_{i+\Delta_2} - \eta)/\sigma^3 \end{cases} \quad (4.9)$$

We consider translations Δ_1, Δ_2 of size less than 5 pixels which seems to be enough to bring the structural information not available from model 2.

As it is a statistical model, both model, computation and texture synthesis are performed exactly in the same way as with models 1, 2 or 3.

Hierarchical Model (model 5)

The model proposed [43] is a general constructive model, but it suffers from the fact that the analysis of this model is still an unsolved problem: given a hierarchical macroscopic texture, we still do not know how to compute the model parameters. This model consists of three types of information:

- an *array of labels* telling, for each pixel, the type of primitive to which it belongs. The array describes the arrangement of primitives in the texture image (if only one primitive exists, this array may be obviously suppressed). It is the representation of the highest level of the texture information
- the various *microscopic models* (1,2 or 3) related to the various existing texture primitives. This part of the model describes the low level texture information (see section 4.1.1)
- an array of vectors (x-y components) giving for each pixel, the coordinates of the x-axis unit vector of the *local coordinate system* of the local texture primitive. These local coordinate systems describe the local orientation of the primitive involved.

Model computation: given a hierarchical texture sample, there is no procedure to compute the model parameters (label array, microscopic models and local coordinate systems) in the literature

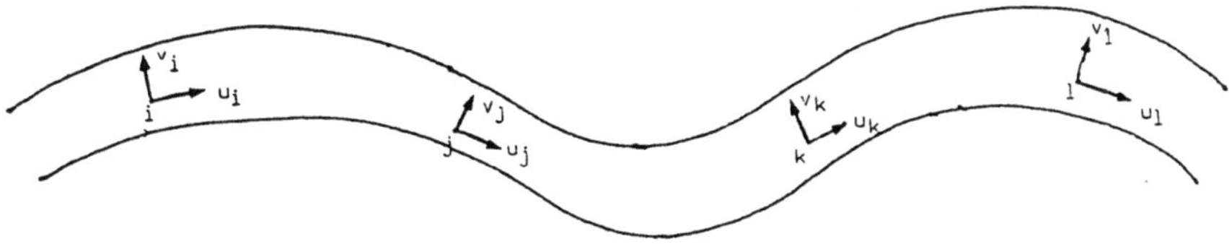

FIGURE 4.4. Visualization of several local coordinate systems on a curved texture primitive

128	128	
A	B	128
C	D	128

TABLE 4.2. Presentation format of plates 1, 2, 4, and 5

Texture synthesis: given a hierarchical model, there is no difficulty in synthesizing each texture primitive one after the other using the procedure previously described. Care must be taken when we compute the grey tone $X_{i+\Delta}$ of the neighbours of each location i in the updating phase. The coordinates of $i + \Delta$ have to be determined using the local coordinate system at location i. More details and an efficient implementation of this algorithm may be found in [7].

Results in the Case of Grey Tone Planar Texture Fields

We first show some results on planar grey tone texture modelling. These are representative of the performances of our models for all the textures studied.

Plates 1 and 2 will be displayed with the presentation format of table 4.2. Each image is subdivided into four 128×128 subimages; subimage A is always a reference natural texture sample on which a texture model was computed. Subimage C presents a *synthetic* image when model 1 (second order spatial averages) was used. Subimage B corresponds to the use of model 3 ($S_{p,q}$ model) and subimage D to the use of model 4 (third order model).

In plate 1 we compare the performances of the models 1, 3, and 4 on a *'rattan'* texture extracted from Brodatz' book [14]. This texture is a weakly structured macroscopic texture.

We may see that model 4 gives better results (same number of iterations and similar errors) than the two others when we compare all the syntheses to the original sample on subimage A. Nevertheless, microscopic models already give a good approximation of the result.

In plate 2, we show similar results in the case of *'bark'* texture also extracted from Brodatz's book. Conclusions are also similar.

Plate 3 gives an example of a hierarchical macroscopic texture sample made of three different microscopic primitives. The lowest part of the image corresponds to the synthesis of seismic texture (first primitive) which is obtained in seismic recordings of underground reflections. It is a texture with substantial linear structures corresponding to seismic horizons. Each pixel intensity at location (x, y) in the image represents the acoustic resistance of an earth element situated at depth y and horizontal distance x. This texture was then twisted using two inter-woven sine waves on the left and right centre parts of the image to simulate torsades. The orientation information was given by the array of vectors described above. A wool sample (second primitive) was woven in a straight manner apart

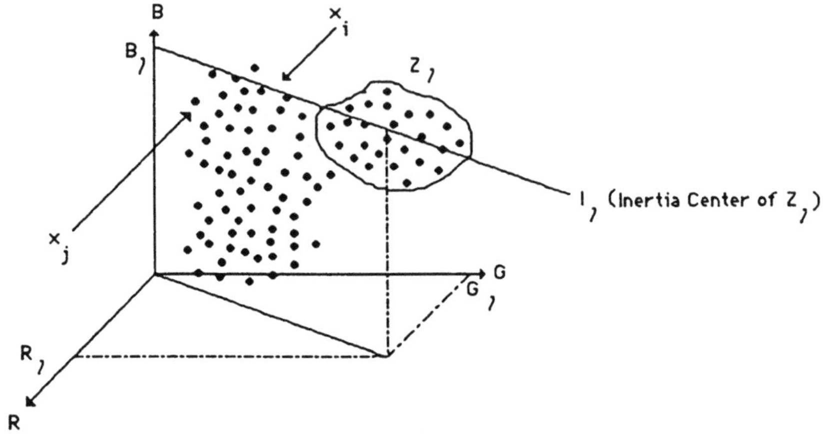

FIGURE 4.5. Colour distribution of the texture sample

from the two torsades. Another fabric (third primitive) was woven along the two main diagonal directions on the centre of the image to produce the last pattern. Once again, this orientation information was given by the local coordinate systems (the third part of the model). The three microscopic models used (model 2) were computed on a planar homogeneous sample (no spatial distortion).

Such a model allows us to conceive a synthetic weaving machine and obtain results of the type shown in plate 3.

Colour Textures

Colour textures are usually represented by the [RGB] components of each pixel. The motivation of the following work is to avoid the growth by a factor 3 of the models and the corresponding synthesis algorithms.

The solution is to replace each pixel by a simple *scalar* code and then create models of the same size as in the black and white case. Synthesis algorithms developed earlier for black and white textures become applicable with only few changes to the colour case (see[41]).

Texture Coding

In this section we propose a procedure to replace a colour texture (described by its [RGB] components) by an image of scalar codes (usually 8 to 12 codes) and by a look-up table giving the [RGB] coordinates of the 8 to 12 codes approximating the colour texture.

The problem is in fact to approximate a great number of colour pixels in a RGB space (see figure 4.5) by a small number of them.

It is a clustering problem: suppose that we allow L different codes, we have to cluster the cloud of the pixel colours shown in figure 4.5 by L clusters and then to attribute the same colour code l, and the same colour I_l (belonging to the domain Z_l) to each pixel of each cluster Z_l. The colours I_l are nothing else but the different elements of the required look-up table. The solution adopted minimizes the mean square coding error:

$$Coderr = \sum_{l=0}^{L-1} \sum_{i \in Z_l} | X_i - I_l |^2 \tag{4.10}$$

where:

l denotes a particular code

$I_l = [R_l, G_l, B_l]^T$, the RGB coordinates of the colour associated to code l

Z_l is the cluster corresponding to all the texture pixels coded by l

X_i is the pixel colour (at location i) represented by its RGB components: R_i, G_i, B_i.

A solution to this fundamental problem of clustering analysis has been proposed by E. Diday. We describe simply the principle of his algorithm. More details, and the demonstration of the convergence of his algorithm may be found in [26].

The algorithm starts from an initial clustering of the cloud in L clusters. The choice of the initialization is left to the user and usually has no crucial effect on the final result.

We propose the following one: Each pixel X_i is transformed into a scalar Y_i such that:

$$Y_i = R_i + 256G_i + 2^{16}B_i \tag{4.11}$$

All Y_i values are then ranked by increasing values in an interval $[Y_{min}, Y_{max}]$ where Y_{min} and Y_{max} are the min and max values of all Y_i obtained. The interval is subdivided in L equal segments and we finally cluster the image by inspecting the Y_i values of each pixel; if Y_i falls in the l-th interval, we assign pixel X_i to the cluster Z_l.

The algorithm proposed by Diday, called the Dynamic Clustering Algorithm, is an iterative procedure which converges usually after 5 to 8 iterations.

We mention beforehand that it is possible to show that the colours I_l which minimize the mean square coding error are simply the inertia centres of each cluster Z_l. Each iteration of the algorithm starts from a clustering and gives a new clustering solution at the output: we compute the L inertia centres of the clusters available at the input and use the L points as poles which attract the closest pixels to obtain new clusters.

The algorithm is briefly:

1. For each cluster Z_l, compute its inertia centre I_l in the RGB space
2. For each point X_i of the cloud, compare its distance to all centres I_l
3. Reassign each X_i to the label l corresponding to the nearest inertia centre I_l
4. Compute the coding error Coderr
5. if Coderr $\geq$ threshold GO TO 1 else END.

Each pixel X_i is coded with one of the L possible codes (an array of $M \times M$ codes is created), and for each code l, (R_l, G_l, B_l) values of the final inertia centre of Z_l are stored. The result is three look-up tables with the colour information used for the approximation of the coded texture. Each pixel X_i is thus approximated by the inertia centre I_l of the cluster to which X_i belongs.

Extension of the Models

Case of model 1 The extension is straightforward as we have simply to compute the various co-occurrences of colour codes: $p_\Delta(l_1, l_2)$ on the coded image, where l_1 and l_2 are the codes of pixels i and $i + \Delta$

Case of models 2 to 5 We should (and could) consider all coefficients of the covariance tensors, but for simplicity we keep only their trace: the RR, GG, and BB covariances which have been proven to contain all the textural information retained by a human observer [41]

As an example, if we consider model 2, we replace the histogram of grey levels by the histogram of codes and we compute the *scalar* autocovariance $M_2(\Delta)$:

$$M_2(\Delta) = \frac{1}{I}\sum_{i=1}^{I}(I_l(X_i) - MEAN)^T(I_l'(X_{i+\Delta}) - MEAN)/\sigma^2 \qquad (4.12)$$

where $I_l(X_i)$ (respectively $I_l'(X_{i+\Delta})$) is the RGB coordinate vector of the code l representing X_i, (respectively $X_{i+\Delta}$), $MEAN$ is the mean value of these vectors over the texture and σ^2, their variance:

$$MEAN = \begin{bmatrix} \frac{1}{I}\sum_{i=1}^{I} R_l(X_i) \\ \frac{1}{I}\sum_{i=1}^{I} G_l(X_i) \\ \frac{1}{I}\sum_{i=1}^{I} B_l(X_i) \end{bmatrix} \qquad (4.13)$$

$$\sigma^2 = \frac{1}{I}\sum_{i=1}^{I}(I_l(X_i) - MEAN)^T(I_l(X_i) - MEAN) \qquad (4.14)$$

Similar considerations also hold for models 3, 4 and 5.

Colour Texture Synthesis

If we take model 1 (second order spatial averages) as an example, the algorithm will produce an image of labels controlled by their second order statistics. From the model parameters we can compute the histogram of labels using an equation of type (4.5). Using this histogram as input, we first produce a white noise image of labels having the histogram as input.

In the second phase, we modify sequentially the label at each pixel location which minimizes the mean square error between the current feature vector $\mathbf{B}^{TX}$ and the input model $\mathbf{B}$. The algorithm compares only L labels (8 to 12) at each location due to the use of the colour coding result which is a benefit for the synthesis performance.

Finally, it remains to load the label image in a frame buffer and load the available look-up table with the model to obtain the synthetic colour texture.

Modelling Results in the Case of Colour Textures

Plates 4 and 5 are displayed with the presentation format of table 4.2. In plate 4 we compare the performances of the models 1, 3 and 4 on a formica texture (imitation of wood). This texture is a weakly structured macroscopic texture. Subpicture A corresponds to the natural sample coded on 3×8 bits. This natural sample was coded using the vector quantization method with only 12 codes from which the models were computed. As for plates 1 and 2 the third order model of subimage D gives the best results.

On plate 5, we show similar results in the case of a woven string wall covering. It is interesting to note that all three models give quite acceptable results though this texture is macroscopic.

Plate 6 shows the synthesis result concerning hierarchical texture obtained by twisting 4 different microscopic textures along torsades. It is in fact possible to imagine a great variety of weaving schemes with this technique.

4.1.3 Textures Lying on 3D Surfaces

We are interested by textures lying on a 3D surface but having no thickness. We do not consider solid textures.

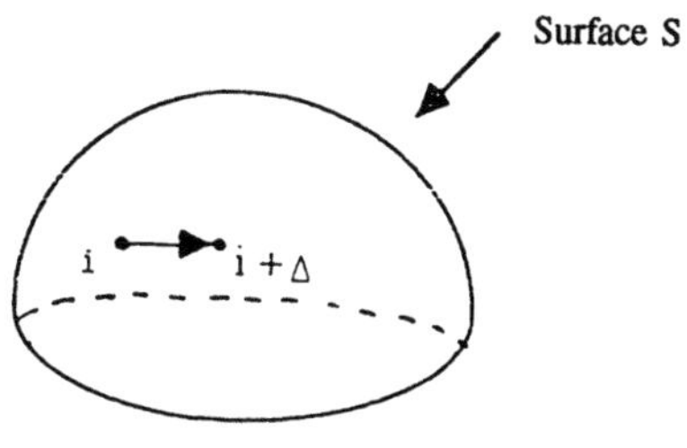

FIGURE 4.6. A 3D surface and a pair $(i, i + \Delta)$ of locations on the surface S

3D Models

Suppose that we dispose of texture drawn on a piece of paper. If we lay it down on various 3D objects, these objects become textured, but it is the same texture. Thus, *a texture model does not have to depend on the geometry of the surface on which it lies.*

Given a 3D object, in order to measure textural properties we have to get rid of shadow effects, lighting conditions, variations of the surface reflectance of the objects (not due to the texture information itself). These effects can be easily added when one synthesizes textured objects, but we will see later on, that given a 3D textured object, it is difficult to eliminate these various effects (the analysis problem is not yet solved !).

How can we model 3D textures?

The interesting feature of the models we defined for planar textures is that they can be easily extended to the 3D case (a plane is nothing else but a particular surface) as they consider various correlations between pixels.

If we take model 1 as an example, its parameters can be computed if for each location i of the 3D surface, we are able to find its translated $i + \Delta$ (see figure 4.6) and measure both colours X_i and $X_{i+\Delta}$.

The problem is that we have to constrain $i + \Delta$ to lie on the surface also, which is not possible for every choice of Δ translations in the 3D space. The fact that the Δ translations (which parameterise the model) must not depend on the geometry led us to a choice of 2D translations lying on the tangent plane to the surface at location i.

The problem is that the 3D surface has to be smooth enough in order for $i+\Delta$ (lying on the tangent plane) to also belong to the 3D surface.We shall see later on, how to extend these considerations to the case of a "not so regular" surface.

Assuming that $i + \Delta$ is on the tangent plane at location i, we still have to define a local coordinate system on the tangent plane to solve this problem completely.

In conclusion, *the key to the extension of the former models to the 3D case is to define and utilize local coordinate systems on the 3D surface [65].*

Local Coordinate Systems (LCS)

Local coordinate systems are defined at each location i of a 3D surface by a triple of 3D orthogonal unit vectors:

$\mathbf{n}_i$ the unit vector carried by the normal to the surface

$\mathbf{u}_i$, $\mathbf{v}_i$ two unit vectors on the tangent plane to the surface

As the only ambiguity concerning these vectors are related to those lying on the tangent plane, the definition is often restricted to the $(\mathbf{u}_i, \mathbf{v}_i)$ vectors lying on the tangent plane.

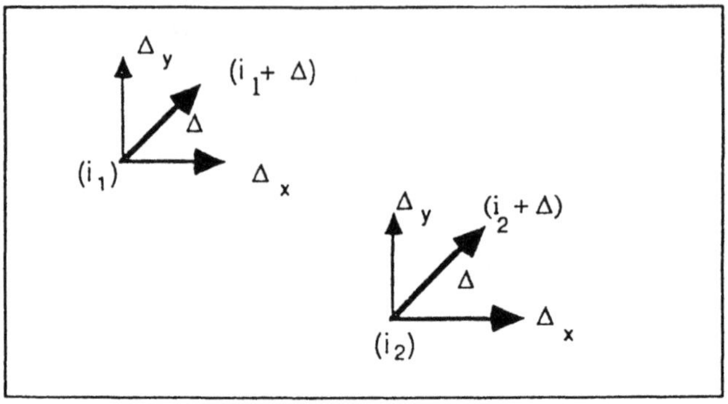

FIGURE 4.7. Parallel local coordinate systems defined on a plane

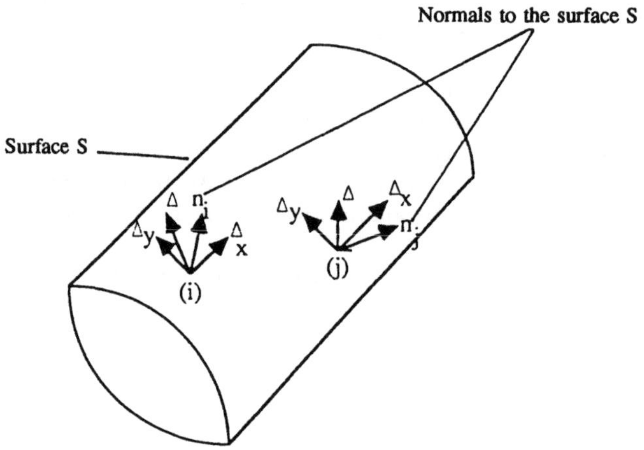

FIGURE 4.8. Local coordinate systems on a 3D surface S

Planar Case

When we compute the various 2D microscopic models while scanning the texture image as shown in figure 4.1, we unconsciously suppose that all the local coordinate systems are parallel to each other (see figure 4.7).

This choice of local coordinate systems corresponds to homogeneous texture. If texture is no longer homogeneous, the local coordinate systems follow the local orientation (distortion) of the texture (see model 5).

3D Case

Suppose that we deal with homogeneous texture lying on the 3D surface.

How must the local coordinate systems be defined ?

It is clear that they are not parallel to each other due to the influence of the surface geometry (see figure 4.8).

Given the 3D surface S, it is possible to compute the normals at each location as well as the tangent plane, but we still do not know how to orient the x-axis Δx on this tangent plane and the unit vector u_i attached to it.

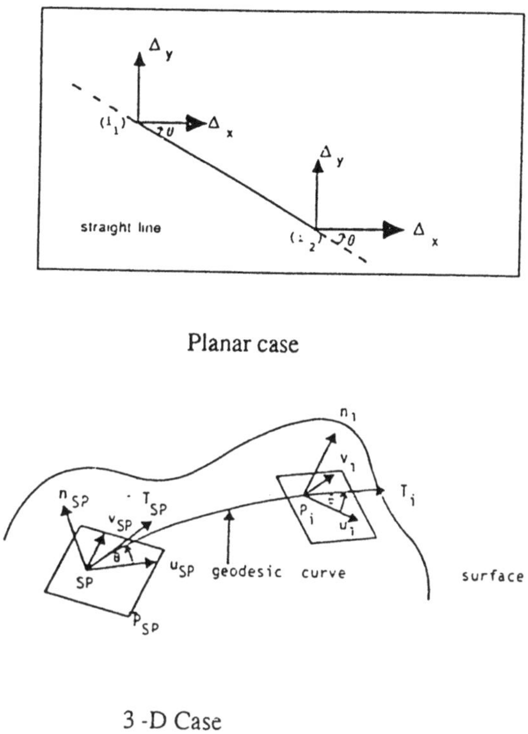

FIGURE 4.9. Parallel transport along geodesics

Determination of the Local Coordinate Systems in the Homogeneous Case

We suppose that homogeneous texture is laid down without any stretching or folding on the 3D surface.

In the planar case, a conceptual way to produce parallel local coordinate systems is to start from a point *SP* with a given local coordinate system and for any other point i we have simply to draw a line segment joining *SP* to i and define the new local coordinate system at location i in such a way that the angle between the segment [*SP*, i] and the local coordinate system remains constant ($= \theta$). The local coordinate system is sliding along the segment [*SP*, i] (shortest path joining *SP* to i). The particularity of the planar case is that whatever the starting point *SP* chosen, the solution obtained is the same for the other locations i.

In the 3D case, we follow a similar procedure [65, 37, 39]: we achieve a parallel transport of the local coordinate systems along the geodesics.

Given an initial position *SP* on the 3D surface and an initial local coordinate system $(\mathbf{u}_{SP}, \mathbf{v}_{SP})$, we compute the shortest path joining *SP*, to i (geodesic) on the 3D surface and we transport the initial local coordinate system along the geodesic in such a way as to keep constant the angle between the x-axis Δx (and the $\mathbf{u}_i$ unit vector carried by it) and the tangent to the geodesic as shown in figure 4.9. Remarks:

- We suppose that we deal with C^2 surfaces without borders so that geodesics have these global properties
- The curvature of the projection of a geodesic on the tangent plane to any 3D surface is always equal to zero. Thus, the geodesic is locally like a "small" segment on the

tangent plane to the surface. Parallel transport along geodesics insures the minimum spatial distortion of the homogeneous texture

- The local coordinate systems obtained *depend* on the starting point and the initial coordinate system. This effect could seem strange at first sight but it is natural. Suppose you have a piece of planar texture that you want to map on a 3D surface; depending on the initial point of contact and the orientation of the planar texture at this point, you will produce different effects when you will map the texture apart from this point. The degrees of freedom available for the construction of the LCS reflect indeed these possibilities. A technique which would not offer these degrees of freedom would not be general.

Non-Homogeneous 3D Textures

The use of geodesics gives a solution for the construction of the LCS. This may be considered as a reference solution about which spatial distortion may be produced (these distortions could be predefined in a plane for example and then be mapped relatively to the reference LCS).

When the type of distortion is of not much importance for the designer, it is possible to use other LCS which are easier to compute than homogeneous LCS constructed with geodesics.

Parametric surfaces If a surface is given by:

$$x = x(u,v) \tag{4.15}$$
$$y = y(u,v) \tag{4.16}$$
$$z = z(u,v) \tag{4.17}$$

and if the u = const (i.e., constant), v = const curves define an orthogonal set, then it is possible to use the following LCS:

$$\Delta x = \left(\frac{\partial x}{\partial u}, \frac{\partial y}{\partial u}, \frac{\partial z}{\partial u}\right) \Big|_{v=v_0}^{u=u_0} \tag{4.18}$$
$$\Delta y = \left(\frac{\partial x}{\partial v}, \frac{\partial y}{\partial v}, \frac{\partial z}{\partial v}\right) \Big|_{v=v_0}^{u=u_0} \tag{4.19}$$
$$\tag{4.20}$$

otherwise, when the set is not orthogonal,

$$\Delta x = \left(\frac{\partial x}{\partial u}, \frac{\partial y}{\partial u}, \frac{\partial z}{\partial u}\right) \Big|_{v=v_0}^{u=u_0} \tag{4.21}$$
$$\Delta y = \Delta x \wedge \mathbf{n}(u_0, v_0) \tag{4.22}$$

n is the normal to the surface.

The corresponding type of mapping consists in cutting a planar texture painted on a piece of paper in small slices in the two main directions (horizontal and vertical) and fixing it along the u=const curves of the 3D surface

Surfaces described by $z =$ const slices These slices are available when surfaces are sensed by a laser sensor for example, or by other scanners. In this case, we define the following LCS:

n normal to the surface (easy to compute)

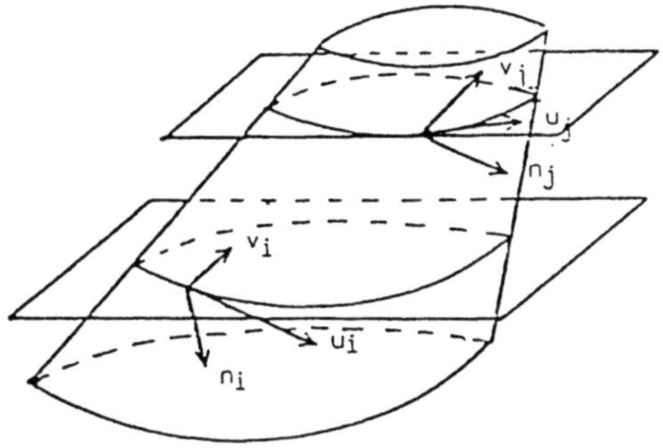

FIGURE 4.10. LCS on a surface described by z = const curves

Δx tangent vector at sample i to the z = const curve (see figure 4.10)

$\Delta y = \Delta x \wedge \mathbf{n}$.

The corresponding type of mapping (physical analogy) consists in fixing the horizontal slices of the planar texture around the surface along the z = const curves.

More generally, the construction of LCS has to describe the type of mapping desired by the designer to texture an object.

Non "Regular" Surfaces

The solution proposed supposes that the 3D surfaces are differentiable and that the tangent plane at each location is a good approximation of the surface neighbourhood up to the size of the domain TC.

Practically, we applied this technique with 3D objects which were not differentiable, simply continuous (C_0 surfaces) and no aberration was even noticed along the C_1 discontinuity curves (see figure 8 as an example).

If problems arose, it would be possible to compute, for each location i, the geodesics starting from i and with the various orientations $\theta = \theta_i$ around the $\theta = 0$ geodesic whose tangent vector at location i is in coincidence with the Δx axis obtained by parallel transport along the global geodesic. Global geodesics, starting from SP and ending at location i, define all $\theta = 0$ local geodesics. The local geodesics define the local mapping between the domain TC (of figure 4.3) defined on the plane and the equivalent directions on the 3D surface (see figure 4.11). The various Δ translations of TC are then measured and the neighbours $i + \Delta$ are obtained as having the same orientation compared to the $\theta = 0$ orientation as in the planar version, and the same *length* (line segment length on the plane TC, and geodesic length on the 3D surface).

Computation of the 3D Texture Models

The solution we adopted is to use a *planar* texture sample and to compute models 1, 2, 3 or 4 using the scanning presented in figure 4.1 (we are in the 2D case). There is *no solution for hierarchical texture.*

If no planar texture sample is available but only an image of 3D texture, the problem is much more difficult to solve (it is of course not solved!) as we have to get rid of the shadow effects, even if we know the 3D geometry. The main difficulty comes afterwards when we have to determine the type of mapping used and the texture models, both types of information being encapsulated in the same texture image.

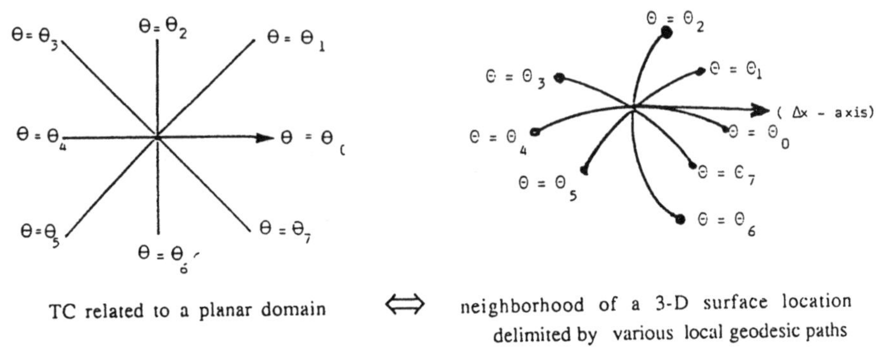

FIGURE 4.11. Mapping proposed to find the neighbours $X_{i+\Delta}$ of any location i in the case of a non regular surface

3D Texture Synthesis

Principle

We suppose that we have a hidden surface algorithm (for example a Z-Buffer algorithm) so that we know for each object available in the scene, its region zone seen on the image screen.

We attach to each textured object a texture model of the type previously described (model 1 to 4) a label image and an LCS array.

The idea (see [37, 63] for more details) is to sample the object surface through the screen image sampling, (back projection of the image sampling) as shown in figure 4.12.

Brief Description

As in the planar case, we first use the model histogram to fill up the image zone corresponding to the visible zone of a chosen object, with white noise codes having a histogram equal to that of the input one.

In the second phase, we consider sequentially each screen pixel i' of the zone and determine the corresponding 3D location i on the 3D surface as shown in figure 4.12. We then compute the equation of the tangent plane P_i to S at location i and determine from the input data the orientation of the LCS at this location (interpolation may be necessary, but usually the software performing the LCS computation uses the same sampling).

Once the LCS is obtained, we compute the 3D positions of all neighbours $i + \Delta$ of location i given by the domain TC (see figure 4.3) corresponding to the various model parameters.

$$i + \Delta = i + k\mathbf{u}_i + m\mathbf{v}_i \qquad \forall k, m \in [-N, N], \text{(integers)}$$

where $\mathbf{u}_i$ and $\mathbf{v}_i$ are the unit vectors on the Δx and Δy axes of the tangent plane.

In order to compute the code of $X_{i+\Delta}$ we project the various locations $i+\Delta$ on $(i+\Delta)'$, on the screen image and read the image code at this location. As we do not consider shadow effects, we suppose that the texture information on the image plane is *equal* to the texture information $X_{i+\Delta}$ on the 3D surface.

Thus, it is possible to update the current feature vector B^{TX} as previously because we have determined the texture labels of all neighbours and perform the optimum updating of the zone labels.

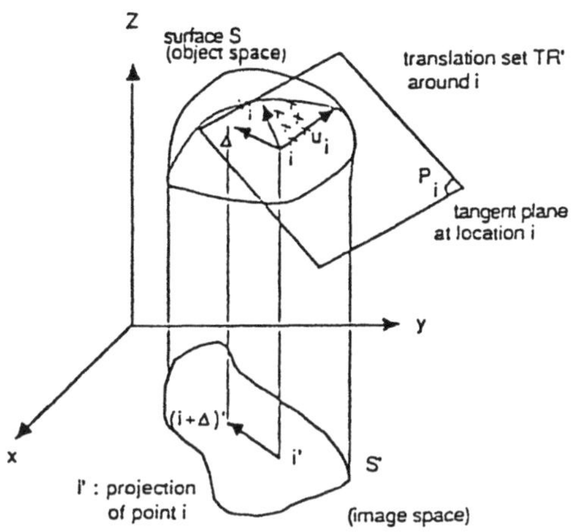

FIGURE 4.12. An object surface S and its observed image S′ projected on the image screen (x-y plane)

Modelling Results in the Case of 3D Surfaces

Plate 7 presents a cast-iron part of a RENAULT car engine. A set of 5000 points was measured on the surface of this object with the aid of a 3D laser sensor developed at INRIA (50 sections of 100 points each measured at z=const positions). The x-y-z coordinates of the points are known with a precision of 5% . The aim of the experiment was to cover this real and not precisely determined object with simulated grey tone SEISMIC texture (why not!). The scalar product of the normals to the surface with the vision axis is displayed in subimage D which gives an idea of the object's complicated shape. Subimage C shows a synthesis result displaying only the texture information mapped on the object using parallel transport along geodesics. The reader may verify that there is no visible distortion of the texture information on the object surface (texture seems to have the same thickness everywhere). Subimages A and B concern two final results when shading was added (the 3D nature of the object appears only at this moment!) with two different starting points for the computation of the LCS.

Plate 8 presents a synthesis result in the microscopic colour case.The texture used is a wool scarf sample on which model 2 was computed (left part of the image). The object is an artificial surface of revolution described by 30 sections of 30 points lying on a portion of a cone, cylinder, paraboloid and sphere concatenated in a shape similar to that of a nail. The object surface is clearly only C^0 but data is given with infinite precision (from analytical formulas). The reader may verify on the right part of the image that the texture is almost not discriminable from the sample, that texture does not seem distorted even along the discontinuity circles separating the various solids.

Plate 9 shows the same nail covered with a hierarchical texture of the type of plate 3 as an example of macroscopic 3D texture synthesis. More details about the synthesis of this image may be found in [7].

Application to Fashion

Our goal was to synthesize the appearance of different types of clothing of different materials on a real mannequin (eventually a human customer). The mannequin was digitised using a laser sensor. Plate 10 showing the variations of the measured depth information.

This information is also presented in plate 11 where the depth information is visualized by grey levels, the closest points appearing as the whitest. We also took a picture with a normal camera of the naked mannequin at approximately the same relative position (see plate 12). We used a threshold technique to separate the body of the mannequin from the background (see plate 13) and had then to match the depth information coming from the laser sensor (plate 11) with the information of the camera picture (plate 13). We used control points to map both data. The 3D texture synthesis was performed on the body of the mannequin with the help of the laser data (see plate 14) and background plus texture body were finally mixed to produce plate 15.

4.2 Special Models for Natural Objects

This part is devoted to special models used to simulate various objects. The first type of models, called "fractals", is very popular and was introduced by B. Mandelbot [66].

4.2.1 Fractals

Aims of Fractal Geometry

Fractal geometry was introduced by B. Mandelbrot [66] to deal with complex functions, for instance continuous but nowhere smooth, that classical geometry fails to describe.

Such functions appear frequently in nature and everyday life, and the usual methods, such as derivation, cannot be used.

Basic Idea

The assumption made is that the phenomena studied keep their complexity at each level of analysis. The very well known example is the measurement of coasts or borderlines between countries. To measure a coast, one can use a rule of length m, and place it along the coast. If the rule has been placed n times, the length $L(m)$ will be:

$$L(m) = n \times m$$

But, as one decreases the length m of the rule, n increases in such a way that:

$$L(m) \to \infty$$

If, instead of using a rule, one tries to measure the length by covering the coast with circles of radius m, and then let m decrease toward 0, the same difficulty is encountered. This problem is solved with the notion of fractal dimension. Several definitions exist for it, and the following one corresponds to Haussdorff dimension. To measure a line, one usually adds lengths; to measure surface, one adds squares of lengths, for volumes cubes of lengths, etc.

A shape is said to be fractal, if, to measure it, one has to add lengths at a power which is not an integer: For many coasts (see figure 4.13), there exists a number D (the fractal dimension of the shape) such that:

$$\lim_{m \to 0}\left(\sum m^{\alpha}\right) = 0 \quad if\ \alpha < D$$

$$\lim_{m \to 0}\left(\sum m^{\alpha}\right) = \infty \quad if\ \alpha > D$$

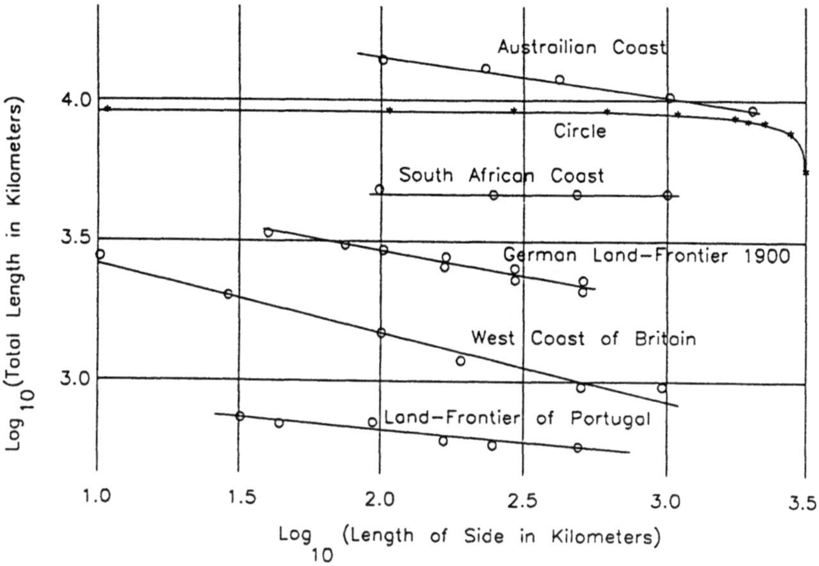

FIGURE 4.13. Lengths of some coasts according to L.F.Richardson

Image Synthesis

Many features make fractal sets of great interest for image synthesis; for instance, the property that as one views the sets at greater magnification more and more structure is revealed, or the fact that only a small number of parameters are needed for their specification, independently of the visual complexity.

Fractals also provide often a convenient way to synthesize some natural objects with significant data compression. Fractal landscapes, clouds or vegetation are quite common.

One problem with fractal techniques is to control precisely the shape of the synthesized object: generally, only a few parameters control the fractal dimension, the size and so on.

Iterated Function Systems

The method of Iterated Function System (IFS), initially developed by M. Barnsley [3], provides a good control and understanding of fractal synthesis.

An IFS is a couple (W, P) where:

$W = \{w_1, \ldots, w_n\}$ is a finite set of mappings, all of them being strict *contractions.*

$P = \{p_1, \ldots, p_n\}$ is a set of probabilities: $\forall i,\ 0 < p_i \leq 1,\ \sum p_i = 1$.

$$w_k(i,j) = \begin{bmatrix} a_1^k & a_2^k \\ a_3^k & a_4^k \end{bmatrix} \begin{bmatrix} i \\ j \end{bmatrix} + \begin{bmatrix} a_5^k \\ a_6^k \end{bmatrix} \qquad \forall k \in \{1, 2, \ldots, n\} \tag{4.23}$$

for each k, w_k is such that $\forall u \in \mathbf{R}^2, \mid w_k(u) \mid \leq s_k \mid u \mid, s_k < 1$

We then consider the following process: let z_0 be any point in $\mathbf{R}^2$, we randomly choose a map w_i (with probability p_i), compute $z_1 = w_i(z_0)$. We repeat this process a great number of times. If all points are plotted after a sufficiently great number of iterations, they will distribute themselves approximately upon a compact set A, called the *attractor* of the IFS.

Thus with every IFS is associated a unique attractor A, and a special probability measure, the p-balanced measure, which is the stationary distribution of the random walk ; the probabilities, through this measure, have an influence on the density of the points of A (the grey levels).

A characterization of A is given by the following theorem:

if (W, P) is an IFS and A its attractor, then:

$$A = \cup_{i=1}^{n} w_i(A) \tag{4.24}$$

Moreover, if L is a compact 2D set of R^2, the "collage theorem" [4] tells us that if there exist n contractions and some $\epsilon > 0$ such that:

$$h(L, \cup_{i=1}^{n} w_i(L)) < \epsilon$$

then:

$$h(L, A) < \frac{\epsilon}{1-s}$$

where A is the attractor of the IFS, h is the Haussdorff metric:

$$h(B, C) = \max[\max_{x \in B}(\min_{y \in C}(d(x, y))), \max_{y \in C}(\min_{x \in B}(d(x, y)))]$$

for any two subsets B and C,where d is a Euclidian distance and s is such:

$$0 < s < \frac{1}{d(w_i(x), w_i(y))} \leq sd(x, y), \qquad \forall x, y$$

It follows that, to find a suitable set of maps to reconstruct A (or $L\ldots$), we need only to make an *approximate* covering of L by continuously distorted smaller copies of itself.

The design of fractal objects with IFS is made relatively simple and intuitive with the use of (4.24). In fact, this method leads to the inverse problem, which can be stated as follows.

Image Analysis

Given a set A, determine an IFS that will approximately generate A.

Some authors have discussed this problem. M.Barnsley has proposed a method based on the moment theory of p-balanced measures that works under certain hypotheses [4].

The method proposed in [59] has the advantage of being general. It is based upon an optimization technique. Given a 2D shape A and a value n of number of contractions, the problem solved determines the set of parameters $\{a_k^l, k = 1, 2, 3, 4, 5, 6;\ l = 1, 2, \ldots, n\}$ by minimizing a criterion related to the error between A and its fractal approximation A_F.

Two criteria have been utilized:

- The first one is the Haussdorff distance between the two shapes. As this criterion is highly non differentiable with respect to the contraction parameters, a generalized gradient technique called a bundle algorithm [58] was used to converge to a set of parameters which minimizes *locally* this criterion. If we start too far from the solution, we may thus not converge to an acceptable solution

- The second one is the simple mean square error between the two images (which thus allows the manipulation of grey tone 2D shapes) which has the advantage of being differentiable with respect to the parameters. The use of a classical gradient technique allows a more comfortable implementation and convergence.

J. Lévy Véhel and A. Gagalowicz [60] have developed a method based upon the ideas of simulated annealing that allows the generation of a shape A plus the grey levels on A as before, but also insures the convergence to a (the) global minimum.

IFS is a powerful method that makes it possible to synthesize very complex shapes very fast; one can approximate natural objects like the natural maple leaf of plate 16, or design complex images (plate 17) with a very few number of parameters: 30 for the maple leaf, 100 for plate 17, which gives a data compression ratio of 25000. Figure 16 is subdivided in 4 subimages. The top left image presents a natural maple leaf sensed by a black and white camera and slightly coloured with a pseudo-colour technique.The top right image displays the natural leaf thresholded to isolate it from the background. The two bottom subimages show the fractal approximation obtained of the natural leaf at two different resolutions.

4.2.2 Graftals

Graftals have been studied by A. R. Smith [80] and applied to the simulation of trees and plants.

General Considerations

It is possible to model complex natural objects like trees and plants using procedural models. An interesting class of these models which handles plant growth is called *graftals*. It is an adaptation to computer graphics of parallel rewriting grammars introduced by Lindemmayer[66]. This name which sounds like fractals was chosen on purpose because graftals share with fractals the property of appearing "self-similar" due to the parallel application of the same syntactic rules to each position of a graftal. In the terminology of Mandelbrot [61] graftals are *subfractals*: they seem self-similar but are not entirely due to a *residue* interfering with the self-similar part. The parallel rewriting grammars are similar to those used in conventional formal language theory, except that rules are applied *simultaneously* on a given word: there is no distinction between terminal and non terminal symbols. All strings are generated from an initial string called *axiom*, and are words of the language of the grammar. These grammars are called pL-systems, where p is the number of nearest neighbours used in the context of a symbol in the application of the various rules (if p is equal to 0, it is a context-free grammar).

In order to apply this technique to trees, Lindenmayer has introduced additional symbols to the alphabet of the language, these brackets {[,]} will carry the branching information of trees.

An example extracted from [16] will make these notions very clear.

Context-Free Grammars

Consider a 0L-system (context-free grammar) with alphabet S = {0,1,[,]}.

Production rules axiom = {0}

$$\begin{array}{rcl} 0 & \rightarrow & 1[0]1[0]0 \\ 1 & \rightarrow & 11 \\ {[} & \rightarrow & {[} \\ {]} & \rightarrow & {]} \end{array}$$

The construction of the various words starts from the axiom, and at each step, *each* symbol is processed according to its specific production rule. If we consider the successive words generated we obtain:

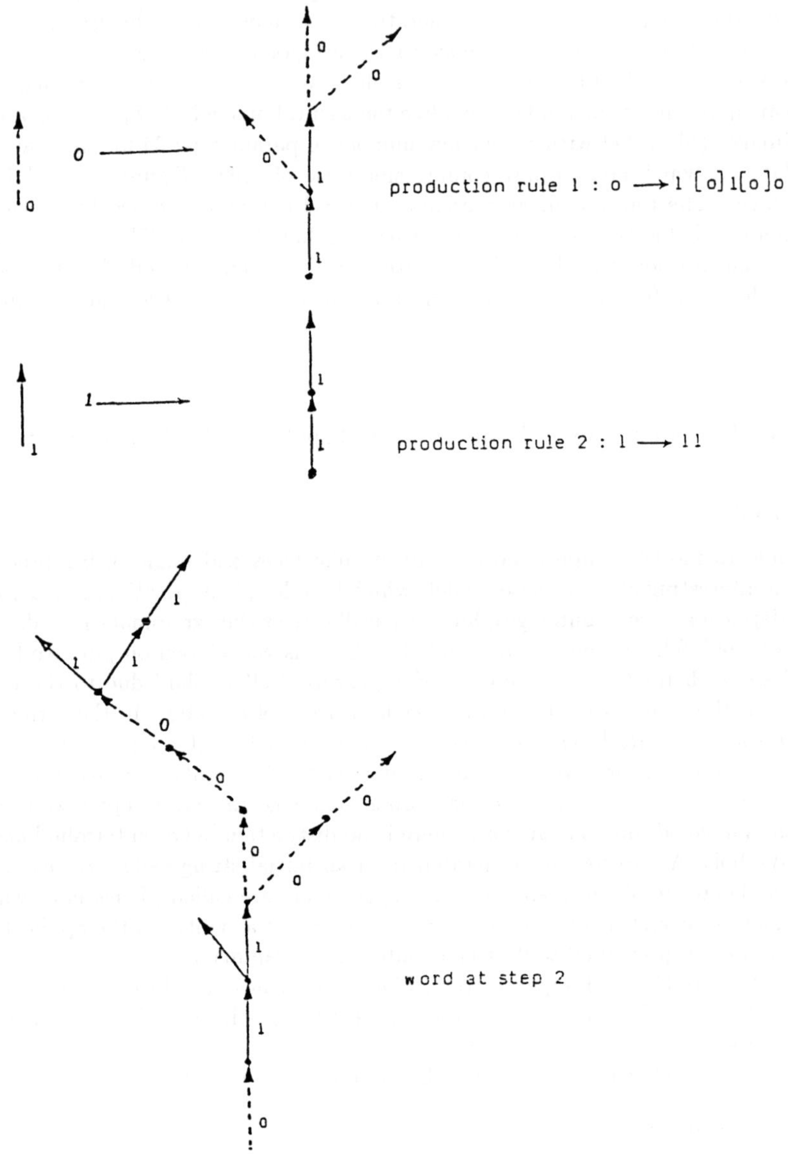

FIGURE 4.14. Graphic representation of a 0L-system (from[16])

step 0: 0 (axiom)
step 1: 1[0]1[0]0
step 2: 11 [1[0]1[0]0] 11 [1[0]1[0]0] 1[0]1[0]0
and so on ...

Each word may be graphically represented as shown in figure 4.14. In his simulation of plants [80], Smith calls the tree associated to a word a *genotype*, and the plant model based upon a genotype, a *phenotype*, which allows for the result visualisation.

An example of phenotype is given in plate 18 where:

- each 0 or 1 is represented by a line segment (or a cylinder)
- each branch end] is considered as a leaf and is represented by a small circle (or sphere).

The program developed by Smith gives control over the colour of stem and leaf, variations of the colour and geometrical attributes with regard to the distance of stems and leaves from the root. Arbitrary sets of angles for the branch formation are also available.

In the former example, we may notice that if T(n) is the order-n word, then the application of the rules is such that:

$$T(n+1) = 1^{2^n}[T(n)]1^{2^n}[T(n)]T(n)$$

If there were be no ones in this recursion, $T(n+1)$ would be a concatenation of 3 n-th order versions of itself; the shape would be self-similar. The occurrence of this special element of the alphabet, not similar to the generic word) is an example of *residue* which shows the sub-fractal characteristic of graftals.

Context-Sensitive Grammars

More realistic plant models are obtained by context-sensitive L-systems. Production rules depend or the nearest neighbours of each symbol. Consider an example of 2L-system often used by Smith for his plant syntheses. The alphabet chosen is (as before) S = {0, 1, [,]}. The two neighbours of each symbol used for the production rules are the nearest one on *the left* and on the *right* of each symbol (another choice could be possible of course), with the constraint that it is neither [, nor].

Production rules axiom = {1}

1) [→ [
2) [→ [
3) 000 → 0
4) 001 → 1[1]
5) 010 → 1
6) 011 → 1
7) 100 → 0
8) 101 → 11
9) 110 → 1
10) 111 → 0

When left or right neighbours of a symbol do not exist (terminal symbols) we suppose that these neighbours are, in fact, *ones.*

Consider as an example the word w = 0 1[1]. To transform the 5 symbols of this word at the next step, we have to replace it by:

$$w' = \frac{1\,0\,1\,[1]\,1}{w}$$

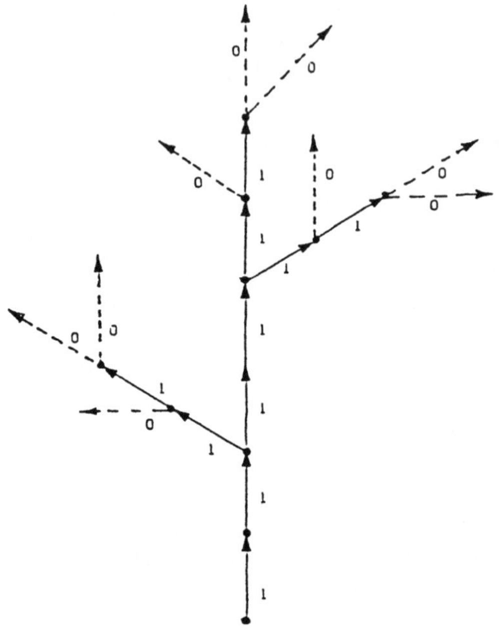

FIGURE 4.15. Step 10 in the generation of the 2-L systems described above

The successive application of rules 8, 6, 1, 10, 2 on w brings:

$$T(w) = 111[0].$$

The various generations of the genotype are in this case:

step 0: 1
step 1: 0
step 2: 11
step 3: 00
step 4: 01[1]
step 5: 111[0]
step 6: 001[11]
step 7: 01[1]1[00]
step 8: 111[0]1[01[1]]
step 9: 001[11] 1[111[0]]
step 10: 01[1] 1[00]0[001[11]]
⋮ ⋮

Similar graphic interpretation of step 10 produces figure 4.15. An example of synthesized plant using a 2L-system after 35 generations is displayed in platee 19.

4.2.3 Particle Systems

A completely different model applied for the simulation of plants and other objects was proposed by Reeves [76] who called it the *particle system.*

Introduction

Particle systems have been investigated by W.T. Reeves et al. [76, 77] since 1983 in order to represent fuzzy objects such as fires, clouds, grass and trees. Traditionally, one can distinguish two main types of data representation: surface-based and volume-based. In the first class, objects are defined by their boundaries; either plane facets or curved surfaces. Volume-based representation defines objects from primitive elements. The CSG (Constructive Solid Geometry) model is a volume-based model where primitive elements are for example a sphere, a cube, a cone, a torus, or a polyhedron. The object is defined by applying boolean operations such as union, intersection and difference to the primitive elements. Voxels is also a volume-based model where primitive elements are elementary volume elements: small cubes parallel to axes. An object is defined as the union of all the volume elements.

Particle systems models lie in the second class. Objects are represented by a multitude of minute primitives named particles. Particle shapes are always very simple, for example: points or spheres can be considered. These particles together represent the volume of an object. At the inverse of classical models, a particle system model is a dynamical model able to represent motion. Over a period of time, particles are generated into the system, move and change state within the system and die from the system.

Particle systems have proven to be powerful modelling techniques because:

- A particle is much simpler than most graphical primitives thus, sophisticated treatments like motion blurr can be applied easily
- A particle model is both a procedural and stochastic model
- The modelling of quite complex and irregular shapes does not need any sophisticated user specification
- Movements and distorsions of objects can be simulated
- The resolution of the generated object can be adapted to the view point.

Two drawbacks must also be specified.

- As a lot of particles have to be generated, some specific rendering techniques have to be considered
- Mixing particle systems with other scene description is often complex.

Particle systems allow the representation of natural objects as opposed to man-made objects. Dynamical objects such as fires, fireworks, torrents, waves or smoke as well as static ones such as grass and trees have already been modelled. Dynamical phenomena are represented by a sequence of images representing the particle evolution as a function of time. Static objects are often defined by the particle trajectories. One can appraise the realism of particle systems modelling on two well known films. In "Star Trek II: The Wrath of Khan", the Genesis Demo sequence shows the application of particle systems to the representation of fire elements [71]. Trees and grass elements were also represented by sophisticated particle systems in the film "The Adventures of André and Wally B." [62].

The remainder of this paper is a summary of W.T. Reeves et al. work [76, 77] where various particle system models and their applications: fires, grass, trees are described. The next paragraph studies the generation process and the following one presents the rendering techniques. Even if these two steps seem to be independent in this presentation; they are usually mixed from a programming point of view because of the number of particles.

Some Models of Particle Systems

Basic Model

In its basic form, a particle system is a collection of particles where each particle is independent of each other. A set of attributes is associated to each particle at its creation. The evolution of a particle system can be described as follows: at each time,

- some particles are added to the system
- some particles are removed from the system
- the existing particles are transformed according to the values of their attributes.

Simple particle attributes can be:

- initial position
- initial velocity (both speed and direction)
- initial size
- initial colour
- initial transparency
- shape
- lifetime.

The number of particles generated and their associated attributes are computed from some global parameters of the system. These parameters can be distributed among two classes: the *generation parameters* and the *transformation parameters.*

Generation parameters: Some of the generation parameters are:

- a position in a three-dimensional space that defines the origin of the particle system
- two angles of rotation about a coordinate system through this origin to define an orientation
- a generation shape which defines a region about its origin into which newly born particles are randomly placed. Common generation shapes are a sphere, a circle or a rectangle
- the mean number of new particles $MeanParts_f$ generated at a frame and its variance $VarParts_f$ such that the number of particles generated at a frame is:

  $$NParts_f = MeanParts_f + Rand() \times VarParts_f$$

  with *Rand* a procedure returning a uniformly distributed random number between -1.0 and +1.0
- the mean initial speed and its variance, with the same computation as for the number of particles
- in the same way, an average value and a maximum variation value are also defined for colour, transparency, size and lifetime
- the particle shape.

One can also increase the system by allowing the previous parameters to change with time

Transformation parameters: Transformation parameters such as the rate of colour change, the rate of transparency change or the rate of size change are also defined.

Some Extensions

The simple model presented above allows the representation of dynamical phenomena such as fires or fireworks (see figure 21). Other applications require more structured models. Until now, each particle was independent of each other. An improvement consists in allowing a particle to generate some new particle systems. One can also use a procedural particle system generation process. Trees have been modelled by recursively generating some new particle systems for each sub-branch.

Realistic effects can also be added by allowing particles movement to be modified by external phenomena such as wind, or the pull of gravity.

Each application requires the definition of a new set of specific parameters. For example, some of the specific global parameters used to model trees are the height of the tree, the mean branch length, the width of the tree.

Particle systems can also be combined with traditional models. For example, in plate 20, the tree trunks are modelled as truncated solid cones and rendered with conventional texture mapping techniques.

Particle Rendering

Let us consider the rendering process associated to a frame. Assuming that the position, geometry and appearance parameters of each particle are known, the rendering of particle systems is quite similar to the rendering of common graphical primitives such as polygons or curved surfaces. Unfortunately, the number of particles is so high (usually between 10,000 and 1,000,000) that traditional rendering techniques cannot be applied. This paragraph describes some approximations that can be made to simplify the rendering algorithms. Two steps of the rendering algorithm are considered: the visible surface determination and the shading process.

Visible Surface Determination

The rendering process differs according to the application. One can distinguish light emitting particles and light reflecting particles. Particle systems simulating fires, fireworks or explosions can be considered in the first class whereas particle systems representing trees or grass lie in the second one.

Light-emitting particles: Considering particles as light-emitting greatly simplifies the rendering process. Each particle is considered as an independent light source. The display algorithm can thus be quite simple: each particle adds a bit of light to the pixels that it covers. The amount of light added and its colour depend on the particle transparency and colour. With this algorithm, no sorting of the particles is needed, the rendering is thus very efficient

Light-reflecting particles: Light–reflecting particles require more sophisticated rendering techniques. Only the algorithms used on plate 21 for the display of carpet of grass and forest are presented here. The forest contains many trees. We will first determine in which order trees have to be displayed. Displaying trees independently implies to assume that the bounding boxes of the trees do not intersect. Under this assumption, we can sort all the trees into a back-to-front order with respect to screen depth. The algorithm consists thus in determining the sort list and then for each tree, beginning with the former one, computing and rendering the tree.

The hidden surface determination of each tree is more complex because some branches of a tree can obscure others; a bucket sort is thus used. The bounding volume of the tree is subdivided into buckets indexed by the eye-space z distance of the particle. Particles are thus displayed according to this order using the painter's algorithm.

The algorithm chosen to render grass is similar except that more accuracy is needed to sort clumps of grass. The non-intersection assumption is not valid any more. Instead of displaying all the buckets associated to a clump at one time, sole the buckets that do not intersect the bounding boxes of the remaining clumps in the eye-space z distance are displayed. Remaining particles are then reassigned to the following buckets.

Shading

Shading fire or firework particle systems is simple because each particle is independent, i.e., cannot play any part in the shading of other particles. On the other hand, motion blur is necessary to solve temporal aliasing and strobing effects. In our application, we can take advantage of the particle simplicity. We can be satisfied with drawing antialiased straight lines instead of points. The straight lines are drawn between the particle positions at the beginning of the frame and halfway through the frame.

Trees and grass particle systems require more sophisticated shading models with ambient, diffuse and specular shading components as well as shadows. An exact shading is again not possible because of the number of particles in the scene. It is thus necessary to approximate the shading. The chosen solution consists in using probabilistic shading models. For example, for grass, the contribution of both the diffuse and ambient components depend on the distance from the top of the clump of grass to the particle in question, decreasing exponentially as the depth increases. Similarly, external shadows are added to trees by considering the plane passing through the light source and the top of the neighbouring trees. This plane cuts the tree in question. Shadows are defined according to this plane: particles above the plane are in full sunlight and particles below the plane are partially in the shade depending on the distance from the plane to the particle.

Conclusion

Some applications of the particle systems have been presented. Both the generation and the rendering process are very application dependent, new parameters and rendering techniques have to be considered for each new application. On the other hand, once the system has been designed, very few user specifications are needed. Several other applications can be investigated such as the modelling of waves [34].

4.2.4 Solid Texturing

Introduction

Traditionally, the problem of texturing objects has been solved using surface texturing methods (including the mapping of a 2D texture on a 3D-object and the procedural generation of a 2D texture over the surface of a 3D object). Surface texturing consists of the evaluation of a function $T(u, v)$ defined over a set of points located on the surface of the object to be textured.

Solid texturing is the generalisation to the 3D space of the concept of surface texturing. Solid texturing also consists of the evaluation of a space function $T(x, y, z)$ that is defined over the entire volume of the object (often in a cube containing it): it is also defined in the interior points of the object.

A space function can be seen as representing a block of solid material: the evaluation of such a function at the visible points of an object's surface will give the texture of that surface (i.e., the texture that could be seen when sculpturing the initial block to obtain the desired object). So it is called "solid texture".

Originally, with surface texturing methods, the value returned by the 2D function was used to determine the reflectance or colour at one point of the surface. Any improvement that has been introduced for surface texturing is also available for solid texturing, such as the use of the returned value to control local reflection, shininess, roughness, or surface normal direction. The latter technique permits the simulation of bumps and wrinkles on geometrically smooth surfaces [9, 11].

Solid texturing methods can be used in exactly the same way, so it is possible to simulate anything that can be simulated using surface texturing methods but without the traditional inconvenience of these methods [74].

Solid Texturing Examples

The concept of "solid texture" has been introduced in 1985 independently by K.Perlin [74] and D.R. Peachey [72]. A wide range of materials and "visual effects" has been simulated using these solid texturing methods. A similar technique has been used by G.Y. Gardner [44] to obtain a "visual simulation of clouds".

1. The 3D equivalents of all the 2D functions that have been used in previous works for procedural generation of 2D textures can be used to generate 3D textures:
   - Sine functions
   - Stochastic functions
   - Functional composition.
2. Traditional texture mapping technique can be included in solid texturing by using the composition of a look-up table function (getting values from the texture map) and a projection function from the 3D space to a 2D space
3. The value returned by the space function can be used to control the colour, intensity or transparency of an object:
   - The use of a "perturbed" sine function allows for the modelling of an impressive realistic marble texture [74] as shown in plate 22
   - The use of short sums of sine functions (each with a different period and phase) permits the simulation of clouds of all kinds [44] as shown in plate 24.
4. The value returned by the space function can be used to control the direction of the surface's normal vector. This allows the simulation of bump textures [74] (See plate 23)
5. The perturbation of the normal direction can be correlated to time: this allows the simulation of waves [74]
6. Both controlling the normal direction and the colour as a function of time permits the simulation of fire [74].

Advantages of Solid Texturing

One of the advantages of solid texturing techniques is that texture becomes independent of the object's shape: texture does not have to be fitted into the object's surface and, so, we do not have to face the traditional problems encountered when using surface texturing techniques (for example, with the poles of a sphere). A second one is that, as for all procedural techniques, the database is extremely small.

Problems with Solid Texturing

Though solid texturing methods are really powerful, there are a certain number of problems that make them not so easy to use. The first problem is that the generated texture depends on the density of sample points on the surface of the objects being textured. The second is that there are no means to automatically determine which functions need to be used to generate a wanted texture: the user has to rely on his own intuition and inspiration for choosing the functions and the values of the parameters that are going to control the texture.

In fact, the problem is that solid texturing only provides the means of a visual simulation of the aspect of some materials and natural elements and that this simulation is not based upon reality. So practice and experience are needed to obtain convincing results.

4.2.5 Structural Modelling of Botanical Structures

Introduction

In the past few years many attempts have been made to simulate the development of plants, trees and botanical structures.

Some interesting results have been obtained using branching process constructions [54], [2] graftals and particle systems [80, 77, 70], combinatorics on trees [28]. The very first model taking account of the botanical structure of plants and trees has been proposed by de Reffye, Edelin, Françon, Jaeger and Puech [22]. It is a parametric model which allows the image generation of a wide variety of plants and trees, permits the integration of time to simulate the growth of plants, and also the incidence of wind, insect attacks and other natural parameters. Their system works using the simulation of the activity of *buds* at different times. These buds can either become a flower and die, or go into sleep, or become an internode, or die, etc.

These different kinds of events can occur according to specific parameters each of which is included in the model. These parameters are also probabilistic: this explains the wide range of plants that have been rendered using this system.

Simulation of the Architecture of Plants

The model takes into account a wide variety of structural parameters such as the order of an axis (see figure 4.16), phyllotaxy (see figure 4.17), and ramification (figure 4.18).

The *order of the axis* of a plant is a parameter relating the branching process to the apical buds of a growth unit. *Phyllotaxy* governs the relative position of a lateral bud with the lateral bud of a previous node. *Ramification* can be either continuous, rythmic or diffuse. It is *continuous* whenever every node of an axis is the root of an axis of greater order, *rhythmic* whenever some nodes are the root of an axis of greater order and *diffuse* whenever these roots are located at random. Some laws are known between botanical species and these parameters. Hence, as explained in [22], the kinds of ramifications are functions of the order of the axis for a given variety. The development trend of a tree may be *orthotropic* or *plagiotropic* (see figure 4.19).

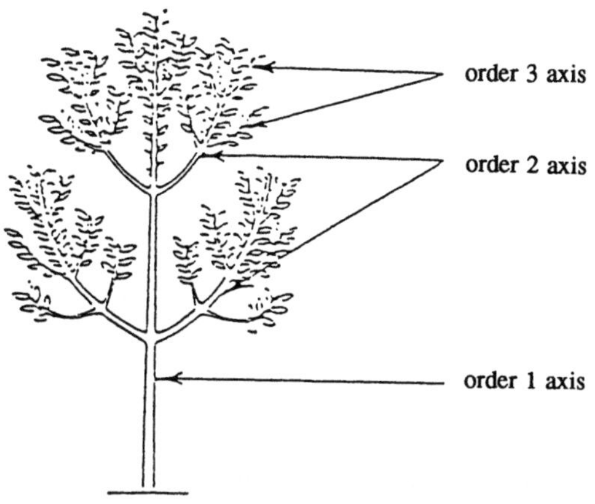

FIGURE 4.16. The notion of order of an axis

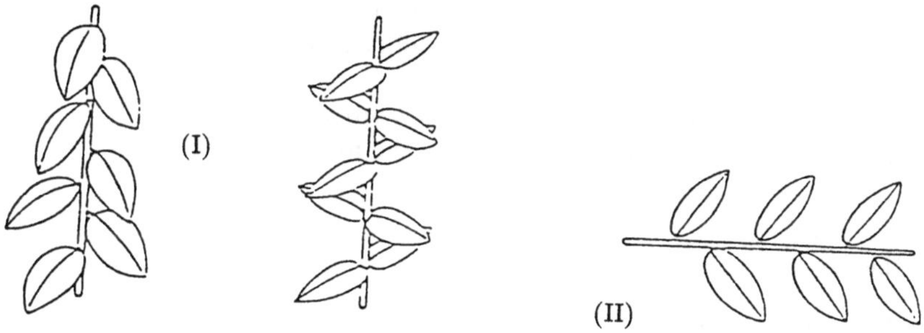

FIGURE 4.17. Phyllotaxy:(I) spiraled (II) distic

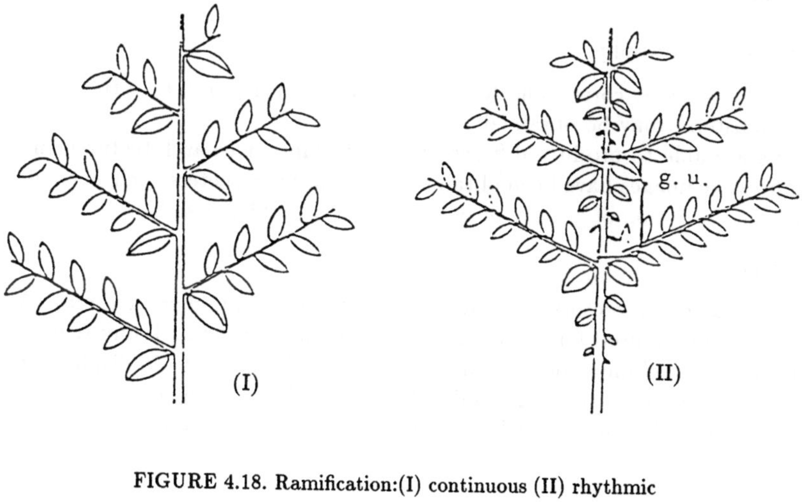

FIGURE 4.18. Ramification:(I) continuous (II) rhythmic

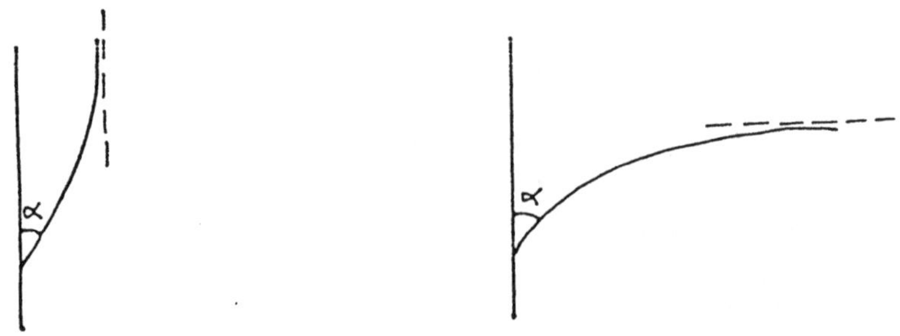

FIGURE 4.19. Development trend of a tree (left: orthotrophy, right: plagiotrophy)

Lastly, the model accounts for tree architecture. Only 23 different architectural models are known [22].

The model proposed by these co-authors is a quantitative (but probabilistic) one. Time is digitized so that the fundamental unit of time is the time taken by the growth of a growth unit, which is supposed to be constant for the axes of a given order (hence axes of different order do not grow at the same speed in general). To each bud are given two probabilities: an *abortion* probability and a *wait* probability (the bud can wait during a certain amount of time without growing).

The ramification process is included in the *non-abort* probability parameter. As explained in [22], growth simulation is obtained including the following parameters:

- age
- growth of the axes
- number of buds at each node
- probability of death, pause, ramification and reiteration
- type of growth unit
- geometry of internodes
- orthotrophy and plagiotrophy.

Each of these parameters may correspond to existing ones in nature or not. In the latter case, interesting "monsters" may be obtained!

These parameters are able to produce different types of plants, tree architectures and growing. Some impressive images have been obtained, and other parameters such as bending of branches under the action of gravity or wind have been implemented.

Visualization and Rendering

The system produces a set of vectors and faces (usually a great number of them). A classical Z-buffer algorithm is used for the elimination of hidden surfaces. Trunk texture may be obtained using the technique developed by Bloomenthal [12]. Plate 25 displays an image of a palm tree constructed using this model (from [22]). Plate 26 shows a "natural" weeping willow obtained by the same model. Both images are impressive for their realism.

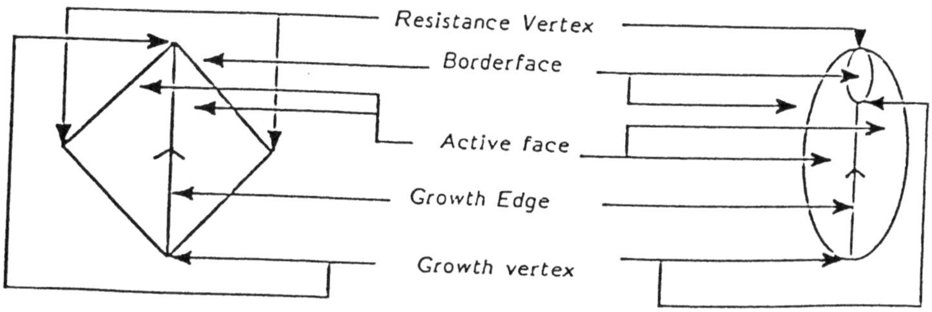

FIGURE 4.20. Elementary modules (growth edges are represented graphically by directed edges, by contrast with resistance edges which are undirected)

4.2.6 Free Surface Modelling by Simulated Evolution

Introduction

We describe a procedural method for modelling free surfaces (free in the sense that, for a surface, the exact position of every point and the precise form of the surface in the neighbourhood of each point is not of primary importance). This method enables the modelling of open or closed planar surfaces, with or without holes, and in particular, the simulation of the evolution of these surfaces. This method [56, 57] has been applied to the synthesis of both static and dynamic images of natural surfaces, movement of articulated objects, etc.

The method is based on principles common to two methods for the simulation of natural phenomena: the simulation of the growth of plants by internodes developed by P. de Reffye [21, 22], and for particle systems presented by W. Reeves [75]. These principles are:

digitization: each model is composed of basic elements (particles, internodes) which lead to digitization in time (denoted h hereafter)

notion of element proper activity: the principle of each method consists in simulating the behaviour of basic elements

characterization of the object evolution: being modelled according to a set of "parameters" associated with the basic elements

priority of topology over geometry: which assumes increasing importance in three-dimensional modelling (see [5, 1, 67, 46] for example).

Particle systems constitute a pointwise model, where the basic element is a point-like particle. De Reffye's model is a linear model where the basic element is a curve-like internode. The method presented here is an extension of the previous methods to the modelling of *surfaces*; it is based upon a surface-like model, called *modular map*, composed with surface elements denoted elementary modules (see figure 4.20). Following a classification proposed by A. Fournier [32], this model is a structural model.

Modular Map

A modular map ([83, 5, 46] for details) represents a set of faces, edges and vertices modelling the topology of subdivision of a surface. Two kinds of vertices and edges (VandE) are defined:

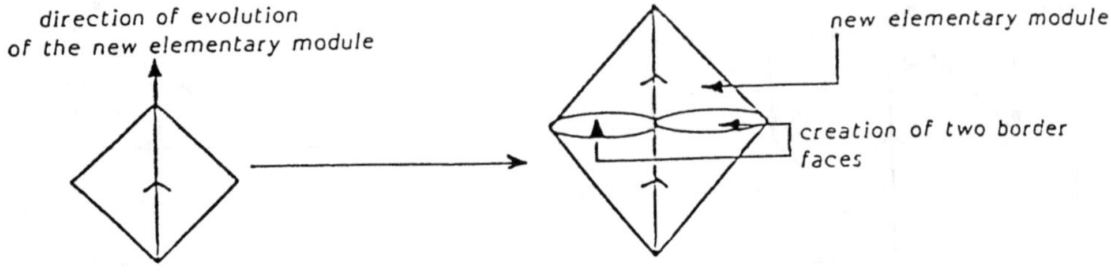

FIGURE 4.21. External ramification of a modular map

growth VandE corresponding to the object skeleton and making up a rooted tree

resistance VandE, which are passive during the map evolution. They constitute the outline of the object in particular.

Vertices and faces of a modular map are partitioned into growth vertices, *active faces* (incident with at least one growth edge) and resistance vertices, border faces (only incident with resistance edges). Every modular map is either reduced to a vertex, which is a growth vertex, or composed of elementary modules. Each elementary module is formed from two distinct triangular actives faces, incident to a common growth edge. Each active face is incident with a growth edge and two resistance edges, two growth vertices and a resistance vertex (see figure 4.20).

Evolution of Modular Maps

The method is as follows: the topology of surface S at time h is modelled as a modular map C_h, defined as a set of vertices, edges, and faces (the faces are simple cycles of edges). One vertex is distinguished in each modular map and called the *initial vertex*. In particular, the modular map C_0, defining the topology of S at h=0 is reduced to the initial vertex. The evolution of the topology of S between times h=0 and h=n is modelled by a sequence of modular maps $(C_0, \dots, C_n)$. Each modular map $C_i (i > 0)$ is determined from the modular map C_{i-1} by an operation on the map. These operations are:

Creation of an initial elementary module; the operation is applicable to a modular map reduced to a growth vertex (in particicular, it applies to the map C_0)

External ramification, corresponding to the creation of a new elementary module in the "interior" of a border face (see figure 4.21)

Internal ramification, or the creation of a new elementary module in the "interior" of an activity face (see figure 4.22)

Deletion of an elementary module

Fusion, which involves the creation of a border face defined by two resistance edges (see figure 4.23)

Disjunction, which involves the deletion of a boundary face (the inverse of the preceeding operation). These last two operations, in particular, permit the representation of the contour of the modelled surface.

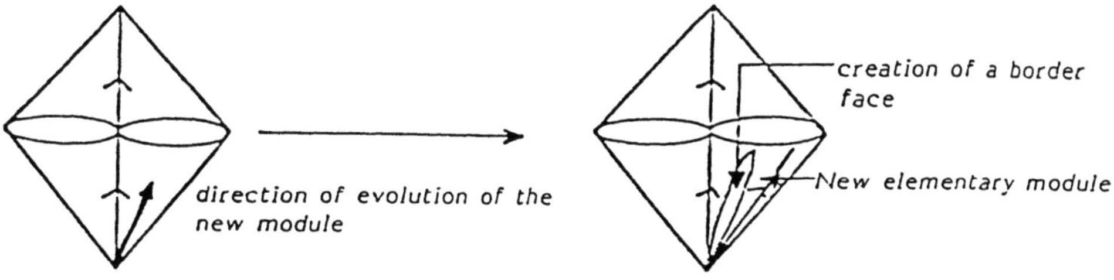

FIGURE 4.22. Internal ramification

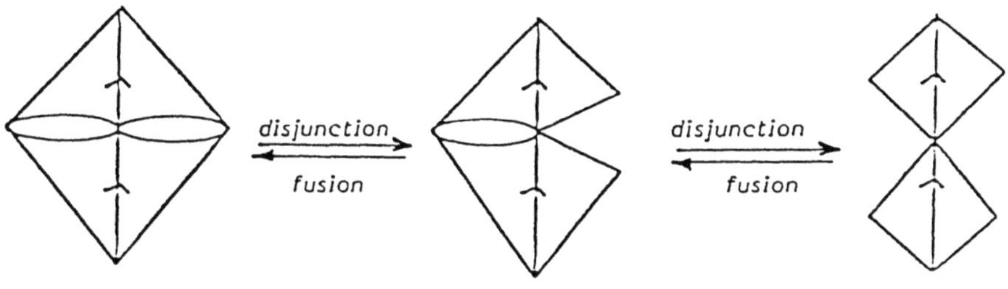

FIGURE 4.23. fusion / disjunction of a modular map

The principle of the characterization of a sequence of modular maps is inspired by graph *grammars*, that is, the characterization requires the definition of:

a finite alphabet (set of "labels")

an initial modular map (C_0, reduced to a vertex)

production rules, composed of a left-hand side, which is a label, and a right-hand side defining a sequence of operations applicable to a modular map.

If several sequences of operations are applicable to a single label, it is possible to have conditions which specify processing (c.f. figure 4.24). These conditions utilize parameters associated to edges and to the initial vertex of the map (these parameters are detailed below).

The principle is as follows. A label is associated with every edge and to the initial vertex. Whenever this label appears as the left-hand side of a production rule, the sequence of operations defined by the right-hand side is applied (i.e., to every edge and to the initial vertex). This is repeated at each time h for every edge or initial vertex.

Characterization of Modular Maps

Before describing the parameters associated with the edges and to the initial vertex of a modular map, it must be pointed out that the growth elements (vertices and edges) form a tree, called the growth tree. This tree is directed: the initial vertex of the map is the root of the tree, and the tree is directed from the root towards the leaves. The growth tree of a modular map is structured along axes; that is, an integer, called the order of the edge, is associated with each edge. An axis is thus defined as a maximal length path in the growth tree (given the directions of the growth edges), with all the edges of the path having the same order.

Various parameters may be associated to the edges and to the initial vertex of a modular map, which permit the characterization of applicable operations.

Given an edge A and the growth edge A′ such that both are incident with the same elementary module, then associated with A are (c.f. figure 4.24):

- the time value h
- the label of A
- the order of A
- the order of A′
- the time of creation of A, each edge (and each vertex, excepting the initial vertex of the map) being created at time h> 0
- the type of A, depending on whether A is incident with the root or with a leaf of the growth tree
- distance relationships between A′ and some other element X of the growth tree; the distance between A′ and X is defined as the number of edges in the unique simple elementary path between them (for example, the distance from A′ to the root of the growth tree, or the origin of the axis incident with A′).

The submersion of the surface in the usual three-dimension space, as well as the aspect (limited to colour) are defined at each instant by real functions associated with the vertices and faces of the modular map (defining the positions of the vertices and faces of the modular map and the colours of its faces). These real functions are initially defined by the user in the production rules. The functions take as arguments the characteristics defined above (time, date of creation, etc.).

The surface modelling method defined here permits the simulation of the evolution of the topology of the surfaces (for growth simulation, for example), as well as the evolution of purely geometric properties (that is, the positions of the vertices of the map, which permits the simulation of the evolution of articulated objects), and the colouring of the surface (for example, the colour variation of a leaf in autumn).

Plate 27 presents the evolution of a maple leaf using this technique and plate 28, a beautiful bell flower.

The sequence of modular maps is obtained by successive applications of external and internal ramifications. The growth tree is composed of two types of axes: an order 1 axis (vertical), and order 2 axes (horizontal). Each clock time produces the creation of a new elementary module whose growth edge extends the axis of order 1, and the creation of elementary modules whose growth edges extend an already existing axis of order 2, or is the origin of a new axis of order 2. Modules whose growth edge extend an axis of order 2 are only created if the length of the axis (number of growth vertices forming the axis) is strictly less than the distance (the number of growth vertices) between the origin of the secondary axis and the root of the tree, which accounts for the halt in growth of the secondary inferior axes.

Acknowledgements:

I would like to thank L. Bourcier, S. Coquillart, F. Cros, J. Lévy Véhel, P. Lienhardt, P. Sander and H. Yahia who helped me in the preparation of these tutorials.

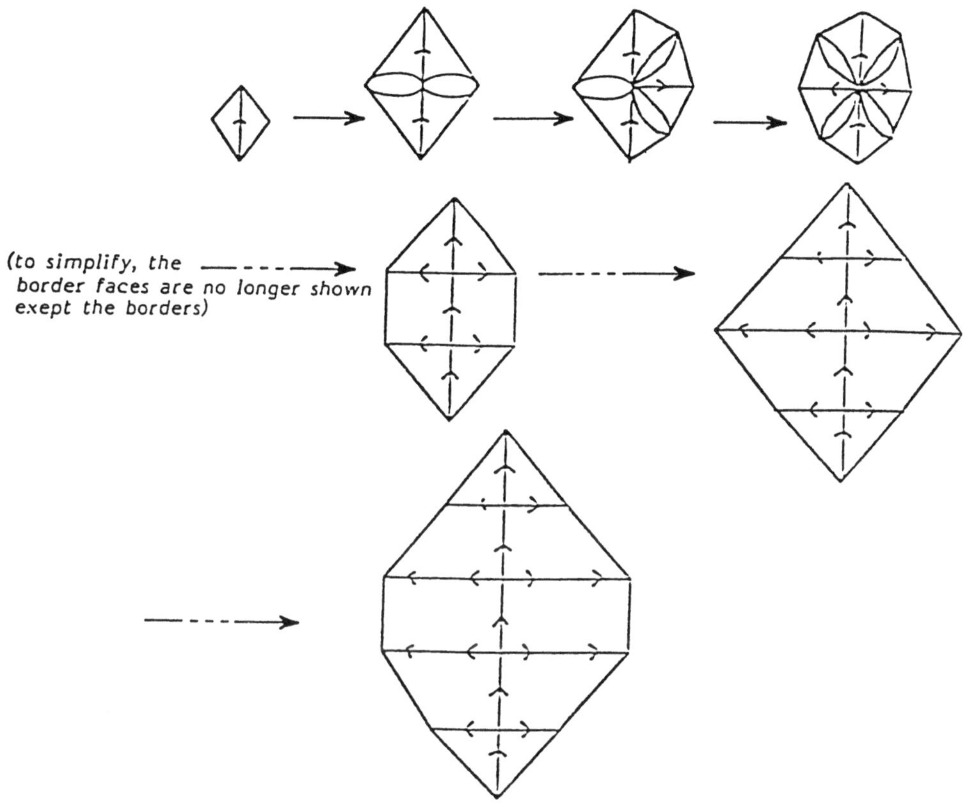

FIGURE 4.24. An example of evolution

4.3 References

[1] S Ansaldi, L de Floriani, and B Faldieno. Geometric Modeling of Solid Objects by using a Face Adjacency Graph Representation. *Computer Graphics (Proc. Siggraph 85)*, 19(3):131–139, 1985.

[2] M Aono and T Kunii. Botanical Tree Image Generation. *IEEE Computer Graphics and Applications*, 4(5):10–34, 1984.

[3] M Barnsley. Making Chaotic Dynamical Systems to Order. Lecture Notes, INRIA Workshop on Fractals, 1986.

[4] M Barnsley, V Ervin, D Hardin, and J Lancaster. Solution of an Inverse Problem for Fractals and Other Sets. *Proc. Nat Acad Sci USA*, 83, 1986.

[5] B Baumgart. A Polyhedron Representation for Computer Vision. *Proc. AFIPS*, 44:589–596, 1975.

[6] J Beck, A Sutter, and R Ivry. Spatial Frequency Channels and Perceptual Grouping in Texture Segregation. *CVGIP*, 37:299–325, 1987.

[7] C Bennis and A Gagalowicz. Hierarchical Textures Synthesis on 3D Surfaces. In *Proceedings of Eurographics-89*. North-Holland, 1989.

[8] J F Blinn. *Computer Display of Curved Surfaces*. PhD thesis, Department of Computer Science, University of Utah, 1978.

[9] J F Blinn. Simulation of Wrinkled Surfaces. *Computer Graphics (Proc. Siggraph 78)*, 12(3):286–292, July 1978.

[10] J F Blinn and M Newell. Texture and Reflection on Computer Generated Images. *Communications of the ACM*, 10, 1976.

[11] J F Blinn and M E Newell. Texture and Reflection, in Computer Generated Images. *Computer Graphics (Proc. Siggraph 76)*, 10(3):542–547, 1976.

[12] J Bloomenthal. Modeling the Mighty Maple. *Computer Graphics (Proc. Siggraph 85)*, 19(3):305–311, 1985.

[13] A C Bovik, M Clark, and W S Geisler. Computational Texture Analysis Using Localized Spatial Filtering. *WCV*, pages 201–206, 1987.

[14] P Brodatz. *Textures: A Photographic Album for Artists and Designers*. Dover, New-York, USA, 1956.

[15] T Caelli and B Julesz. On Perceptual Analysis Underlying Visual Texture Discrimination (Parts 1 and 2). *Biological Cybernetics*, 28 and 29:167–175 and 201–214, 1978.

[16] E Catmull. *A Subdivision Algorithm for Computer Display of Curved Surfaces*. PhD thesis, Department of Computer Science, University of Utah, 1974.

[17] E Catmull and A R Smith. 3D Transformation of Images in Scanline Order. *Computer Graphics (Proc. Siggraph 80)*, 14, 1980.

[18] S Chang, L S Davis, S M Dunn, J O Eklundh, and A Rosenfeld. Texture Discrimination by Projective Invariants. *PRL*, 5:337–342, 1987.

[19] M Clark, A C Bovik, and W S Geisler. Texture Segmentation Using Gabor Modulation/Demodulation. *PRL*, 6:261–267, 1987.

[20] F C Crow. Summed-area Tables for Texture Mapping. *Computer Graphics (Proc. Siggraph 84)*, 18(3), 1984.

[21] P de Reffye. *Modelisation de l'Architecture des Arbres par des Processus Stochastiques : Simulation Spatiale des Modèles Tropicaux sous l'Effet de la Pesanteur. Application au Coffea Robusta.* PhD thesis, Université de Paris Sud, Orsay, 1979.

[22] P de Reffye, C Edelin, J Françon, M Jaeger, and C Puech. Plant Models Faithful to Botanical Structure and Development. *Computer Graphics (Proc. Siggraph 88)*, 22(3), July 1988.

[23] H Derin and H Elliot. Modeling and Segmentation of Noisy and Textured Images using Gibbs Random Fields. *PAMI*, 9:39–55, 1987.

[24] H Derin and C S Won. A Parallel Image Segmentation Algorithm using Relaxation with Varying Neighbourhoods and its Mapping to Array Processors. *CVGIP*, 40:54–78, 1987.

[25] P A Devijver and M M Dekesel. Learning the Parameters of a Hidden Markov Random Field Image Model : A Simple Example. *PRTA*, pages 141–163, 1987.

[26] E Diday and J C Simon. Clustering Analysis, Communications and Cybernetics. In K S Fu, editor, *Digital Pattern Recognition*, volume 10, pages 47–94. Springer-Verlag.

[27] K B Eom and R L Kashyap. Texture and Intensity Edge Detection with Random Fields Models. *WCV*, pages 29–34, 1987.

[28] G Eyrolles, J Françon, and G Viennot. Combinatoire pour la Synthèse d'Images Réalistes de Plantes. *Actes du deuxième colloque Image, CESTA*, pages 648–652, 1986.

[29] J Fairfield. Finding Edges and Skeletons in Textured Images by Comparison of Diffused Histograms. *WCV*, pages 321–323, 1987.

[30] Z Fan and F S Cohen. Textured Image Segmentation as a Multiple Hypothesis Test. *WCV*, pages 234–236, 1987.

[31] E A Feibush, M Levoy, and R L Cook. Synthetic Texturing using Digital Filters. *Computer Graphics (Proc. Siggraph 80)*, 14(3), 1980.

[32] A Fournier. The Modeling of Natural Phenomena. Siggraph-87 Course notes, number 16, 1987.

[33] A Fournier, D Fussell, and L Carpenter. Computer Rendering of Stochastic Models. *Communications of the ACM*, 16(6), 1982.

[34] A Fournier and W T Reeves. A Simple Model of Ocean Waves. *Computer Graphics (Proc. Siggraph 86)*, 20(3):75–84, 1986.

[35] A Gagalowicz. A New Method for Texture Fields Synthesis; Some Applications to the Study of Human Vision. *IEEE Trans. On Pattern Analysis and Machinery Intelligence*, 3(5):502–533, 1981.

[36] A Gagalowicz. *Vers un Modèle de Textures.* PhD thesis, Université de Paris 6, 1983.

[37] A Gagalowicz and S D Ma. Synthesis of Natural Textures on 3D Surfaces. In *7th ICPR Conference, Montreal, Canada*, 1984.

[38] A Gagalowicz and S D Ma. Sequential Synthesis of Natural Textures. *Computer Vision, Graphics and Image Processing*, 30:289–315, 1985.

[39] A Gagalowicz and S D Ma. Model Driven Synthesis of Natural Textures for 3D Scenes. *Computers and Graphics*, 10(2):161–170, 1986.

[40] A Gagalowicz, S D Ma, and C Tournier-Lasserve. New Model for Homogeneous Textures. In *Proceedings of the 4th Scandinavian Conference on Image Analysis, Trondheim, Norway*, 1985.

[41] A Gagalowicz, S D Ma, and C Tournier-Lasserve. Efficient Models For Colour Textures. In *Proceedings of the 8th International Conference (ICPR 86), Paris*, volume 1, pages 409–411, 1986.

[42] A Gagalowicz, S D Ma, and C Tournier-Lasserve. Third Order Model for Non Homogeneous Natural Textures. In *Proceedings of the 8th International Conference (ICPR 86), Paris*, volume 1, pages 409–411, 1986.

[43] A Gagalowicz, S D Ma, and C Tournier-Lasserve. Un Modele pour Textures Macroscopiques. In *Second Colloque Image du CESTA, Nice*, volume 2, pages 690–697, 1986.

[44] G Y Gardner. Visual Simulation of Clouds. *Computer Graphics (Proc. Siggraph 85)*, 19(3):297–303, 1985.

[45] D Geman, S Geman, and C Graffigne. Locating Texture and Object Boundaries. *PRTA*, pages 165–177, 1987.

[46] L Guibas and J Stolfi. Primitives for the Manipulation of General Subdivisions and the Computation of Voronoi Diagrams. *ACM Transactions on Graphics*, 4(2):74–123, 1985.

[47] D C He, L Wang, and J Guibert. Texture Feature Extraction. *PRL*, 6:269–273, 1987.

[48] F Heitz and H Maitre. Application of Autoregressive Models to Fine Arts Painting Analysis. *SP*, 63:1–4, 1987.

[49] M Iizuka. Quantitive Evaluation of Similar Images with Quasi-Gray Levels. *CVGIP*, 38:342–360, 1987.

[50] B Julesz. Visual Pattern Discrimination. *IRE Trans. on Information Theory*, It-8:84–92, 1962.

[51] B Julesz and et al. Inability of Humans to Discriminate between Visual Textures that Agree in Second Order Statistics Revisited. *Perception*, 2:391–405, 1973.

[52] B Julesz and et al. Visual Discrimination of Textures with Identical Third Order Statistics. *Biological Cybernetics*, SMC-8:796–804, 1978.

[53] M Kass and W Witkin. Analyzing Oriented Patterns. *CVGIP*, 37:362–385, 1987.

[54] Y Kawagachi. A Morphological Study of the Form of Nature Computer Graphics. *Computer Graphics (Proc. Siggraph 82)*, 16(3):223–232, July 1982.

[55] A S K Lau. Knowledge-based and Statistical Techniques Applied to Textural Image Classifications. *PRL*, 6:95–100, 1987.

[56] P Leinhardt. *Modelisation et Evolution de Surfaces Libres.* PhD thesis, Université Louis Pasteur, 1987.

[57] P Leinhardt and J Françon. Vegetal Leaves Images Synthesis. MARI 87, 3rd week of l'Image Electronique, Paris (La Villette), 18-22 May, 1987.

[58] C Lemaréchal, J J Strodiot, and A Bihian. On a Bundle Algorithm for Nonsmooth Optimization. In Mangasarian, Meyer, and Robinson, editors, *Non-Linear Programming 4*, pages 245–282. Academic Press, 1981.

[59] J Lévy Véhel and A Gagalowicz. Shape Approximation by a Fractal Model. In *Proceedings Eurographics-87*. North-Holland, 1987.

[60] J Lévy Véhel and A Gagalowicz. Shape Approximation of 2D Objects. In *Proceedings Eurographics-88*. North-Holland, 1988.

[61] A Lindenmayer. Mathematical Models for Cellular Interactions in Development, Part 1 and 2. *Journal of Theoretical Biology*, 18:280–315, 1968.

[62] Lucasfilm Ltd. *The Adventures of André and Wally B (film)*. Lucasfilm Ltd, August 1984.

[63] S D Ma. *Modélisation et Synthèse de Textures, Applications à l'Infographie.* PhD thesis, Université de Paris VI, 1986.

[64] S D Ma and A Gagalowicz. A Parallel Method For Natural Textures Synthesis. In *7th ICPR Conference, Montreal, Canada*, 1984.

[65] S D Ma and A Gagalowicz. Local Coordinate Systems on 3D Surfaces. In *Proceedings of the Eurographics-85*. North-Holland, 1985.

[66] B Mandelbrot. *The Fractal Geometry of Nature.* W H Freeman and Company, San Francisco, 1983.

[67] M Mantyla. Computational Topology: a Study of Topological Manipulations and Interrogations in Computer Graphics and Geometric Modeling. *Acta Polytechnica Scandinava*, 37, 1983.

[68] A Norton, A P Rockwood, and L S Skolmolski. Clamping: A Method of Antialiasing Textured Surfaces by Bandwidth Limitation in Object Space. *Computer Graphics (Proc. Siggraph 82)*, 16(3), 1982.

[69] E Oja and J Parkkinen. Texture Subspaces. *PRTA*, pages 21–33, 1987.

[70] P E Oppenheimer. Real Time Design and Animation of Fractal Plants and Trees. *Computer Graphics (Proc. Siggraph 86)*, 20(3):55–64, July 1986.

[71] Paramount. Genesis Demo from Star Trek II: The Wrath of Khan. Film, 1982. in Siggraph Video Review Number 11.

[72] D R Peachey. Solid Texturing of Complex Surfaces. *Computer Graphics (Proc. Siggraph 85)*, 19(3):279–286, July 1985.

[73] D R Peachey. Modeling Waves and Surf. *Computer Graphics (Proc. Siggraph 86)*, 20(3), July 1986.

[74] K Perlin. An Image Synthesizer. *Computer Graphics (Proc. Siggraph 85)*, 19(3):287–296, 1985.

[75] W T Reeves. Particle System: A Technique for Modelling a Class of Fuzzy Objects. *ACM Transactions on Graphics*, 2(2):91–108, 1983.

[76] W T Reeves. Particle Systems - A Technique for Modelling a Class of Fuzzy Objects. *Computer Graphics (Proc. Siggraph 83)*, 17(3), 1983.

[77] W T Reeves and R Blau. Approximate and Probabilistic Algorithms for Shading and Rendering Structured Particle Systems. *Computer Graphics (Proc. Siggraph 85)*, 19(3):313–322, July 1985.

[78] R A Rensik. On the Visual Discrimination of Self-similar Random Textures. *WCV*, pages 240–242, 1987.

[79] S J Roan, J K Aggarwal, and W N Martin. Multiple Resolution Imagery and Texture Analysis. *PR*, 20:17–31, 1987.

[80] A R Smith. Plants, Fractals and Formal Langages. *Computer Graphics (Proc. Siggraph 84)*, 18(3):1–10, 1984.

[81] K A Stevens and A Brookes. Detecting Structure by Symbolic Constructions on Tokens. *CVGIP*, 37:238–260, 1987.

[82] D Terzopoulos, J Platt, A Barrand, and K Fleischer. Elastically Deformable Models. *Computer Graphics (Proc. Siggraph 87)*, 31(3):205–214, 1987.

[83] W Tutte. *Encyclopaedia of Mathematics and Its Applications*, volume 21, chapter Graph Theory. Addison-Wesley, 1984.

[84] R Vistnes. Computer Texture Analysis and Segmentation. *WCV*, pages 231–233, 1987.

[85] H Voorhees and T Poggio. Detecting Blobs as Textons in Natural Images. *IUW*, pages 892–899, 1987.

[86] H Voorhees and T Poggio. Detecting Textons and Texture Boundaries in Natural Images. *ICCV*, pages 250–258, 1987.

[87] J Weil. The Synthesis of Cloth Objects. *Computer Graphics (Proc. Siggraph 86)*, 20(4):49–54, 1986.

[88] L Williams. Pyramidal Parametrics. *Computer Graphics (Proc. Siggraph 82)*, 16(3), 1982.

[89] G Wywill, B Wywill, and C McPheeters. Solid Texturing of Soft Objects. *IEEE Computer Graphics and Applications*, 12:20–26, 1987.

[90] H Yamada and T Kasvand. Transparent Object Extraction from Regular Textured Backgrounds by using Binary Parallel Operations. *CVGIP*, 40:41–53, 1987.

5 Developments in Ray-Tracing

Christian Bouville and Kadi Bouatouch

ABSTRACT
After a brief recall of the basic principles of ray-tracing, the traditional and distributed approaches will be presented with the related illumination models. This will include an introduction to relevant topics such as photometry and optics, Monte-Carlo integration, stochastic sampling and a discussion on implementation issues. Then, the fundamental problem of ray object intersection will be tackled. First, intersection algorithms for various type of objects will be described including simple objects (solid primitives, polygons,...) as well as more complex objects (bicubic patches, fractals, CSG modelled object,...). Then, after a discussion on bounding volumes, the case of large and complex collections of primitives is considered by examining the two basic classes of organizing structure (i.e. hierarchical data structure and space-partitioning) and their related traversal algorithms. Eventually, the parallel processing issue will be discussed.

5.1 Introduction

The ray-tracing rendering method gives the highest possible level of realism that can be attained in computer graphics. This approach provides a very general framework in which to solve a wide range of problems simply and elegantly. Because of this simplicity, ray-tracing was used in the early stages of computer graphics ([2], [26]) but only since 1980 has the technique been applied to realistic image synthesis [34], [57].

The simplicity of ray-tracing is attractive for many programmers interested in image synthesis but they are often discouraged by the slowness of computations. As a matter of fact, reasonable computation times can only be obtained with powerful computers and efficient algorithms. As far as hardware is concerned, parallel processing is certainly the answer to the problem of computing power.

Many improvements have been made to ray-tracing since 1980, both as regards realism and efficiency, and these topics will be discussed below. The first part of this document, entitled photometry, deals with rendering and the second part, entitled computational geometry, specifically addresses the ray/object intersection problems.

5.2 Photometry

5.2.1 Background

Principles

In image synthesis, a ray is cast in a given direction to evaluate the light energy coming from this direction. Because the eye responds to radiance, this light energy will be characterized by the radiance of the first surface met by the ray. As rays are traced in the opposite direction to the propagation of light, ray-tracing is a backward reconstruction process. Backward ray-tracing is preferred to forward ray-tracing for reasons of efficiency although this leads to some difficulties as will be seen in the last paragraph.

To generate a picture, the viewing frustum is sampled by casting rays from the eye position. If aliasing problems are not taken into account, one ray per pixel is sufficient

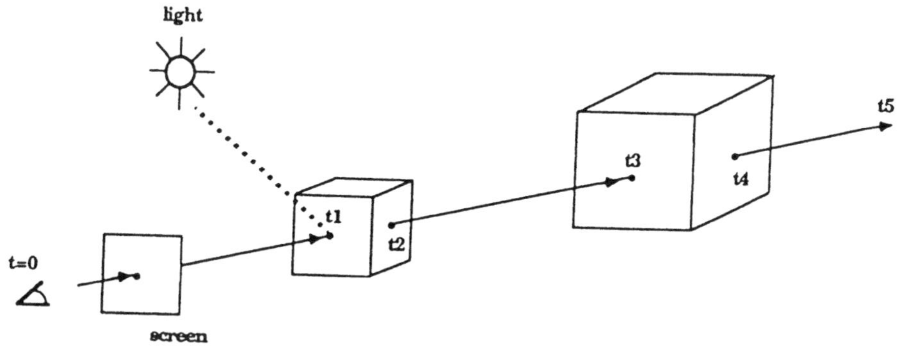

FIGURE 5.1. Ray Casting

to evaluate the intensity of a pixel (figure 5.1). When such a ray intersects objects, pixel intensity is proportional to the radiance of the surface closest to the eye position. In general, the computation of this radiance is a difficult problem because the illumination may come from light sources as well as from the light reflected by other objects in the environment. In the following paragraphs, the photometric algorithms we propose use rays as tools to evaluate illumination in complex environments. As other rays will be necessary to compute the radiance, rays cast from the eye position are called first-generation, or pixel, rays. Note that backward ray-tracing is totally view-dependent.

Principles of Light Energy Computations

Let us consider a small surface dS around a given point **P**. The light energy Q leaving this surface in a given direction consists of reflected and transmitted incident light. As this incident light comes from light source illumination (direct lighting) and from light reflected by other objects (indirect lighting), Q can be broken down as follows:

$$Q = Q_{RD} + Q_{RI} + Q_{TD} + Q_{TI} \tag{5.1}$$

where Q_{RD}, Q_{TD} are respectively the reflected and transmitted light energy produced by direct lighting, and Q_{RI}, Q_{TI} are respectively the reflected and transmitted light energy coming from indirect lighting (figure 5.2).

To compute the indirect lighting effect, namely Q_{RI} and Q_{TI}, second-generation rays are necessary to determine the illumination coming from the environment. When one of these rays intersects an object, equation (5.1) is applied again which implies the shooting of third-generation rays and so on. Consequently, photometric algorithms based on ray-tracing techniques are inherently recursive.

Generally speaking, two levels need to be considered in the mechanisms of light transport: a local and a global level. A local illumination model describes how an incident light interacts locally with a surface, i.e., how a single ray of light is reflected, transmitted, absorbed and scattered when it strikes a single surface. On the other hand, a global illumination model tries to provide a complete description of light propagation within the environment. The ray-tracing approach clearly falls into the category of global illumination models.

As for direct lighting effects, the evaluation of Q_{RD} can be performed using well-known local illumination models ([57], [44]). However, physics-based approaches are preferable because they can be enhanced as necessary to model a broad range of phenomena with

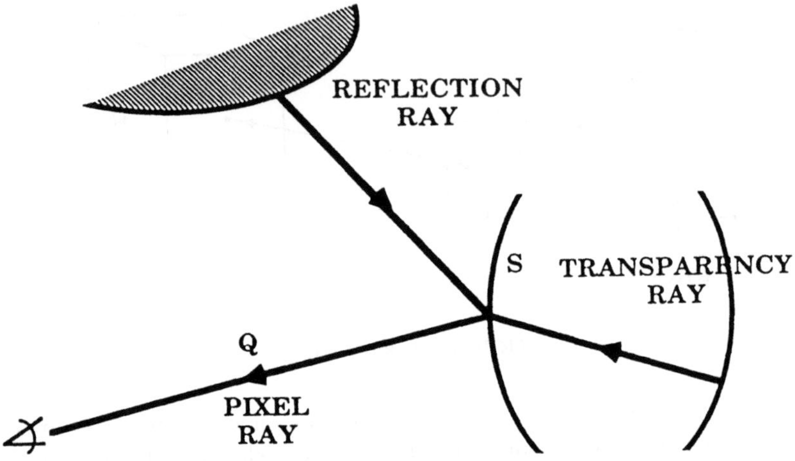

FIGURE 5.2. Light Emitted from a Surface

a rigorous mathematical formulation, which will be helpful in advanced ray-tracing algorithms. Moreover, such an approach can be applied to local illumination models (direct lighting) as well as to global illumination models, and this ensures the homogeneity of the photometric computations.

When computing Q_{RD}, rays are cast toward light sources to evaluate light source illumination. These rays are called shadow rays or shadow feelers because they can be occluded by an opaque object, thus creating shadow or penumbra effects. This occurs when not all the shadow rays cast toward a given light source are occluded, i.e., when these rays are close to the boundary of the occluding object.

Except for shadow rays, which always have a finite length, other rays are infinite when they do not intersect any object in the scene. In this case, the light energy that they transport comes from the remote environment. The illumination flowing from the remote environment can be uniform (the sky for example) or direction-dependent if reflection maps are used [34]. A reflection map specifies a common method of generating approximate reflections of remote objects but it can also be used for transparency or direct viewing (landscape seen through a window for example). With this technique, the remote objects or digitized pictures of the remote environment are projected on a suitably-large bounding volume and the radiance at each sample point is stored in a lookup table called a reflectance map. The light energy flowing in infinite rays can thus be obtained by intersecting the bounding volume and by interpolating the relevant radiance values from the reflectance map (figure 5.3).

Ray-tracing is basically a point-sampling technique and is, therefore, prone to aliasing problems. Unlike the usual rendering techniques, aliasing affects the whole rendering process and not only image plane sampling. However, these shortcomings can be removed by using an appropriate sampling strategy as will be seen below.

5.2.2 Physical Analysis of Reflection and Transmission

The modelling of reflection on rough surfaces is an important issue in the field of photometry and this subject has been widely discussed in the relevant literature. Cook and

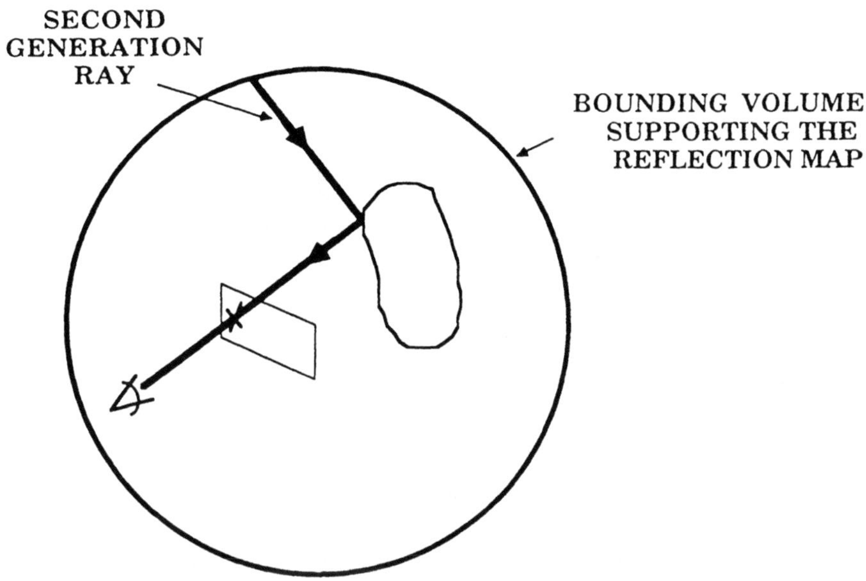

FIGURE 5.3. Bounding volume with reflectance map

Torrance's paper referenced in [18] is a good introduction to these problems and the local reflectance model described below is based largely on their work.

Light transmission phenomena are generally more difficult to capture except in the simple but frequent case of a perfectly smooth surface and homogeneous media [12]. These conditions will be assumed to be present for our purposes. However, a light transmission model for slightly rough surfaces can be empirically derived from a reflectance model as described in [28].

Reflection on Rough Surfaces

Let us consider a small area around a point **P** on a surface and assume that this surface is illuminated by a small extended light source seen under a solid angle $d\Omega$ from P (see figure 5.4). The radiance dL_r of this surface in viewing direction **V** can be expressed as follows:

$$dL_r = RL_i(\mathbf{N}.\mathbf{L})d\Omega \tag{5.2}$$

where R is the bidirectional reflectance and L_i is the radiance of the light source. $(\mathbf{N}.\mathbf{L})$ is the dot product of vectors **N** and **L**. The solid angle $d\Omega$ is defined by:

$$d\Omega = \cos\theta dS/r^2$$

where dS is the area of the light source, θ is the angle $(\mathbf{L},\mathbf{N}')$, $\mathbf{N}'$ being the normal of the light source surface, and r is the distance between **P** and the light source. dS may also be considered as a small area of a large extended source or, more generally, any radiating surface. The term $\cos\theta dS$ is called the projected area and the product $L_i\cos\theta dS$ is called radiant intensity.

In this formula, R characterises the reflectivity of the surface. A more detailed analysis reveals that the reflected light can be broken down into two components. The first one is

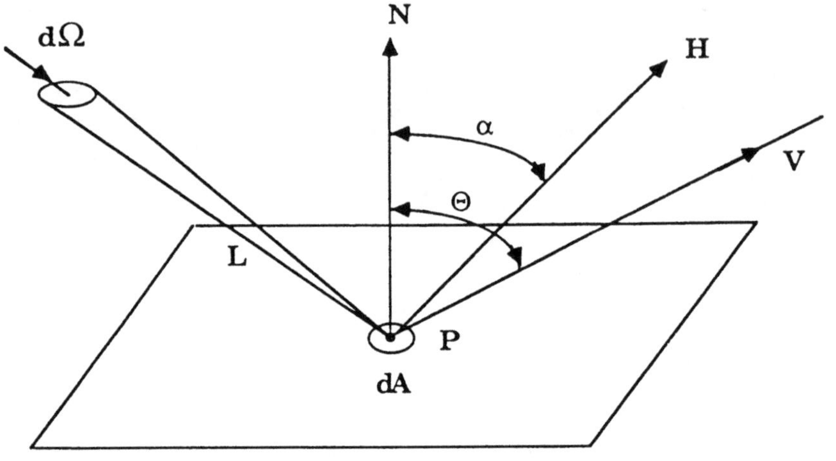

FIGURE 5.4. Reflection on Rough Surfaces

produced by a diffuse reflection effect and thus complies with Lambert's Law, i.e., that the radiance of reflected light is the same in any direction. A surface having this property would look unpolished and no reflections would show up. On the other hand, the second component is due to specular reflection and reflected light is therefore concentrated around a privileged direction. This component gives the surface a shiny smooth appearance. These two components may have a different origin as in the case of plastics which consist of coloured pigments embedded in a transparent substrate. Therefore the light reflected directly from the surface produced the specular components whereas the light that penetrates the object interacts with the coloured pigments and produces the diffuse component (figure 5.5). The plastic appearance is due to the fact that the specularly-reflected light undergoes almost no colour shift, which creates white specular highlights (if the light sources are white). This breakdown of reflected light can be formalized by the following equation:

$$R = sR_s + dR_d \qquad \text{with} \qquad r + d = 1$$

where s and d are the proportions of specular and diffuse light respectively. R_s and R_d are the bidirectional reflectances that characterize the specular and diffuse components respectively.

As the diffuse component complies with the Lambert's law, R_d is constant for a given surface. R_s can be expanded as follows:

$$R_s = \frac{FDG}{4\pi(\mathbf{N.L})(\mathbf{N.V})} \tag{5.3}$$

where F is the coefficient of reflection, D is a distribution function and G is a geometrical factor. Let us analyze these three terms in detail.

The distribution function D characterizes the roughness of the surface. A rough surface can be assumed to consist of a multitude of random-spread mirror-like microfacets, with a normal distributed according to a probability density function $D(\alpha)$ in which α is the angle ($\mathbf{H}$,$\mathbf{N}$), $\mathbf{H}$ being the microfacet normal (see figure 5.4). In other words, $D(\alpha)$ is the proportion of microfacets whose normal forms an angle α with $\mathbf{N}$. Many distributions have been proposed in the literature ([8], [18]). Beckman's law is particularly interesting

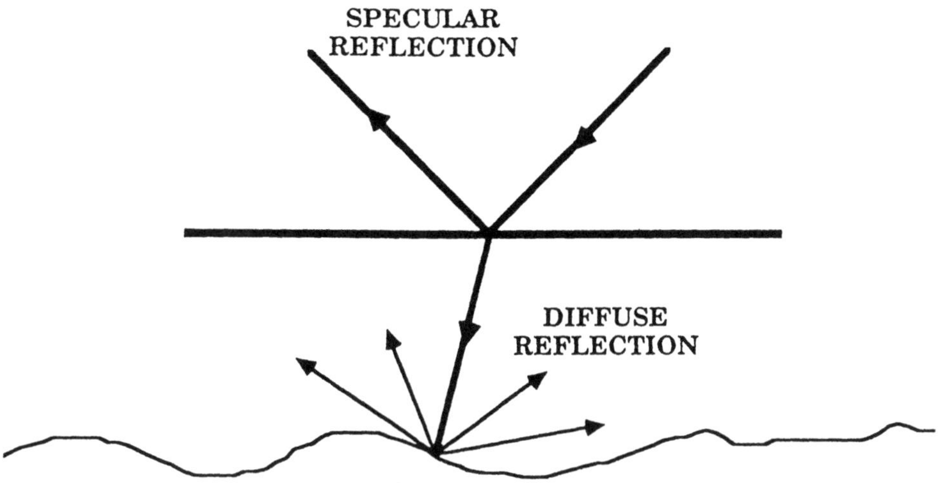

FIGURE 5.5. Diffuse and specular reflection

because it verifies the properties of a probability density function, which is helpful for distributed ray-tracing as shown below. In addition, there is no arbitrary constant. It can be expressed as follows:

$$D(\alpha) = \frac{exp(-tg^2\alpha/m^2)}{m^2\cos^4\alpha} \tag{5.4}$$

where m characterizes the roughness of the material. The smoother the surface, the smaller is m. The computation of this formula is not as difficult as it seems because $\cos\alpha = (\mathbf{H}.\mathbf{N})$ and $tg^2\alpha = (1/\cos^2\alpha) - 1$.

The reflection coefficient, F, characterizes the reflective properties of the mirror-like microfacets. F depends on the refraction index of the material, n_s, and on the angle of incidence with respect to microfacet normals, $\theta' = (\mathbf{L}, \mathbf{H})$, according to Fresnel's formulae:

$$\begin{aligned} F(\theta', n_s) &= (\mid A \mid^2 + \mid B \mid^2)/2 \quad \text{with} \\ A &= \frac{n_s\cos\theta' - \cos\theta''}{n_s\cos\theta' + \cos\theta''} \\ B &= \frac{\cos\theta' - n_s\cos\theta''}{\cos\theta' + n_s\cos\theta''} \end{aligned} \tag{5.5}$$

and θ'' is defined by: $\sin\theta'' = \sin\theta'/n_s$

Only microfacets reflecting light toward direction $\mathbf{V}$ are relevant. This being so, their normal must coincide with the unit angular bisector of $\mathbf{H}$ and $\mathbf{L}$ (see figure 5.4). Therefore we have:

$$\theta' = (\mathbf{L}, \mathbf{H}) = (\mathbf{H}, \mathbf{V})$$

This formula is important if we are to obtain a correct rendering of reflection on glass because the dependence of F on θ' is much more pronounced for dielectric material than for metals. Fast computation of this formula can be easily performed through table look-up and interpolation. For metals, n_s becomes a complex in which the imaginary part characterizes the absorption effect. As n_s depends on wavelength, the reflectance spectrum, and consequently the colour of the reflected light, depends on θ'. Note that at grazing incidence ($\theta' = \pi/2$), $F = 1$ for any value of n_s, which means that there is no colour shift under this incidence for any material.

The geometrical factor G accounts for the shadowing and masking effect between microfacets. Its value is given by the following expression:

$$G = min[1, \frac{2(\mathbf{N.H})(\mathbf{N.V})}{(\mathbf{V.H})}, \frac{2(\mathbf{N.H})(\mathbf{N.L})}{(\mathbf{V.H})}]$$

Now that we know more about specular reflection, let us turn to the relationship between diffuse and specular reflection. A rough estimate of R_d can be obtained by the following expression:

$$R_d = F_0/\pi$$

where F_0 is the reflection coefficient at normal incidence (i.e., $F(\theta, n_d)$ for $\theta = 0$, n_d being the refraction index of the diffusely reflecting material). Remember that the index of refraction n_d is different from n_s for composite material (plastic materials for example). As the index of refraction is wavelength-dependent, each component thus has a different colour.

Transmission

In this paper we are considering only the case of homogeneous media and smooth surfaces. In this situation, an incident ray is refracted (without scattering) at the boundary between the air and the transparent medium in accordance with Snell's law (see figure 5.6):

$$\sin\theta'' = \sin\theta'/n_s$$

where n_s is the refraction index of the transparent medium. Remember that this formula has been used above for computing the reflection coefficient $F(\theta', n_s)$.

For a given direction of observation $\mathbf{V}$, let us call L_t the radiance component due to the transmitted light and L_i the radiance of the surface intersected by the incident ray as given by Snell's law (i.e., the radiance of the surface indirectly seen by the observer). According to Fresnel's formulae, this gives us:

$$L_t = s[1 - F(\theta', n_s)]L_i$$

Remember that s is the proportion of specular light. Note that any incident ray always undergoes two refractions before traversing a transparent object. Therefore, if the refraction index is n_s for the first refraction, it will become $1/n_s$ for the next refraction. If n_s (or $1/n_s$) is greater than 1, total reflection occurs when $\sin\theta'' > 1/n_s$ (or $\sin\theta'' > n_s$) and the transmitted component vanishes.

The transmitted light undergoes additional attenuation due to internal absorption within the transparent medium. This attenuation is given by Bouguer's law:

$$e = \exp(-\beta l)$$

where l is the length of the ray path within the transparent medium and β, called the extinction coefficient, is a characteristic of the material and highly dependent on wavelength.

5.2.3 Simple Ray-tracing

In the simple ray-tracing approach, only point light sources are considered and the scattering of light is not accounted for in computing indirect illumination effects. In the following paragraph, a local reflection and transmission model will be derived from the above theory. The complete photometric algorithms will then be presented.

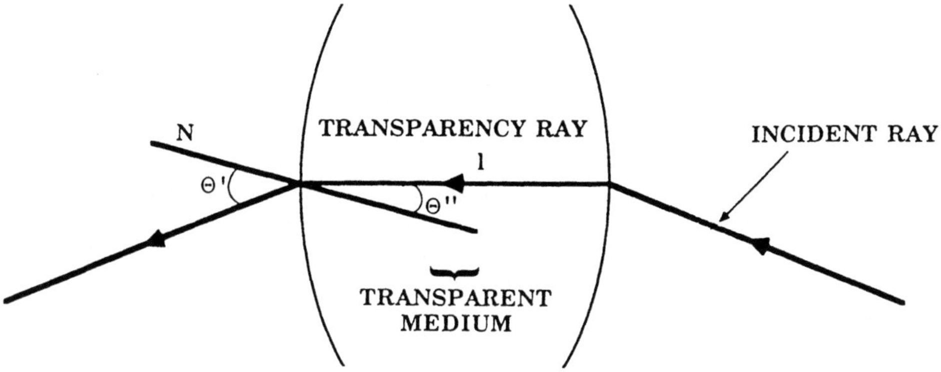

FIGURE 5.6. Refraction — Snell's law

Reflection

Direct Lighting

In case of virtually point light sources, equation (5.2) in 5.2.2 becomes:

$$L_{rd} = R(\mathbf{N}.\mathbf{L})L_s S/d^2$$

where L_s is the radiance of the source, S its projected area and d the distance from the source to $\mathbf{P}$ (see figure 5.4). Note that the radiant intensity, which is equal to $L_s S$ in this case, is independent of direction. L_{rd} is the radiance of the reflection light due to direct lighting. Remember that the bidirectional reflectance R encompasses both diffuse and specular terms.

Only one shadow ray is cast, which precludes the rendering of penumbra. A difficulty arises when a light source illuminates an object through a transparent object or through a perfect mirror, which amounts to direct illumination. This problem will be discussed below. However, simple solutions often suffice such as creating a virtual image of light sources for illumination through mirror reflection. For illumination through transparent objects, the refraction effects can be ignored and only attenuation effects are taken into account, assuming that there is a straight line ray path from $\mathbf{P}$ to the light source. Another solution consists of using a restricted form of forward ray-tracing, i.e., by tracing rays from the light sources to the observer [4].

Indirect Lighting

As far as specular reflection is concerned, surfaces are assumed to be perfectly smooth, which precludes the rendering of scattering effects. In this case, the distribution function $D(\alpha)$ is a Dirac's distribution and only one direction need be sampled. This direction L_0 corresponds to $\alpha = 0$, that is the direction of perfect mirror reflection (see figure 5.4):

$$(\mathbf{L}_0, \mathbf{N}) = (\mathbf{N}, \mathbf{V}) = \theta = \theta'$$

and $\mathbf{L}_0$ in the same plane as $(\mathbf{N},\mathbf{V})$.

This being so, one ray is shot from P in direction $\mathbf{L}_0$ to evaluate the incident light L_i. This ray is called the reflection ray. The reflected light is obtained by integrating equation (5.2), which leads to:

$$L_{ris} = sF(\theta', n_s)L_i$$

where L_{ris} is the radiance of the specular light due to indirect lighting effects and L_i is the radiance of the surface intersected by the reflection ray.

The diffuse component resulting from the indirect lighting effect cannot be captured precisely because simple ray-tracing does not account for all light transport mechanisms. Unlike specular reflection, diffusely reflected light could originate from any and all directions, therefore there is no particular direction to sample. The problem of almost direct lighting effects that has been discussed in 5.2.3 falls into the same category of difficulties. Furthermore, it appears that in most cases, much of the incident light originates from diffuse reflection on the surrounding objects. This global diffuse illumination effect is often called ambient lighting. The radiosity approach is well suited to simulating this phenomenon [15]. In simple ray-tracing, global diffuse illumination effects can be roughly approximated by the following formula:

$$L_{rid} = dR_d E_{ia}$$

where L_{rid} is the radiance of the diffuse component due to global diffuse illumination and where E_{ia}, which is the irradiance producing this lighting effect, is given an arbitrary value.

Transmission

As scattering effects are ignored, the theoretical results given in 5.2.2 apply here. Transmitted light comes from the direction of incidence by Snell's law. This being so, a ray is cast in this direction to evaluate incident light, viz. the radiance L_i of the surface intersected by this ray. These rays are called transmission rays. The radiance L_{ti} of the transmitted light component is thus given by the following equation:

$$L_{ti} = e\, s[1 - F(\theta', n_s)]L_i$$

Due to the basic assumptions on scattering effects and light sources, no light can be transmitted directly from a light source, therefore the Q_{td} component is always zero (see 5.2.2). Note however that the scattering effect can be modelled using an empirical approach, as described in [28].

Photometric Algorithm

The total radiance L_c at point P in direction V is obtained by summing the different components of the above illumination model. Equation (5.1) from 5.2.2 then becomes:

$$L_c = L_{rd} + L_{ris} + L_{rid} + L_{ti}$$

The recursivity in the computations of L_{ris} and L_{rid} can be represented by a binary tree, with nodes corresponding to ray/surface intersections and branches corresponding to transmission and reflection rays. Figure 5.7 illustrates the building of such a tree. In principle, the tree building process stops when there are no more ray/surface intersections. The photometric computations may then begin, starting from the leaves up to the root node. Practically speaking as the ray/surface intersections require extensive computation times, the tree building process must be limited so as to reduce the number of rays.

A straightforward solution consists of strictly limiting the overall number of rays, the depth of the tree and the number of successive reflections within the same transparent object (to avoid costly internal reflection cycles). A less radical solution is based on the fact that rays do not contribute equally to the resulting root radiance. Therefore the spawning of a ray can be abandoned if its contribution to the root radiance is deemed

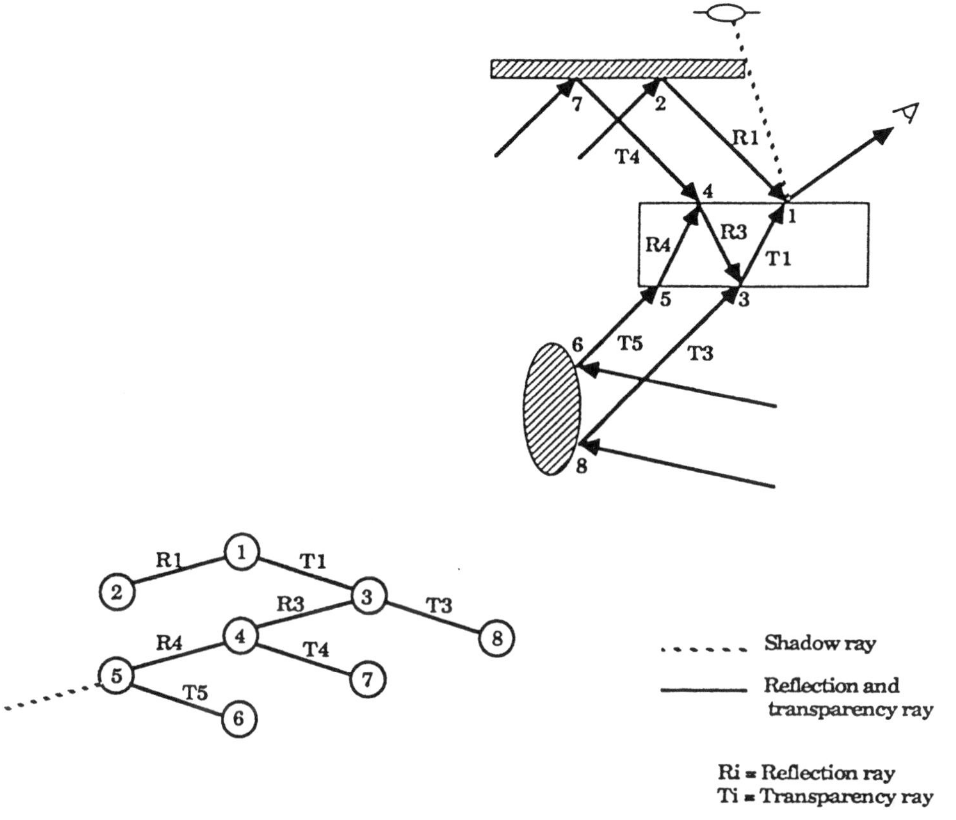

FIGURE 5.7. Example of Tree Building

sufficiently small. As the exact contribution of a ray is not known when building the tree, an upper-bound can be estimated, assuming that this ray intersects a surface with the largest possible radiance in the scene. This very effective technique is named adaptive tree-depth control [28].

The Handling of Colours

The exact rendering of colours requires the modelling of all emission, reflection and transmission characteristics in terms of spectrum. All radiances involved in the photometric computations must then be represented by their spectral density. Eventually, the RGB values corresponding to a particular colour monitor are obtained by integration of the resulting radiance spectrum, an example being the Red component where we have:

$$R = \int_{\mu m}^{\mu M} L_c(\mu) r(\mu) d\mu$$

where $L_c(\mu)$ is the spectral density of the radiance, $r(\mu)$ is the matching function corresponding to the red phosphor of the monitor, and μm and μM are the bounds of the visible spectrum. Consequently, depending on sampling density, the computing load may be very heavy. However, a judicious choice of the wavelength for samples makes it possible to reduce sampling density (see [41]).

As the specular reflectance and transmittance spectra of many materials are almost flat (glass, aluminium, steel, etc.), satisfactory results can be obtained at less cost using an RGB representation for every wavelength-dependent variable. This simplification is broadly used in computer graphics, but its extension to the global illumination model raises a number of difficulties. For example, the generalized RGB representation yields poor results with a major colour shift in specular reflection (copper or brass for example). As global diffuse reflection is roughly modelled in simple ray-tracing, the impairments that are due to the generalized RGB representation are much less visible on diffuse reflection.

A difficulty arises when dealing with the reflection coefficient $F(\theta', n_s)$ because it also depends on the angle of incidence θ' (remember that n_s is wavelength-dependent). As the computation of $F(\theta', n_s)$ by Fresnel's formulae is quite complex, Cook and Torrance have proposed an interesting approximation method [18]. In this approximation, the "colour" of a material is represented by the RGB values of $F(\theta', n_s)$ at normal incidence ($\theta' = 0$), that is R_0, G_0 and B_0. Let us call F_θ the value of $F(\theta', n_s)$ for an average value of n_s. The RGB values of $F(\theta', n_s)$ for any incidence θ', that is R_θ, G_θ, B_θ, can be approximated by means of the following formula:

$$R_\theta = R_0 + (R_{\alpha/2} - R_0)\frac{\max(0, F_\theta - F_0)}{F_{\alpha/2} - F_0}$$

Remember that $F_{\alpha/2} = R_{\alpha/2} = 1$ according to the laws of reflection at grazing incidence (see 5.2.2). The computation of F_θ can be simplified through lookup tables and interpolation. The same formula can be easily extended to compute $F(\theta', n_s)$ at any wavelength, given reflectance spectra at normal incidence.

Antialiasing

With simple ray-tracing, the main drawback arises from the inherent difficulties of this approach when required to cope with aliasing problems. There is no both simple and general solution to these problems. The most commonly used method consists of locally supersampling and filtering according to certain criteria such as edge detection or contrast thresholding ([57],[42]). The resulting improvement is noticeable, but impairments remain in the case of small objects (slivers) or texture mappings for example. Amantides has suggested modelling rays by cones instead of straight lines so as to do away with point sampling inadequacies [1]. However, his solution makes intersection problems even more critical and not all photometric phenomena are easily modelled with cone-tracing (e.g. light scattering). So far, the only really effective and general solution that exists consists of using adaptive stochastic sampling and distributed ray-tracing, which will be discussed below.

5.2.4 Distributed Ray-tracing

Background

The limitations of simple ray-tracing come from its over-simplified sampling strategy. We have seen that some improvements can be made to this approach so as to cope with the critical problem of aliasing at pixel level. However, these techniques cannot be easily extended to deal with scattering effects or motion blur for example. A certain amount of supersampling is necessary in these cases. The remaining problem is how to control the sampling process and how to efficiently compute all the integrals that are involved in the filtering process.

Generally speaking, the pixel value can be expressed as a multiple integral based on the nesting of different filtering functions. Let us illustrate this by considering the case

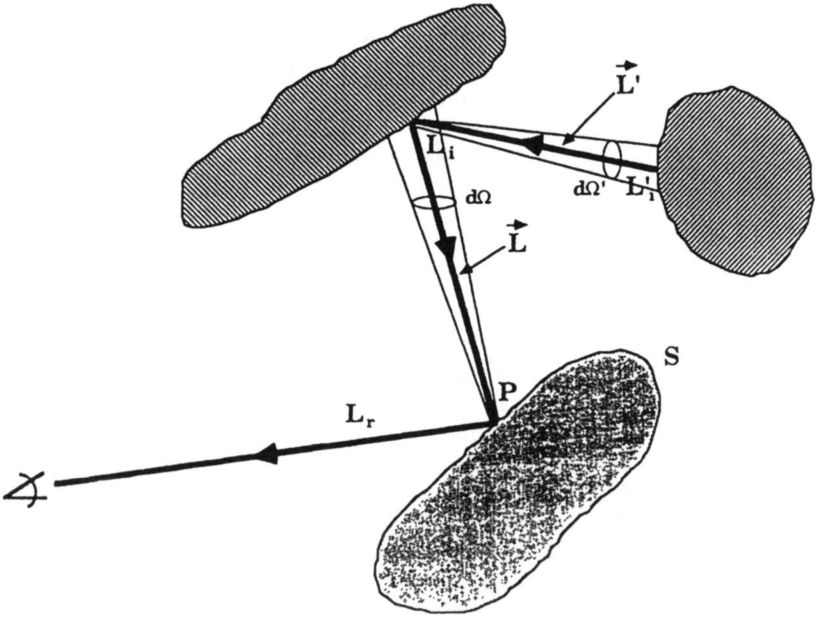

FIGURE 5.8. Multiple Scattered reflections

of multiple scattered reflection (see figure 5.8). From equation (5.2) in 5.2.2, the radiance L_r of the surface seen by the observer in a direction $\mathbf{V}$ is given by:

$$L_r = \int\int RL_i(\mathbf{N}.\mathbf{L})\, d\Omega$$

In this equation, radiance L_i at a given point of the radiating surface is also expressed as a similar integral:

$$L_r = \int\int\int\int RR'L'_i(\mathbf{N}.\mathbf{L})\,(\mathbf{N}'.\mathbf{L}')\, d\Omega\, d\Omega'$$

and so on and so forth. The more dimensions we add to the integral, the more accurate the solution. This process is no more than an iterative solution to the global integral equation describing the energy balance between emission, reflection and absorption [32]. The same analysis can be generalized to account for antialiasing at pixel level, motion blur, RGB value computations, depth of field effects, etc.

The fundamental idea of distributed ray-tracing involves using a Monte-Carlo approach to cope with the complexity of the integration problem. According to this method, the multiple integral that yields the pixel value will be approximated by sampling and averaging the integrand. The sampling and filtering process is thus extended from 2D (i.e., the image plane) to N-Dimensional space, which amounts to a generalized anti-aliasing procedure. As seen from the previous example, the modelling of scattering effects introduces variables which define sampling directions and, consequently, rays. Therefore, as far as these variables are concerned, the sampling process is distributed all along the ray-tracing process, hence the name distributed ray-tracing [13].

A certain amount of oversampling at pixel level is vital in order to obtain a correct result, however the strength of the method lies in the fact that sampling density depends directly on image content rather than on the number of variables. This being so, once the cost of oversampling is accepted, additional effects can be added without entailing any noticeable change in computing time.

To illustrate the photometric algorithm that results from this approach, let us consider a simple example (see figure 5.9). First, given a sample in the image plane, a pixel ray is traced. Let us assume that this ray intersects a specular reflecting surface. In this case, a reflection ray is cast in another direction to sample incoming light. From the same point, a shadow ray can be cast toward a given point of light source surface. This ray-tracing process can be handled in the same way as simple ray-tracing, i.e., with a tree-building step followed by a computation of the root radiance through backward tree traversal. From this point of view, distributed ray-tracing can be seen as an enhancement of simple ray-tracing. However, in the distributed approach, the resulting image plane radiance value represents only a sample of the integrand and not the integral itself as in the case of simple ray-tracing. Furthermore, weighting terms may be included to account for aspects other than photometry such as image plane low-pass filtering or time filtering.

Several image plane samples are thus collected as described above, changing reflection and shadow ray direction and, possibly, the location of moving objects if the time variable is accounted for (motion blur) or the wavelength at which all photometric computations will be done if the RGB values computations are included [13]. When a sufficient number of image plane samples has been collected, the pixel value can be obtained by simply averaging these samples:

This example highlights a number of important issues:

- what sampling process to use to avoid the highly-visible artefacts due to uniform sampling
- how to control the sampling process and estimate whether or not enough samples have been collected
- how to select the samples so as to obtain a sufficiently accurate solution with the lowest possible sampling density.

These are questions which we shall try to answer below.

Principles

Theory

As we have already seen, the pixel value, or more precisely the value of one of its RGB components, can be expressed as a multiple integral as follows:

$$C = \int L(X)dX \qquad \text{with } X \in R^n$$

The integrand $L(X)$ is the image plane radiance function. Actually, this integral is the result of iterating several simple or double integrals, each one corresponding to a particular "effect". In all these component integrals, the integrand can be broken down into the product of two functions: one including all the terms that depend on the luminous environment and another that depends only on geometry (and some modelling parameters). This latter function, which can be seen as a weighting as well as a filtering function, is called the Component Weighting Function (CWF). Taking equation (5.2) in 5.2.2 as an example, the term L_i falls into the first category whereas the terms $(\mathbf{N}.\mathbf{L})$ R are only geometry-dependent. Therefore, we may write:

$$L(X) = F(X)P(X)$$

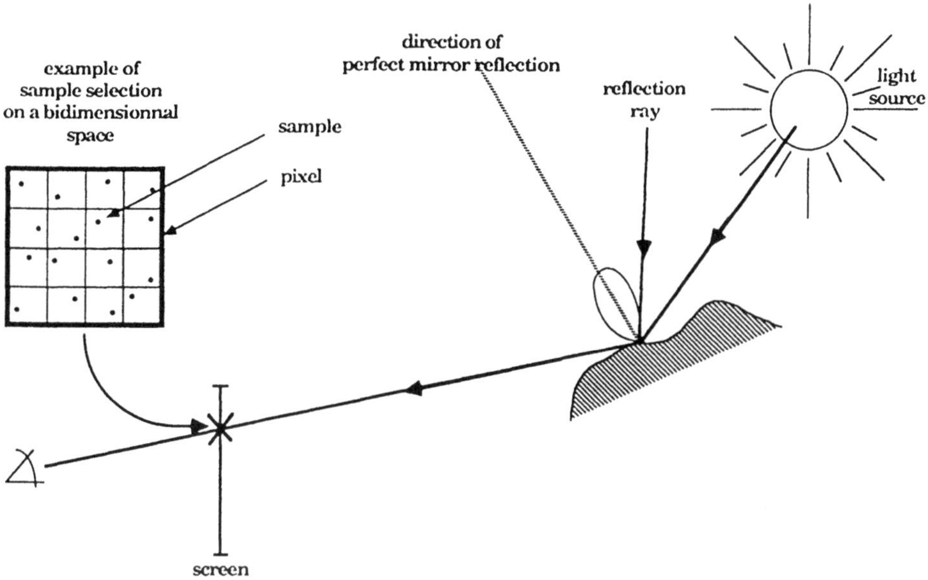

FIGURE 5.9. Schematic of the photometric algorithm

where $F(X)$, called the lighting function, depends on the luminous environment, and $P(X)$, the overall weighting function, is the product of all the CWFs.

In the Monte-Carlo approach $P(X)$ can be considered as a probability function provided that:

$$P(X) \geq 0 \text{ for } X \in R^n \text{ and } \int P(X)dX = 1$$

In general, these conditions are easily obtained through mathematical manipulations. Given these restrictions, the pixel value C can be seen as the expected value of $F(X)$ where X is a N-dimensional random variable with $P(X)$ as a probability density function.

Let $\{X_i\}$ be a sample of size n of the random variable X (*having the probability density function P(X)*). A statistical estimate of C can thus be obtained by simply averaging the $F(x_i)$s where x_i is the observed value of X_i. According to the estimation theory, the sampling process may stop when the confidence interval is sufficiently small. The width of this interval is:

$$\delta = 2t_\epsilon S/\sqrt{n}$$

where S is the observed variance. Given a probability threshold ϵ, t_ϵ must be such that:

$$\epsilon = 2[1 - \theta(t_\epsilon)]$$

θ being the Gaussian distribution function and ϵ the test threshold, i.e., the probability for the estimated value outside the confidence interval.

Sampling Strategy

According to the above theoretical analysis, the samples should be selected in such a way that they follow a probability distribution defined by $P(X)$. In this case, more samples will be collected where $P(X)$ is large, i.e., in the most important regions. This form of sampling, called importance sampling, reduces the variance of the lighting function samples (the $F(x_i)$s) and, consequently, improves the convergence of the method.

Such a sample distribution can be obtained using the well-known stratification technique [28]. With this technique, the R^n space is divided into m sub-regions and the number of samples in each sub-region is made proportional to the probability of this region. In general, this partition of R^n is performed in such a way that all regions have same probability, in which case an equal number of samples is collected in each region. At least one sample must be taken from each region and, if the need for additional samples is indicated, one new sample must be collected in each region.

Seen from the standpoint of probability, the variables of different CWFs are statistically independent. This means that the sub-spaces defined by the variables of each CWF can be considered separately, which produces 1D or 2D sub-regions. Even for 2D space, the computing cost of partitioning into equal probability regions may be heavy if the CWFs are too complex. Furthermore, pre-computing is not always possible because the CWFs may depend on parameters that can only be determined during the rendering process. This means that the CWFs must be chosen sensibly in order to avoid unnecessary complications. When defining the CWFs, it is not necessary to maintain all the weighting terms. In fact, the main goal of CWFs is to obtain a sample distribution suitable for most situations. Consequently, the CWF may include only weighting terms which have the greatest influence in most cases. In 5.2.4 we give an example of CWF definition for the computation of the illumination produced by a spherical light source. The case of scattered reflection (blurred reflection) and colour computations is dealt with in [13].

In the same reference, there is a description of techniques which further improve the sample distribution.

One drawback of stratified sampling is that several samples may be collected in the same sub-region, in which case importance sampling can no longer be used. The best sample distribution, i.e., in which there is only one sample in each region, cannot always be obtained with stratified sampling because the number of samples is not known a priori. However, as we shall see below, the use of stochastic sampling techniques makes it possible to reduce aliasing effects and prevents the collection of samples that lie near each other, which leads to a better estimate of the overall integral.

To avoid this drawback, J Kajiya has proposed an adaptive sampling technique in which sub-regions or cells are subdivided recursively [32]. Whenever a new sample is needed, a cell is first selected and then divided into new cells. The old sample from the original cell lies in one of these new cells. The new sample is chosen to lie in the opposite cell. To account for importance sampling, cell division is performed in such a way as to ensure that the new cells have equal probability. As the cell size may vary, the algorithm must keep track of this data and, if necessary, the samples are weighted according to these sizes.

Example of CWF Definition and Partitioning

Point light sources are certainly the easiest case to handle but they create non-realistic sharp shadow boundaries. Real shadows always have penumbra at the edges because diffused light sources may be partially occluded by an opaque object. The penumbra effect can therefore only be rendered if the non-occluded light source illumination is known at shadow boundaries. From equation (5.2) in 5.2.2, the radiance of reflected light at a given point **P** of a surface is:

$$L_r = \int\int_{\Omega} RL_i\,(\mathbf{N}.\mathbf{L})\,d\Omega$$

where Ω is the non-occluded solid angle from which the light source is seen. For the sake of simplicity, the light source will be assumed to be spherical with a constant radiance L_i

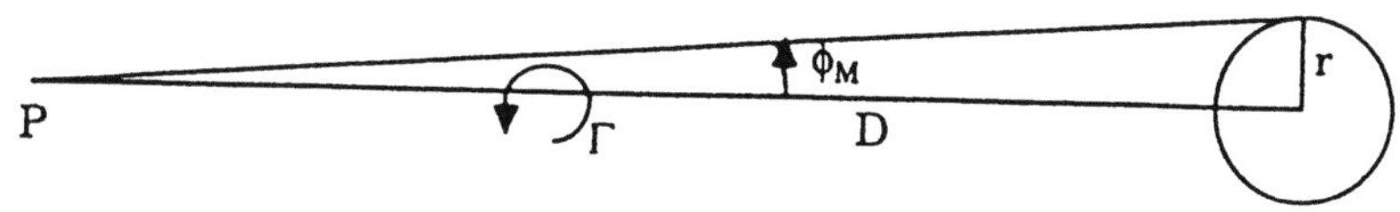

FIGURE 5.10. Light source bounding cone

on its surface. If the light source is far enough away from **P**, the solid angle from which the light source is seen can be assumed to be bounded by a right circular cone the axis of which is the line joining **P** to the centre of the sphere and the base of which is the disk of radius r normal to $\mathbf{L_0}$ (see figure 5.10). This gives us:

$$L_r = \int_0^{\Phi M} \int_0^{2\alpha} RL_i\,(\mathbf{N.L})\,h(L) sin\Phi\, d\Gamma\, d\Phi$$

where $\Phi M = \sin^{-1}(r/D) \approx r/D$, r is the radius of the sphere and D is the distance between **P** and the centre of the sphere. *h(L)* is a boolean function with a value of 0 when the light source is hidden from direction **L**, and 1 in all other cases.

CWF Definition. According to the Monte-Carlo approach, this integral will be computed by sampling the sub-space defined by the variables Φ and Γ with a sample distribution governed by the CWF. As explained above, the CWF may include only weighting terms which have the greatest influence in most cases. The bi-directional reflectance R does not vary rapidly unless specular reflection outweighs diffuse reflection. Therefore, as such situations are infrequent for light source illumination, R can be discarded. The term **(N.L)** can also be excluded because its effect only becomes important at grazing incidence, a situation for which the light source illumination is almost zero. The CWF will therefore be:

$$P(\Phi, \Gamma) = k \sin \Phi$$

where the constant k is selected so that:

$$\int_0^{\Phi M} \int_0^{2\alpha} P(\Phi, \Gamma)\, d\Gamma\, d\Phi = 1$$

This condition leads to:

$$1/k = 2\pi(1 - \cos \Phi M) \approx \pi r^2/D^2$$

and the function to be sampled will be:

$$F(\Phi, \Gamma) = R\,(\mathbf{N.L})\,h(L)/k \tag{5.6}$$

Partitioning. Due to axial symmetry, $P(\Phi, \Gamma)$ does not depend on Γ and the Γ values will therefore be uniformly distributed between 0 and 2π. At first sight, the sinΦ distribution function seems to place more importance on samples with large Φ values but it has to be remembered that the directions defined by the Φ and Γ values at constant Φ are more sparse when Φ increases, owing to the uniform distribution of Γ.

In fact, this CWF leads to uniform sampling of the whole solid angle Ω because $d\Omega = P(\Phi, \Gamma) d\Phi d\Gamma$. Therefore, in this case, dividing the (Φ, Γ) sub-space into equal probability sub-regions amounts to dividing the solid angle Ω into equal parts. As the distribution of Γ is uniform, the problem will be to find a set $[\Phi_i]$ such that:

$$\int_{\Phi_{i-1}}^{\Phi_i} \sin \Phi\, d\Phi = A \quad \text{for } i = 0, \ldots n$$

with

$$\Phi_0 = 0 \text{ and } \Phi_n = \Phi M.$$

n is the number of sub-regions for each dimension and A is given by:

$$A = 1/n \int_0^{\Phi M} \sin \Phi d\Phi = (1 - cos\Phi M)/n \approx r^2/2nD^2$$

Note that A depends on r/D and, consequently, the Φ_i's can only be computed during the rendering process. From the above equations, it can be easily shown that:

$$\cos \Phi_{j-1} - \cos \Phi_j = A \quad \text{for } j = 1, \ldots, n$$

By summing these equations from $j = 1$ to i, we obtain:

$$\cos \Phi_i = 1 - iA$$

Using small angle approximation for $\cos \Phi_i$, we have:

$$\Phi_i \approx a_i r/D \text{ with } a_i = \sqrt{i/n} \qquad (i = 0, \ldots n)$$

Therefore, once the a_i sequence is computed, partitioning can be performed for any value of r/D with only $n - 2$ multiplications.

Rendering. The function $F(\Phi, \Gamma)$ defined by equation (5.6) in 5.2.4 represents a sample of the reflected light energy produced by a light source illumination. Each time such a sample is needed during the rendering process, the following operations will be applied:

- select a sub-region at random (see [13])
- compute k and the sub-region limits
- select Φ and Γ (see Stochastic Sampling below)
- ray-trace in direction $L(\Phi, \Gamma)$ to test $h(L)$
- compute $F(\Phi, \Gamma)$.

End of Procedure Test

Adaptive sampling raises a difficult question: how can we decide whether enough samples have been collected? This problem requires a great deal of attention because the End of Procedure Test greatly influences computing time and picture quality.

The confidence interval test mentioned in 5.2.4 is only valid for large collections, which is seldom the case. If the random variable $F(X)$ follows a Gaussian distribution law, a collection of any size can be considered [45]. Lee et al. suggest using the mean quadratic error as a measure of the difference between the estimate and the true value. In their test, sampling stops when the estimate of the mean quadratic error is below a certain threshold, which leads to a X^2 test [39]. This test can be further refined with the use of stochastic sampling techniques [20]. But again, this test assumes that $F(X)$ follows a Gaussian distribution.

Mitchell[42] points out that variance is a poor measure of visual perception of local variation and he suggests the use of contrast as a measure of sample differences, i.e.:

$$C = \frac{I_{max} - I_{min}}{I_{max} + I_{min}}$$

Three separate contrasts are then computed for each primary colour and are tested against three different thresholds to account for the human eye's known response to noise as a function of colour. In his prototype system, red, green and blue thresholds are set to 0.4, 0.3 and 0.6 respectively.

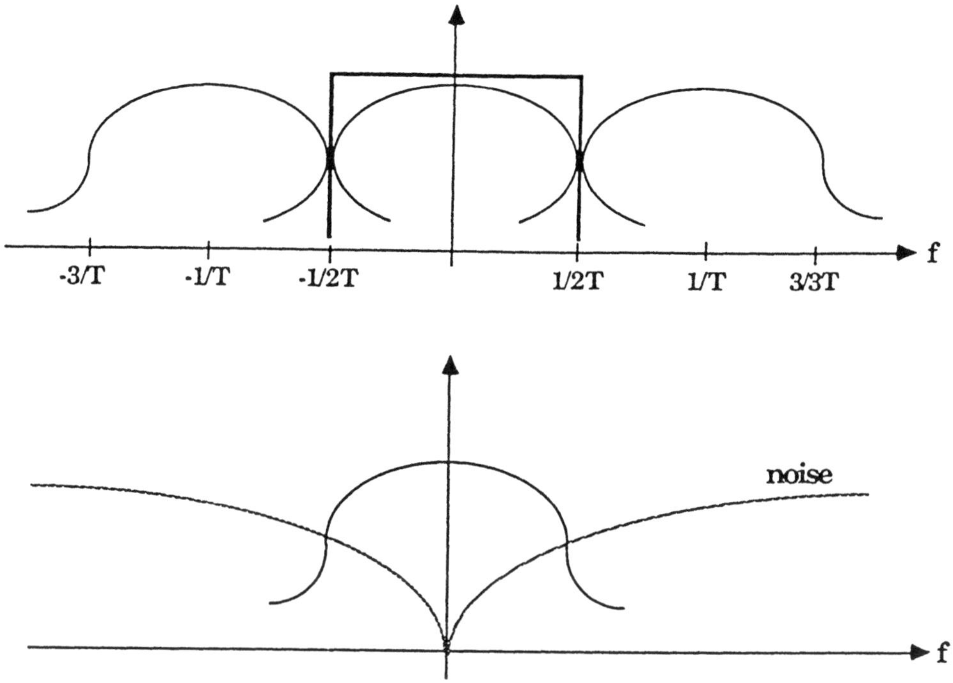

FIGURE 5.11. Regular Sampling (above) - Stochastic Sampling (below)

Stochastic Sampling

In distributed ray-tracing, the sampling process must be carefully chosen so that defects from sampling are barely visible. The main drawback of regular sampling is that it can create highly visible false patterns (staircase for example). Stochastic sampling is not prone to these defects because it uses irregular sampling patterns. The resulting aliasing error gives rise to a featureless noise which is much less conspicuous than false patterns. Figure 5.11 illustrates this by comparing the effect of both sampling methods in the frequency domain.

In consequence, the defects associated with stochastic sampling can be characterized by the power signal to noise ratio (SNR) and it can be shown that the SNR is proportional to the average sampling rate. Therefore, for constant signal, the average squared reconstruction error is proportional to the inverse of the sampling rate. Note that the SNR can also be used as an end of procedure criterion [20]. In addition, the noise spectrum is an important feature because the human visual system is more sensitive to low/mid frequency noise. Therefore, the tradeoff between low, mid and high frequency noise is an important issue in stochastic sampling. Two forms of stochastic sampling pattern have been proposed in the literature: the Poisson-disk sampling pattern and the jitter pattern ([17], [20], [42]).

The Poisson-disk distribution is a Poisson distribution with a minimum-distance constraint between points ([20], [42]). In one-dimensional space, the location of one sample depends on the previous sample on the basis of:

$$X_k = X_{k-1} + L_k$$

where the random variable L_k has the following probability density function:

$$P(L) = \begin{bmatrix} \beta.exp(-\beta(L - L_m)) & \text{if } L > L_m \\ 0 & \text{elsewhere} \end{bmatrix}$$

In this formula, L_m represents the minimum distance between samples. The average sampling rate is given by:

$$F_e = \beta/(1 + \beta.L_m)$$

The Poisson-disk sampling pattern was originally used to model the distribution of receptors in the human eye. The generation of Poisson sampling patterns is inherently global. If the sampling density needs to be increased, a whole new pattern must be generated, which is a drawback in adaptive sampling.

A jitter sampling pattern is simply obtained by jittering the sample points of a regular pattern. In one-dimensional space, this gives us:

$$X_k = Y_k + B_k$$

where Y_k is a sample from the original regular sampling pattern and B_k is a random variable. Jitter sampling can create false patterns if certain conditions on the probability distribution of B_k are not respected. Such defects can be avoided using uniform probability distribution within a limited interval. The width of the interval, i.e., the amplitude of the jitter, is then chosen to be equal to the spacing between the samples of the associated regular sampling pattern.

Unlike Poisson-disk patterns, the generation of jitter pattern is local since samples are independent. This being so, jitter pattern sampling lends itself quite well to adaptive sampling. A recursive subdivision algorithm similar to the cell subdivision algorithm described in 5.4.2 can be used for this purpose [20]. Another interesting feature of jitter patterns is that image plane samples can be shared by different pixels during the filtering step (see 5.2.4).

Given these properties and the fact that jitter patterns are easiest to generate, jitter pattern sampling is often used. However, D Mitchell has proposed a method called the point-diffusion algorithm which makes it possible to generate Poisson-disk patterns at low computing cost [42]. He also demonstrated the superiority of this pattern with regard to the visibility of aliasing noise.

Filtering

The filtering of image plane samples is an important issue in distributed ray-tracing and stochastic sampling. As we have seen, the use of these two techniques produces a certain amount of oversampling in the image plane. The resulting signal therefore needs to be resampled to fit into the uniform array of pixels that makes up the digital image. However, little practical information exists in the literature concerning the problem of resampling a signal from non-uniform samples.

The goal of the image plane filter is to bandlimit the oversampled signal so that resampling does not introduce aliasing. Practical filters are windowed so that the filter support is small and they should be positive since image samples are non-negative. Among all possible filters, the raised cosine filter exhibits reasonable low-pass behavior and is positive:

$$R(x,y) = \frac{1}{W}\cos(\frac{2\pi r}{W} + 1) \qquad \text{with } r < W,$$

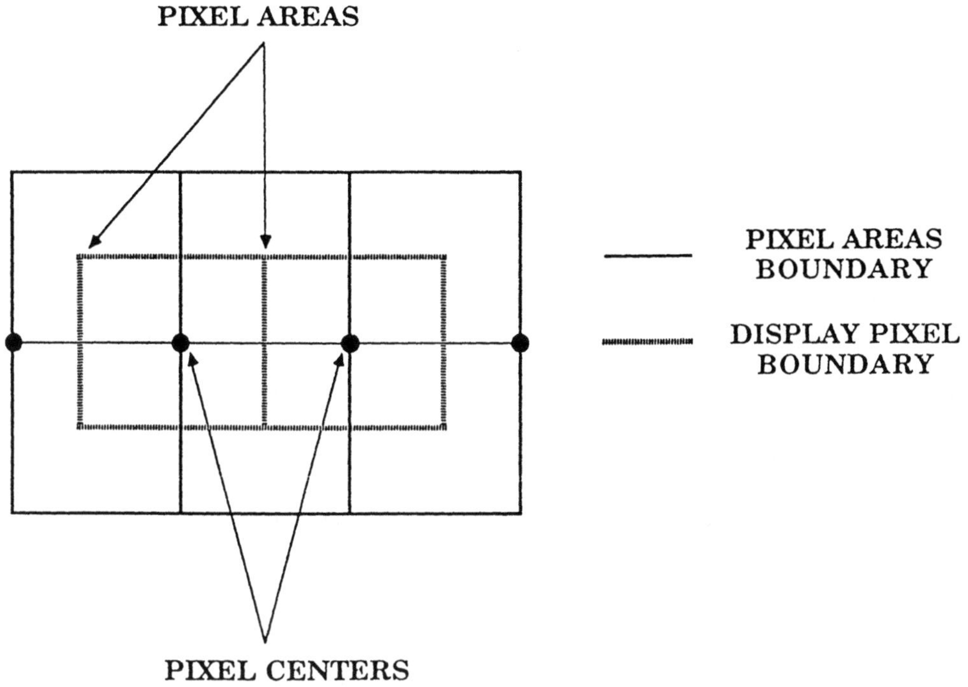

FIGURE 5.12. Pixels and Pixel Area Layout

where $r = \sqrt{(x^2 + y^2)}$ is the distance from the centre of the filter and W the radius of the filter. The filter width can be adjusted according to the signal to noise ratio so as to attenuate noise more effectively. Alternatively, an adaptive filtering technique, such as Wiener filtering, can also be used [20].

If image plane filtering is included in the global Monte-Carlo integration process, this filter is no more than a particular CWF controlling sample distribution in the image plane. Once partitioning into equal probability sub-regions has been performed, the filter value will be assumed to be constant for all samples lying in the same sub-region ([17], [45]). However, in this case, image plane samples cannot be shared by different pixels since sample distribution is specific to a pixel. This method will be called implicit filtering.

Another solution consists of uniformly distributing the image plane samples and filtering the samples at a final stage. This method will be called explicit filtering. This solution amounts to including the image plane filtering function in the $F(X)$ term of the integrand rather than in the weighting term $P(X)$ (see 5.2.4). In this case, the sharing of samples by different pixels is possible under some conditions as regards the stochastic sampling pattern (see 5.2.4). Such a method is described in reference [13]. In this method, samples are first collected in pixel areas that are rectangles of the same size as the pixel but their vertices coincide with pixel centres. As a filter width of twice the pixel size is generally sufficient, a pixel can be computed by considering only the four pixel areas that have the pixel centre as their vertex (see figure 5.12). As the sampling density is allowed to vary in each pixel area, samples are pre-filtered with respect to the four neighbouring pixels, i.e., the corresponding partial weighted sums are computed. This being so, any processed pixel area is characterized by the same amount of data. In this way, each sample is shared by four pixels and storage requirements are kept to a minimum.

To allow for local changes in the sampling rate, explicit filters are normalized. An example of a one-dimensional weighted-average filter is given by:

$$G(X) = \frac{\sum_k F(x_k) r(x - x_k)}{\sum_k r(x - x_k)}$$

where $r(x)$ is the filter function. Small defects may remain after the application of this filter because adaptive sampling may create dense clumps of supersamples that overwhelm the influence of nearby base samples. To cope with this problem, D Mitchell recommends the use of a multi-stage filter that consists of applying in succession three different box weighted-average filters with ever-narrowing low-pass cutoff until the proper bandwith for the display is reached. The box filter widths he uses are successively $\frac{1}{4}W, \frac{1}{2}W, \frac{1}{2}W$, and W where W is the width of a pixel (the $\frac{1}{2}W$ filter is applied twice). This results in a piecewise cubic filter with very low signal leakage above the low-pass cutoff.

Note that in implicit filtering, filter normalization is not necessary because the number of samples is the same in each image plane sub-region. This method is also less sensitive to sample-density fluctuations because importance sampling is designed to concentrate samples near the pixel centre, which reduces the effects of small local changes within the pixel surface.

5.2.5 Enhancements

At this point, it is interesting to come back to the case of indirect illumination discussed in 5.2.3 (see figure 5.13). With simple ray-tracing, only one ray is shot to account for the indirect lighting effect, which produces path R. This being so, the lighting effect caused by path R' is not taken into account although it is often important. The distributed ray-tracing approach will easily solve the problem for specular reflection provided that the light source is sufficiently large to be detected at the base sampling rate. J Kajiya has shown that the distributed ray-tracing approach can be extended to all cases of illumination [32]. However, such an approach cannot deal efficiently with diffuse reflection because there is no privileged direction to assist the importance sampling mechanism.

Four mechanisms of light transport can be identified, depending on the origin of the incoming light and the characteristic of the outgoing light: diffuse to diffuse, specular to diffuse, diffuse to specular and specular to specular. The ray-tracing approach is well suited to the diffuse-to-specular and to the specular-to-specular light transport mechanism whereas the diffuse-to- diffuse mechanism is better handled by the radiosity approach. From this analysis, J Wallace et al. propose a two-pass solution to the rendering equation. A first pass is based on the radiosity approach, with extensions to deal with the effects of diffuse transmission, and specular-to-diffuse reflection and transmission. The second pass uses a variant of the distributed ray-tracing algorithm [55].

5.3 Computational Geometry

5.3.1 Intersection Computation: Principles

In the ray tracing technique, computation of the intersection between a ray and the scene is very time-consuming. In this section, the way to perform this intersection, for complex scenes, is described. This intersection must be simplified and in addition its cost must be reduced by using a hierarchy of bounding volumes. We give a method of building an efficient hierarchy. In the sequel, the scene is supposed to be expressed in the

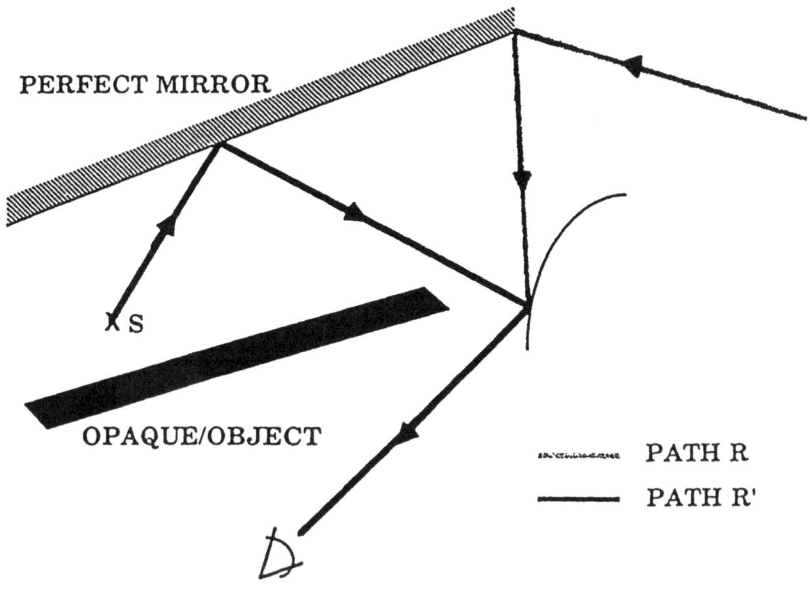

FIGURE 5.13. Indirect illumination

eye coordinates system (ECS). It may have two kinds of representation: either a set of independent objects or a CSG tree (Constructive Solid Geometry) which is a binary tree whose leaves are primitive objects like sphere, cylinder, cone and whose nodes are boolean operators like union, intersection and difference. These two representations may be mixed to give a third one. The purpose is to intersect a scene by a ray whose equation is:

$$\mathbf{P} = \mathbf{P}_0 + t.\mathbf{D}$$

where: $\mathbf{P}_0$ is the ray origin, $\mathbf{D} = (dx, dy, dz)$ is the direction vector of the ray and t is a scalar ranging over [0,1].

The intersection result is then a set of t values corresponding to the intersection points. Only the closest point to the ray origin, is used to compute the lights contribution and to shoot secondary rays.

To simplify the intersection computing, each object is described in a local coordinates system (LCS) are shown in figure 5.14.

Two transformation matrices are then associated with each object: the first one allows the transformation of a point in the eye coordinates system onto a point in the local coordinates system, and the second one allows the inverse transformation. Ray-object intersection is performed in the LCS. To do that, the ray is transformed into the LCS. This simplifies both the computations of the ray-object intersection and that of the normal. Since t is a scalar, its value is not affected by this transformation. To compute the closest intersection point, the smallest value of t is substituted in the ray equation expressed in the ECS. The transformation LCS–ECS is then not necessary. As for the normal calculation, it is performed on the LCS, then it is transformed onto the ECS.

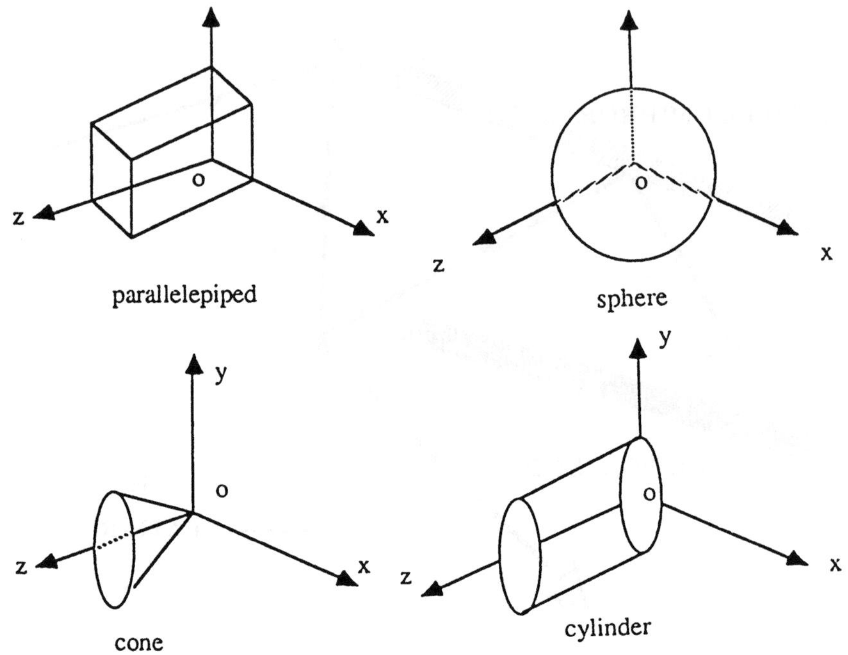

FIGURE 5.14. Local Coordinates Systems

5.3.2 Bounding Volumes

To reduce the amount of ray-object intersections, it is absolutely necessary to use a hierarchical data structure. With this latter, some computations may be avoided, such as the intersection of a ray with the objects which do not lie on the ray's path. This data structure is a tree of bounding volumes. Bounding volumes are simple geometric objects which fit around the objects. They are chosen to be simple to intersect with a ray, such as spheres or parallelepipeds that have faces perpendicular to the axes. These bounding volumes are combined into a tree by picking some of them and surrounding them with another bounding volume. This process is repeated recursively until a bounding volume is generated that surrounds the whole scene (figure 5.15).

Many different trees can be built for a given scene and may considerably influence the rendering time of even a simple image. Thus it is very important to find a way to choose a tree that reduces this time. Trying to construct manually a tree is very tedious and not efficient. A better method consists in dividing the scene into halves along one axis and surround each half with a bounding volume. This process is applied recursively on each half. This method does not account efficiently for the property of neighbouring of the objects (spatial coherence).

A better method has been proposed by Goldsmith and Salmon [25]. The strategy used is a heuristic tree search. Objects are added successively and the tree is searched to find a suitable insertion point for each new node. Since not all nodes of the tree can be considered as a point for insertion, the search must follow only few paths. The choice of subtrees to search from a given node is determined by the smallest increase in surface area of the node's bounding volume that would occur if the new node was to be inserted as a child of it. During the search process, two or more children of a node may have the same increase

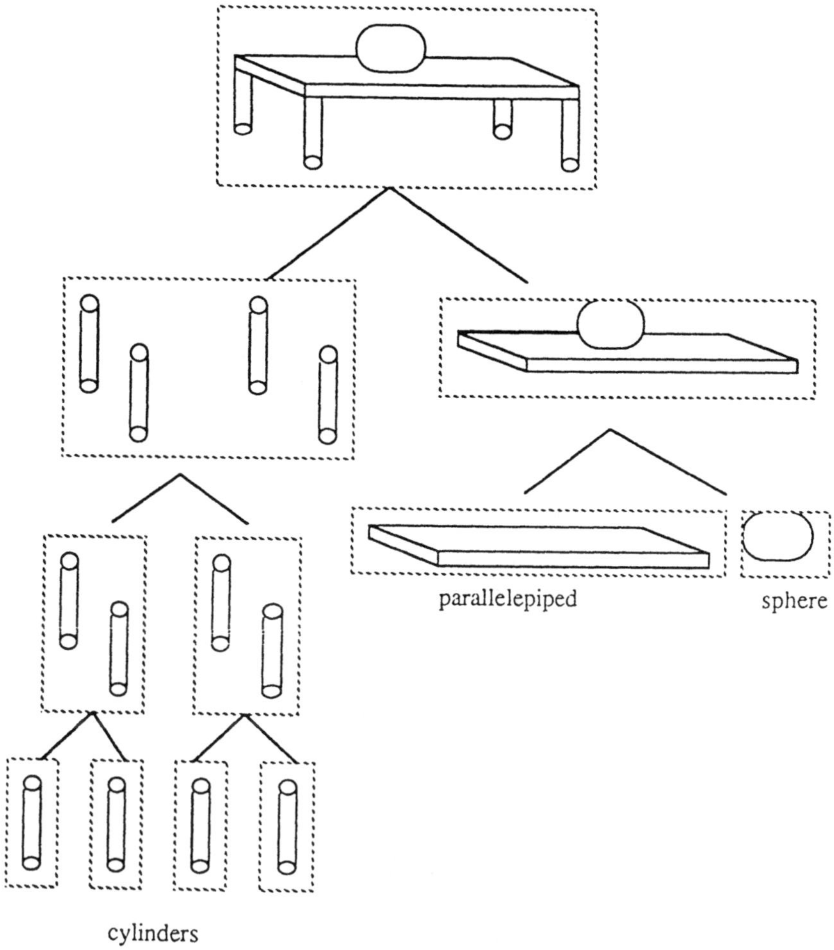

FIGURE 5.15. Hierarchy of Bounding Volumes: Binary Tree

in bounding volume surface area after adding the new node. In that case the relevant subtrees can be searched to find the best location for insertion. In practice, this may occur only at the first two or three levels of the hierarchy.

During the search, at each level of the tree, the node is considered as a prospective child of each node that will be searched (figure 5.16). The tree is evaluated with the proposed insertion and the location with the smallest increase in tree cost is saved. When the search reaches a leaf node, the new node and the leaf node are proposed as children of a new non leaf node (figure 5.16).

If the scene is represented by a CSG tree, the process of construction of the hierarchy is the following. Firstly, the data structure of each leaf of the CSG tree is extended by adding to it the bounding volume of the leaf. Then the tree is searched from the leaves to the root in order to compute the bounding volumes of the non leaf nodes. These bounding volumes are in their turn added to the data structure of the associated nodes. Their evaluation depends on the boolean operator associated with the nodes (figure 5.17). The bounding volume of the root bounds the whole scene.

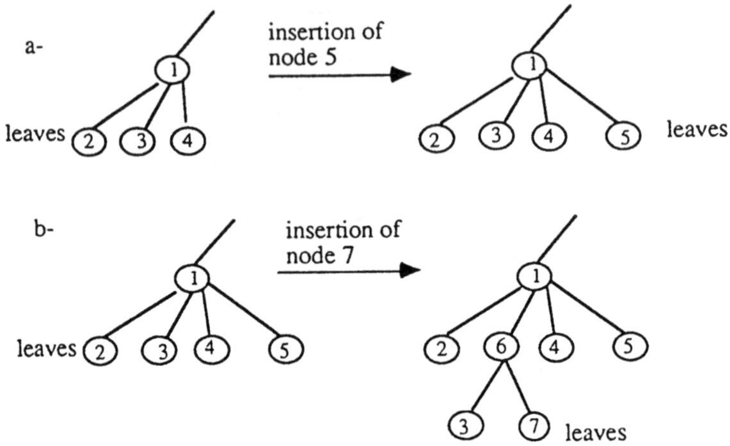

FIGURE 5.16. Hierarchy of Bounding Volumes which is a Quadtree

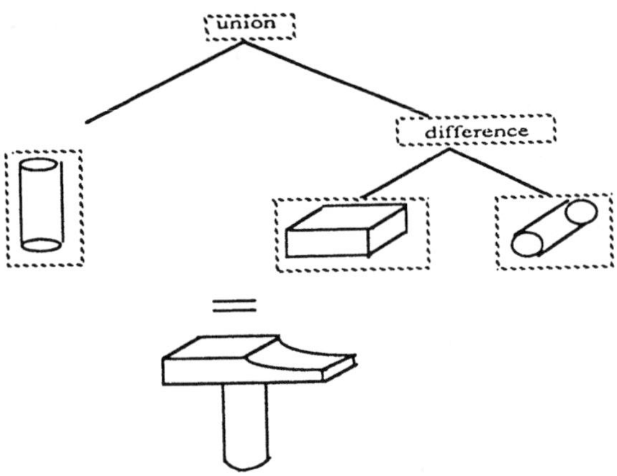

FIGURE 5.17. CSG Tree and Hierarchy of Bounding Volumes

Once the hierarchy of bounding volumes is built, the ray-scene intersection test is performed as follows. The hierarchy is searched from the root to the leaves. During this search, at a node N, the associated bounding volume is checked for an intersection with the current ray. If this bounding volume is intersected by the ray, those of the children nodes are in their turn checked for an intersection. This process is repeated recursively and terminates at the leaf nodes. On the other hand, if the bounding volume of the node N is not intersected by the ray, the associated subtree is left out, that is, it is not searched.

The Different Kinds of Bounding Volumes

The choice of a good bounding volume is of prime necessity. We shall see later what is the compromise that allows a good choice. Several kinds of bounding volumes exist: sphere, parallelepiped, polyhedron, ellipsoid and statistical volumes.

parallelepiped For the sake of speed up, the faces of this bounding volume are perpendicular to the axes of the eye coordinates system. Its perspective projection onto the screen plane is often used to filter the primary rays (starting at the eye)

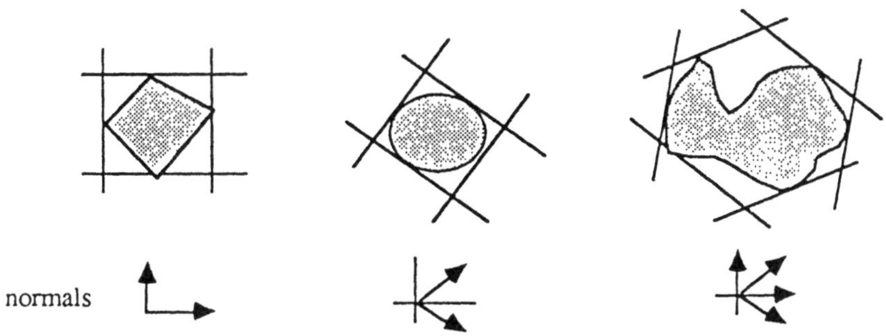

FIGURE 5.18. Polyhedral Bounding Volumes

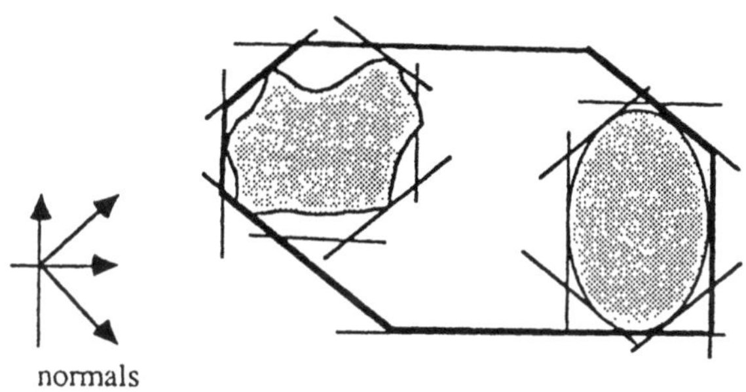

FIGURE 5.19. Polyhedral Bounding Volume of Two Objects

sphere and ellipsoid They may be used to filter the reflected and refracted rays and those directed to the light sources

polyhedron It has been proposed by Kay and Kajiya [35]. The objects are bounded by polyhedra whose sizes may be different but whose face normals have constant direction vectors (figure 5.18). These direction vectors as well as the number of faces are chosen by the user before the synthesis phase For example, we can choose polyhedra whose face normals are:

$\mathbf{N1} = [\sqrt{3}/3, \sqrt{3}/3, \sqrt{3}/3]$
$\mathbf{N1} = [-\sqrt{3}/3, \sqrt{3}/3, \sqrt{3}/3]$
$\mathbf{N1} = [-\sqrt{3}/3, -\sqrt{3}/3, \sqrt{3}/3]$
$\mathbf{N1} = [\sqrt{3}/3, -\sqrt{3}/3, \sqrt{3}/3]$

As shown in figure 5.19, it is easy to build a hierarchy with these bounding volumes.

statistical bounding volumes They bound surfaces defined by stochastic models as fractals [21]. The fractal which models reliefs, is generated by randomly and recursively subdividing a triangle into four new non coplanar triangles. For this type of surface, two bounding volumes have been proposed in the literature: the cheese-cake of Kajiya [31] and the ellipsoid of Bouville [11]. The first one is given by figure 5.20. This is a prism which contains the original triangle or the one created at a certain level of subdivision The intersection between a ray and this bounding volume does not require a lot of computations. On the other hand, its drawback is that it does not account for the exact statistical properties of a fractal generated by a recursive subdivision of a triangle. Indeed, the variance of a point lying on a triangle

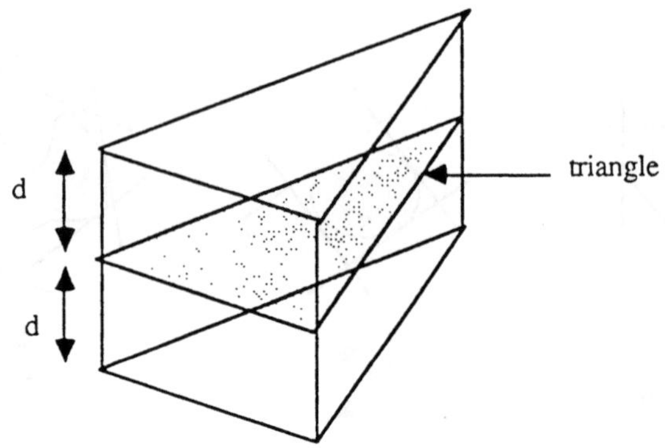

FIGURE 5.20. Statistical Bounding Volume of Kajiya

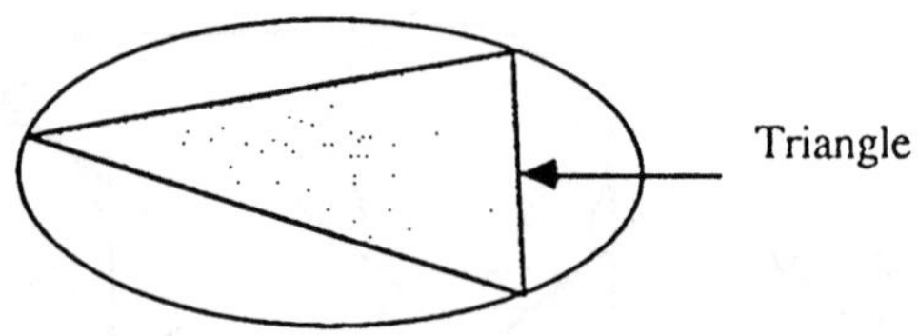

FIGURE 5.21. Ellipsoid of Bouville

decreases near the vertices whereas the height above the triangle remains constant in the case of the prism of Kajiya. As shown in Figure 5.20, this constant is named d and is chosen such that P is close to one. P is the probability, for which all the triangles created by a recursive subdivision of triangle T, are bounded by the prism associated to T Kajiya proposes:

$$d = \beta.\mu$$

and

$$\mu = (1\sqrt{3}/3)^{2H}.V_H$$

where V_H is a constant depending on the dimension H of the fractal and 1 is the length of a side of the triangle. If β is equal to 3 then P takes the value 0.9974.

As for the ellipsoid of Bouville, it is shown by Figure 5.21. This bounding volume is evaluated by intersecting an ellipsoid (which contains the vertices of the triangle) and an infinite height prism subtended by the triangle. It accounts for the previous remarks since the variance is zero at the vertices and maximum at the centroid of the triangle. This kind of volume is consequently more efficient than the one of Kajiya. Since a fractal surface is generated stochastically, it may extend beyond its bounding volume. The probability for which this case can occur is named overflow probability. Moreover Bouville's bounding volume is small, moreover this probability is important. The overflow probability varies at each point of the triangle and reaches a maximum value at the midpoint of each of the edges. If one fixes, for these three points, a maximum overflow probability p_m the bounding volume must contain a displacement whose amplitude is equal to:

$$D = d_m.P$$

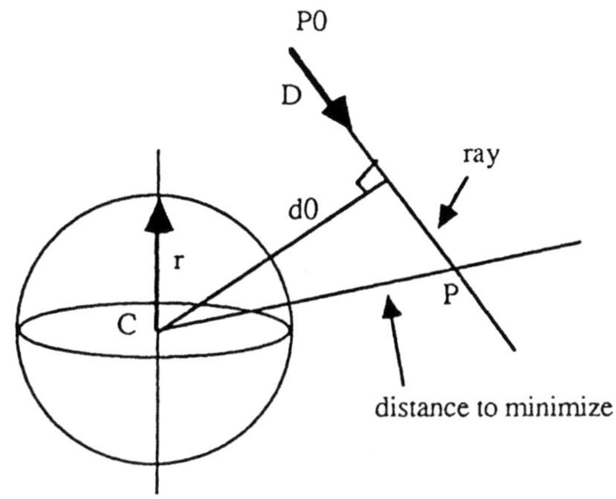

FIGURE 5.22. Intersection Test for a Spherical Bounding Volume

where $d_m = 2\beta.k^2.2^{iH}/\sqrt{3}$, β depends on p_m, k is a scaling factor of the fractal, H is the dimension of the fractal and i the subdivision level.

If $p_m = 10^{-5}, \beta = 3.123$ and if $p_m = 10^{-6}, \beta = 3.46$

Intersection Test

The intersection test consists in checking a bounding volume for an intersection with a ray. In the following, we give the way to perform it for different types of bounding volumes.

sphere The square of the orthogonal distance d_0^2 between the centre of the sphere and the ray is first computed. If d_0^2 is smaller than or equal to the square of the radius of the sphere, then the ray intersects the sphere, otherwise it does not intersect it (figure 5.22) Let $\mathbf{C}$ be the centre of the sphere and let $\mathbf{P} = \mathbf{P}_0 + t.\mathbf{D}$ be the ray equation. d_0 is evaluated by minimizing the distance between $\mathbf{C}$ and a point $\mathbf{P}$ in the ray. This gives:

$$d^2 = \| \mathbf{P}_0 + t.\mathbf{D} - \mathbf{C} \|^2 = \| \mathbf{P}_0 - \mathbf{C} \|^2 + 2t.(\mathbf{P}_0 - \mathbf{C}).\mathbf{D} + t^2. \| \mathbf{D} \|^2 \qquad (5.7)$$

By setting to 0 the derivative of d^2, we obtain:

$$t = ((\mathbf{P}_0 - \mathbf{C}).\mathbf{D} / \| \mathbf{D} \|^2) = -(\mathbf{P}_0 - \mathbf{C}).\mathbf{D}$$

The substitution of the value of t in equation (5.7) yields:

$$d_0^2 = \| \mathbf{P}_0 - \mathbf{C} \|^2 - ((\mathbf{P} - \mathbf{C}).\mathbf{D})^2$$

The evaluation of d_0^2 requires only one dot product and some floating point operations

parallelepiped The faces of the parallelepiped are perpendicular to the axes of the eye coordinates system. First, the intersections between the ray and the faces $x = x_1$ and $x = x_2$ are computed. Two values of t are then obtained:

$$t_1 = (x_1 - x_0)/dx \textit{ and } t_2 = (x_2 - x_0)/dx$$

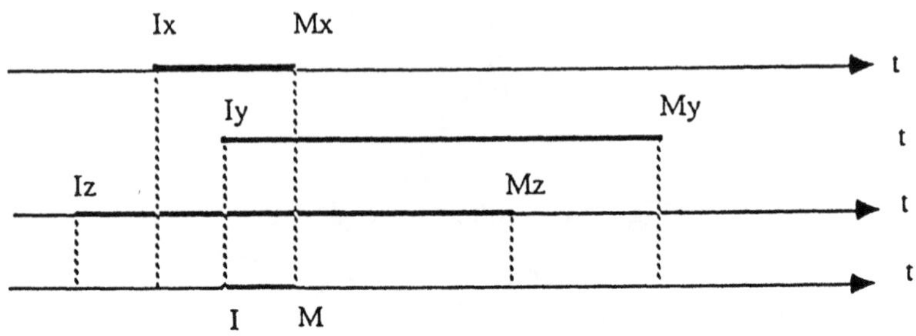

FIGURE 5.23. Intersection with a Parallelepiped

One intersection interval is then determined:

$$[Ix, Mx] = [\min(t_1, t_2), \max(t_1, t_2)]$$

The same processing is applied to the face perpendicular to the y and z axes. The result is then an intersection interval given by:

$$[I, M] = [\max(Ix, Iy, Iz), \min(Mx, My, Mz)]$$

If I is smaller than or equal to M then the ray intersects the parallelepiped bounding volume, otherwise it does not intersect it (Figure 5.23)

polyhedron The intersection test is similar to the previous one, except that the faces are not perpendicular to the axes of the eye coordinate system.
Let **N** be the normal of a face such that:
$\mathbf{N} = [A, B, C]$
and
$Ax + By + Cz + d = 0$ the equation of the plane containing the face.

The value of t corresponding to the intersection between the ray and this face is computed by substituting the ray equation in that of the plane:

$$t = -\frac{(d + \mathbf{N}.\mathbf{P}_0)}{\mathbf{N}.\mathbf{D}}$$

Choosing a Good Bounding Volume

The choice of a good bounding volume is of prime importance. It must offer a good compromise between the calculation complexity of the intersection test and the size of the bounding volume. The smaller the size is, better the bounding volume fits the object; this makes the intersection test more efficient. But such a bounding volume involves more expensive computations. A good compromise minimizes the cost function [56] given by:

$$C = v.V + i.I$$

where

- C is the total cost function
- v is the number of times the bounding volume is intersected to perform an intersection test

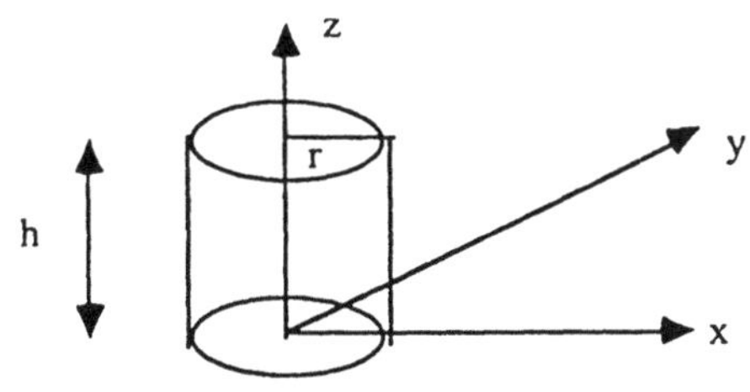

FIGURE 5.24. Cylinder and its LCS

- V is the cost for an intersection with a bounding volume
- i is the number of times the object is intersected
- I is the cost of intersection with an object.

For an object, v and I are constant. On the other hand, V and i may be modified in order to reduce the cost function C. A reduction of the bounding volume complexity yields a decrease of V and an increase of i. Whereas, an increase of this complexity yields an increase of V and a decrease of v. We think that the parallelepiped offers a good compromise. A good alternative is to choose Kay and Kajiya's polyhedron.

5.3.3 Intersection with Simple Objects and Composite Objects

This subsection deals with the intersection calculation between a ray and simple objects (sphere, parallelepiped...) or composite objects created by combining simple objects with boolean operators.

Simple Objects

Sphere

Given the notations of section 5.3.2, the intersection points are solutions of the following equation:

$$\| \mathbf{P}_0 - \mathbf{C} \|^2 + 2t.(\mathbf{P}_0 - \mathbf{C}).\mathbf{D} + t^2. \| \mathbf{D} \|^2 = r^2 \tag{5.8}$$

If the intersection calculation is performed in the local coordinates system of the sphere, equation (5.8) becomes:

$$\| \mathbf{P}_0 \|^2 + 2t.\mathbf{P}_0.\mathbf{D} + t^2. \| \mathbf{D} \|^2 = r^2$$

Parallelepiped

The way to compute the ray-parallelepiped intersection has been shown in section 5.3.2. Let $[I, M]$ be the interval intersection. If I is smaller than or equal to M then the intersection exists and in addition, I and M are the values of the parameter t corresponding to the intersection points. Otherwise it does not exist.

Cylinder

The intersection is performed in the local coordinates system of the cylinder (figure 5.24).

The cylinder is supposed to be the result of the intersection between an infinite height cylinder and the subspace delimited by two planes whose equations are $z = 0$ and $z = h$. The intersection process is the following. The intersection between the ray and the infinite height cylinder is first performed. This yields a first interval $[t_1, t_2]$. The intersection with

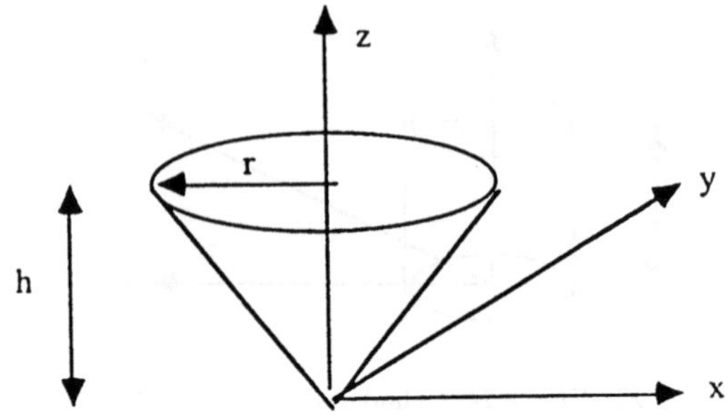

FIGURE 5.25. Cone and its Local Coordinates System

the two planes gives a second interval $[t_3, t_4]$. The final intersection interval $[I, M]$ results from the combination of these two intervals (as for the parallelepiped).

Obtaining (t_1, t_2) The equation of the infinite height cylinder is:

$$x^2 + y^2 = r^2$$

Substituting the ray equation in this equation we obtain:

$$t^2.(dx^2 + dy^2) + 2t.(x_0.dx + y_0.dy) + (x_0^2 + y_0^2 - r^2) = 0$$

Solving the equation gives the interval $[t_1, t_2]$

Obtaining (t_3, t_4) Let A and B the two values of t resulting from the intersection with the two planes:

$$A = -z_0/dz \qquad \text{and} \qquad B = (h - z_0)/z_0$$

We get:

$$t_3 = \min(A, B) \qquad \text{and} \qquad t_4 = \max(A, B)$$

Cone

The intersection is performed in the LCS of the cone (figure 5.25).

The cone is supposed to be the result of the intersection between an infinite height cone and the subspace delimited by two planes, the equations of which are $z = 0$ and $z = h$. The intersection between the ray and the infinite height cone is first performed. The equation of this cone is given by:

$$h^2.(x^2 + y^2) - r^2.z^2 = 0$$

Substituting the ray equation in this equation yields an interval $[t_1, t_2]$. Then the planes are in their turn intersected to give a second interval $[t_3, t_4]$ such that:

$$t_3 = \min(A, B) \qquad \text{and} \qquad t_4 = \max(A, B)$$

where $A = -z_0/dz$ and $B = (h - z_0)/dz$.

The final interval is the combination of these two intervals (as for the cylinder).

Polygon

Several ray-polygon intersection methods have been proposed in the literature. Only three of them are presented in this subsection. For all these methods, the intersection process consists of two steps:

1. the goal of the first step is to perform the intersection between the ray and the plane containing the polygon

2. the second step tests if the resulting point is inside or outside the polygon.

a - Haines's Method

Described in [27].

Ray-Plane Intersection The plane is defined by the following equation:

$$a.x + b.y + c.z + d = 0$$

The unit normal vector of the plane is given by $\mathbf{N} = [a, b, c]$ and the ray equation is similar to the one given previously:

$$\mathbf{P} = \mathbf{P}_0 + t.\mathbf{D} \qquad \text{or} \qquad [x, y, z] = [x_0, y_0, z_0] + t.[dx, dy, dz]$$

The ray-polygon intersection is given by:

$$t = \frac{-(\mathbf{N}.\mathbf{P}_0 - \mathbf{D})}{\mathbf{N}.\mathbf{D}}$$

Inside - Outside Test After calculating the ray-plane intersection, the next step is to determine if the intersection point is inside the polygon. Several methods have been proposed to solve this problem. Berlin [7] gives a good overview of some techniques. Haines's method is a modified version of Berlin's one and is faster. This method consists in shooting a ray in a certain direction and counting the number of line segments crossed. If the number of crossing is odd, the point is inside the polygon; else it is outside. Haines's method handles the special case where the ray hits a vertex in the polygon.

Define a polygon as a set of vertices:
Polygon $= \{[x_i, y_i, z_i],\ i \leq N\}$
Let $\mathbf{I}$ be the intersection point between the ray and the plane containing the polygon. The first step is to transform the problem to a two dimensional plane where a point is specified by a coordinates pair (u, v). It consists in simply projecting the polygon onto the plane defined by two coordinates chose among x, y, and z. Choosing which coordinate to throw away is simply defined: throw away the coordinate whose corresponding plane equation is of greatest magnitude. For example, for a polygon with normal $\mathbf{N}$ = [0, -6, 2] the y coordinate would be thrown away and the x and z coordinates would be assigned to u and v. The coordinate having the largest magnitude is called the dominant coordinate.

Once the polygon has been projected onto the (u, v) plane, the inside-outside test can be simply formed. Translate the polygon such that the intersection point $\mathbf{I} = (u_i, v_i)$ is at the origin. Label the new vertices as (u', v'). Consider a ray starting from the origin and proceeding along the $+u$ axis. Each edge of the polygon is tested against

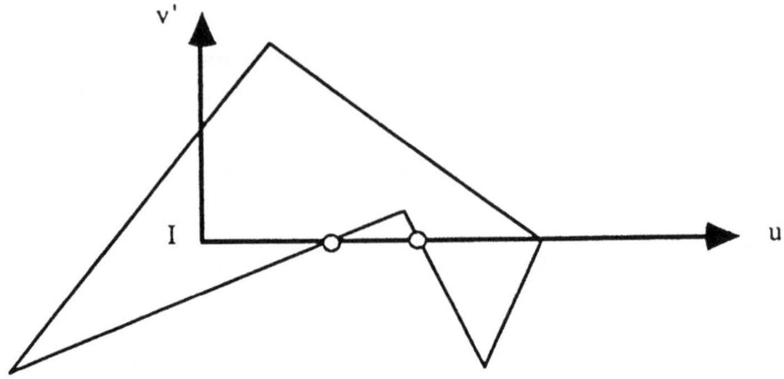

FIGURE 5.26. Polygon Inside-Outside Test

the ray. If a number of edges crossed by the ray is odd, the point **I** is inside the polygon, else it is outside (figure 5.26). A special case may occur when the ray passes through one or more vertices. These vertices have a v' coordinate equal to zero and are to be considered on the $+v'$ side of the plane. In this way no vertices lie on the ray itself. This latter has to be redefined such that it remains close to the original ray but does not pass through any vertex.

The test algorithm is the following. First, the edges are checked for an intersection with the u' axis. For those edges that do cross (this occurs when the endpoints have v' coordinates of different sign), the vertices are checked to see if both endpoints are on the $+u'$ part of the plane. If so, the $+u'$ axis must be crossed, else if one endpoint has a negative coordinate, then the exact location of the intersection between the ray and u' axis must be found. If the u' coordinate of this location is positive, then the edge indeed crosses the $+u'$ axis. Haines's method is very efficient because most edges can trivially be rejected or accepted.

b - Snyder's Method

Described in [49], Snyder's method concerns the ray-triangle intersection. It will be extended, here, to a polygon. Let $\mathbf{P}_i$ be the vertices of a triangle and let $\mathbf{N}_i$ be the associated normals which are used for normal interpolation across the triangle. The normal to the triangle is given by:

$$\mathbf{N} = (\mathbf{P}_1 x \mathbf{P}_0) \times (\mathbf{P}_2 - \mathbf{P}_0)$$

A point **P** lying on the triangle plane satisfies:

$$\mathbf{P}.\mathbf{N} + d = 0 \qquad \text{where} \qquad d = -\mathbf{P}_0.\mathbf{N}$$

An index i_0 is computed to be equal either to 0 if $|\ \mathbf{N}_x\ |$ is maximum or to 1 if $|\ N_y\ |$ is maximum or to 2 if $|\ N_z\ |$ is maximum.

To intersect a ray $\mathbf{P} = \mathbf{O} + t.\mathbf{D}$ with a triangle, first compute the t parameter of the intersection between the ray and the triangle plane:

$$t = \frac{(d - \mathbf{N}.\mathbf{O})}{\mathbf{N}.\mathbf{D}}$$

Let i_1 and $i_2 (i_1, i_2 \in \{0, 1, 2\})$ be two unequal indices different from i_0. Compute the i_1 and i_2 components of the intersection point **I**, by:

$$I_{i1} = O_{i1} + t.D_{i1} \qquad \text{and} \qquad I_{i2} = 0_{i2} + t.D_{i2}$$

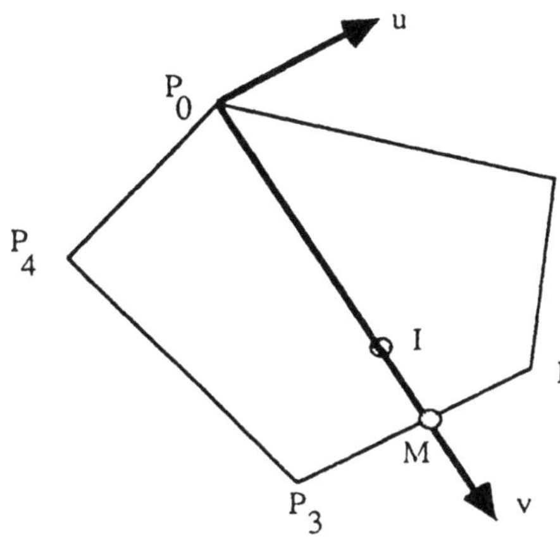

FIGURE 5.27. Ray-Polygon Intersection: Marchal's Method

The inside-outside test can be performed by computing scalars β_0, β_1 and β_2:

$$\beta_i = \frac{[(\mathbf{P}_{i+2} - \mathbf{P}_{i+1}) \times (\mathbf{I} - \mathbf{P}_{i+1})]_{i0}}{[N]_{i0}}$$

where addition in subscripts is modulo 3.

The β_i are the barycentric coordinates of the point where the ray intersects the triangle plane. Only the i_0 component of the cross product is computed; the value of I_{i0} is therefore unnecessary. **I** is inside the triangle if and only if $0 \leq \beta \leq 1$ for $i \in \{0, 1, 2\}$. The interpolated normal at point **I** is given by:

$$\mathbf{N}' = \beta_0.\mathbf{N}_0 + \beta_1.\mathbf{N}_1 + \beta_2.\mathbf{N}_2.$$

Snyder's method can be easily extended up to polygons. The main idea is to consider a polygon as a union of triangles. If the intersection point I is inside one triangle, then the intersection test is terminated and consequently it is not necessary to treat the remaining triangles. Otherwise, if **I** lies outside all the triangles, it is then outside the polygon.

c - Marchal's Method

Described in [40], and illustrated by figure 5.27.

I is the intersection point between ray and the polygon plane. Once **I** has been computed, all the vertices $\mathbf{P}_i$ are transformed to the two dimensional coordinates system (u, v) whose origin is the vertex $\mathbf{P}_0$. The plane of this coordinates system is the polygon plane. The inside-outside test determines if an edge $\mathbf{P}_i\mathbf{P}_{i+1}$ intersects the v axis at a point **M** (this may occur when the u components of $\mathbf{P}_i$ and $\mathbf{P}_{i+1}$ have different signs). If so, and if $\mathbf{P}_0\mathbf{I} < \mathbf{P}_0\mathbf{M}$ then **I** is inside the polygon, else it is outside. On the other hand, if none of the edges intersect the v axis, then **I** lies outside the polygon.

The interpolated normal at point **I** is given by:

$$\mathbf{N}_I = (\mathbf{P}_0\mathbf{I}/\mathbf{P}_0\mathbf{M}).\mathbf{N}_M + (1 - \mathbf{P}_0\mathbf{I}/\mathbf{P}_0\mathbf{M}).\mathbf{N}_0$$

where the normal $\mathbf{N}_M$ at point **M** is given by:

$$\mathbf{N}_M = (\mathbf{P}_i\mathbf{M}/\mathbf{P}_i\mathbf{P}_{i+1}).\mathbf{N}_{i+1} + (1 - \mathbf{P}_i\mathbf{M}/\mathbf{P}_i\mathbf{P}_{i+1}).\mathbf{N}_i$$

and $\mathbf{N}_i, \mathbf{N}_{i+1}$ are the normals at point $\mathbf{P}_i$ and $\mathbf{P}_{i+1}$. $\mathbf{P}_i\mathbf{P}_{i+1}$ is the intersected edge.

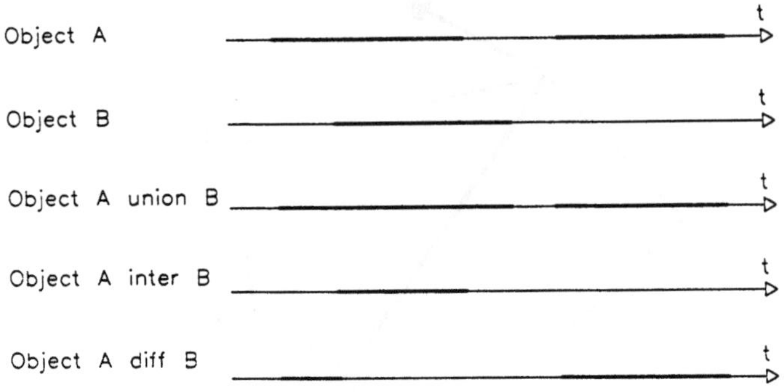

FIGURE 5.28. Ray Intersection with a Composite Object

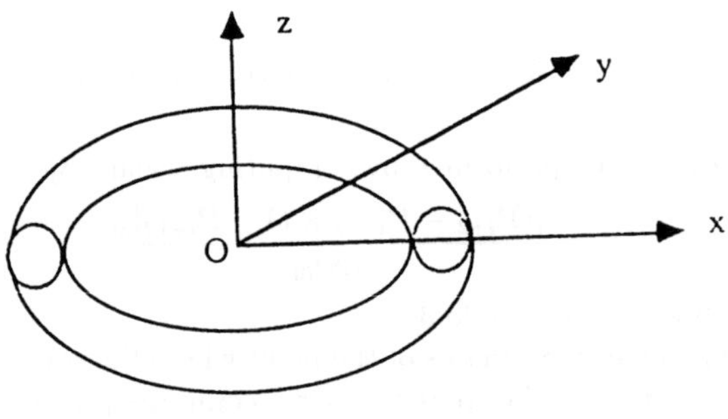

FIGURE 5.29. Torus and its LCS

Composite Objects

A composite object may be created by performing set operations (union, difference, intersection) on simple or on other composite objects. A CSG tree is an example of a composite object. The ray-object intersection results in a list of intervals as shown in figure 5.28. In this example, two objects are combined with each set operator. The intersection result is a list of two intervals, the length of which depends on the set operation used.

5.3.4 Intersection of Complex Objects

Only the following complex objects are considered: torus, surface of revolution, prism, algebraic surface, bicubic surfaces and fractals.

Torus

The local coordinates system associated with a torus is shown in figure 5.29. The torus equation is:

$$(x^2 + y^2 + z^2 + 1 - r^2) - 4.(x^2 + y^2) = 0$$

when $r(r \in [0, 1])$ represents the radius of the toric part of the torus. By substituting the ray equation in the torus question, we get a fourth degree polynomial equation.

$$a.t^4 + b.t^3 + c.t^2 + d.t + e = 0$$

which can be solved by algebraic or numerical methods.

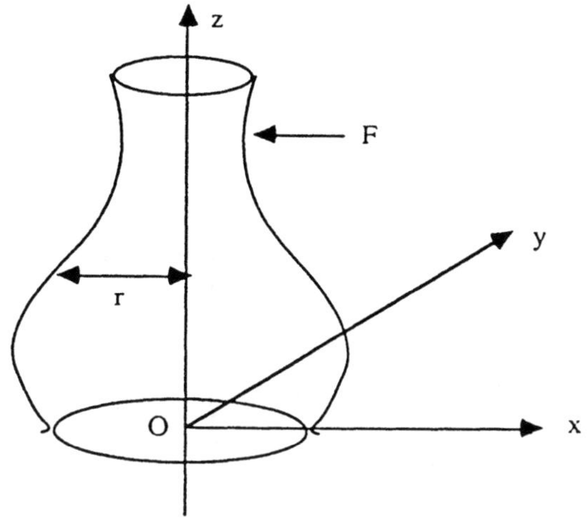

FIGURE 5.30. Surface of Revolution and its LCS

Surface of Revolution

The local coordinates system of a surface of revolution is shown in figure 5.30.

This kind of surface may be generated by rotating a curve F round an axis. A first intersection method has been proposed by Kajiya [31]. It consists in using in the "strip tree" method [5]. The advantage of this method is that it may process any curve F. On the other hand, its drawback is both its important memory requirement and its expensive processing. Bouville's method [12] seems faster and in addition requires less memory. But this method supposes that F and r (see Figure 5.30) are cubic B-spline curves depending on one parameter u. In this method the surface of revolution is then defined by the following system:

$$x^2 + y^2 = r^2(u) \quad (5.9)$$

$$z = F(u). \quad (5.10)$$

By substituting the ray equation in the previous system, we get a sixth degree polynomial equation

$$a.F^2(u) - r^2(u) + b.F(u) + c = 0$$

where a, b and c are constants.

The normal $\mathbf{N}$ at the intersection point $[x_i, y_i, z_i]$ is equal to

$$[x_i, y_i, \frac{-r(u_i).r_u(u_i)}{F_u(u_i)}]$$

where r_u and F_u represent the derivatives of r and F with respect to u, and u_i is the value of u corresponding to $[x_i, y_i, z_i]$.

Prism

It has been introduced by Kajiya [31] and is shown in Figure 5.31. A prism is defined by:

- an axis
- a base plane

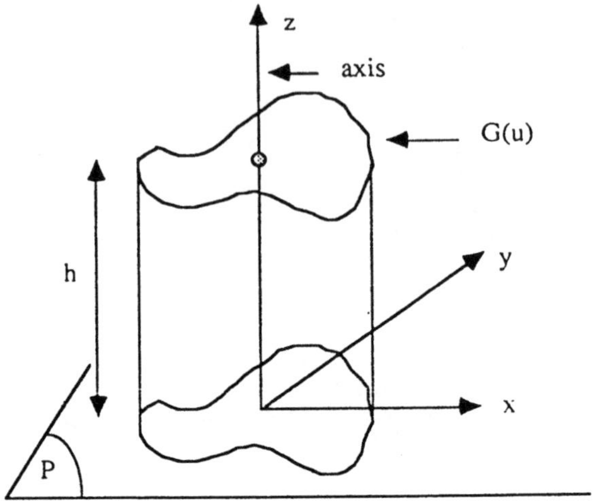

FIGURE 5.31. Prism and its LCS

- a curve G delimiting a crossed section perpendicular to the axis
- a height h.

If G is any curve, Kajiya proposes the "strip tree" method to compute the ray-prism intersection. We suggest a faster method which assumes that G is a cubic B-spline curve defined by

$$x = G_x(u) \qquad \text{and} \qquad y = G_y(u)$$

In addition, the z component of a point on the prism, is bounded by 0 and h. We obtain then the following system:

$$x_o + t.dx = G_x(u)$$
$$y_o + t.dy = G_y(u)$$

A third degree polynomial equation is derived from this system:

$$G_x(u) + a.G_y(u) + c = 0$$

The normal at the intersection point $[x_i, y_i, z_i]$ is given by

$$\mathbf{N} = [x_i, \frac{-x_i.x_u(u_i)}{y_u(u_i)}, 0]$$

where x_u and y_u are the derivatives of x and y with respect to u.

Algebraic Surfaces

An algebraic surface is defined by:

$$S(x,y,z) = \sum_{i=0}^{l} \sum_{j=0}^{m} \sum_{k=0}^{n} a_{ij}.x_i y_j z_k \tag{5.11}$$

The substitution in $S(x,y,z)$ of the ray equation, gives a polynomial equation $S^*(t)$ whose degree is $d = l + m + n$:

$$S^*(t) = \sum_{i=0}^{d} a_i.t^i$$

$S*(t)$ may be solved with non linear programming techniques, such as the one of Laguerre or Bairstow. These techniques are iterative and converge only if they start from an initial value of t close to the exact root. To find a good initial value of t, one must isolate the roots be recursively subdividing the range of t into two equal sized subintervals, and by seeing if the resulting subintervals contain at least one root. This process terminates when the width of an interval is less than a given threshold. Several root isolation methods are proposed in the literature. Only two of them are discussed in this subsection:

- interval methods [36], [37], [43]
- Collins's method [16].

Interval Method

An interval is defined by an ordered pair of real numbers $[a, b]$ with $a < b$. The interval method allows arithmetic operations to be performed on intervals using the operators $+, -, *$ and $/$. Let op be an operator; we have then:

$$[a, b] op [c, d] = \{x \ op \ y, \text{ such that } x \in [a, b] \text{ and } y \in [c, d]\}$$

except that we do not define $[a, b]/[c, d]$ in case $0 \in [c, d]$.

These operations can be performed algebraically using the endpoints of the intervals, as shown in the following:

$$\begin{aligned} [a,b] + [c,d] &= [a+c, b+d] \\ [a,b] - [c,d] &= [a-c, b-d] \\ [a,b] * [c,d] &= [\min(a*c, a*d, b*c, b*d), \max(a*c, a*d, b*c, b*d)] \\ [a,b] \,/\, [c,d] &= [a,b] * [1/d, 1/c] \text{ provided that } 0 \notin [c,d] \end{aligned}$$

Division by an interval containing 0 may be defined as:

$$1/[a,b] = \begin{cases} [1/b, +\infty] & \text{if } a = 0, \\ [-\infty, 1/a] & \text{if } b = 0, \\ [-\infty, 1/a] \cup [1/b, +\infty] & \text{if } a \le 0 \le b, \\ [1/b, 1/a] & \text{if } a > 0 \text{ or if } b < 0 \end{cases}$$

Let $f(x_1, \ldots, x_n)$ be a rational function, and let F be the corresponding interval rational function. If for each $i, 1 \le i \le n$, x_i ranges over $[a_i, b_i]$ then

$$F([a_1, b_1], \ldots, [a_n, b_n]) \supset \{f(x_1, \ldots, x_n)$$

$$\text{such that } x_i \in [a_i, b_i], 1 \le i \le n\} = \text{range of } f$$

Let us see now how the interval method can be used to solve the polynomial equation (5.11). First, the range T of variable t, is determined by intersecting the ray with the bounding volume of the surface. After that, the method checks the possibility of the interval T (and its subintervals) to contain the value 0. This is done by interval evaluation of the polynomial equation (5.11). If this evaluation contains 0, then there is some chance for the polynomial to have real zeros. In this case T is subdivided into two subintervals and the process is repeated for the subintervals in a recursive fashion. The recursion terminates when the width of the current subinterval is smaller than a threshold (in case of isolation) or when it can be treated as a single point which is a real root of the polynomial.

The same technique can be used to isolate or to find the solutions of a system of non linear equations [43]. For the sake of simplicity, consider a system of two polynomial equations where the two unknowns are u and v ranging respectively over $U = [u_1, u_2]$ and $V = [v_1, v_2]$:

$$f(u,v) = 0 \quad \text{with} \quad (u,v) \in I \tag{5.12}$$

$$g(u,v) = 0 \quad \text{with} \quad (u,v) \in I \tag{5.13}$$

where $I = [u_1, u_2] \times [v_1, v_2]$

The method checks the possibility for a solution to lie within the entire domain of the 2D interval I. This is done by interval evaluation of the functions $f(u,v)$ and $g(u,v)$. If both the evaluations contain 0, then there is some chance for the solution to exist. If so, I is subdivided into 2D subintervals and the process is repeated recursively as pointed out above.

Collins's Method

Let $P(x) = \sum_{i=0}^{n} a_i.x^i$

Descartes' rule states that the number of sign variations $var(a_n, a_{n-1}, \ldots, a_0)$ exceeds the number of positive zeros, multiplicities counted, by an even non-negative integer. Hence if $var(P)$ is equal to 0, P has exactly no positive roots, and if $var(P)$ is equal to 1, P has exactly one positive root. A surprising theorem which Uspensky attributes to Vincent in 1886 [54], shows that after a finite number of transformations

$$P'(x) = P(x+1) \qquad \text{and} \qquad P^*(x) = (x+1)^n \, P(1/(x+1))$$

one arrives at a polynomial having sign variation 1 or 0. Collins and Loos have modified Uspensky's method to make it faster. It is described by the following algorithm written in Pseudo-Pascal.

```
procedure real_root_isolation(P: polynomial;
                              var L:list_of_intervals);

var
  bound: real;
  B : polynomial;
  L': list_of_intervals;

begin

  { bound the positive roots of P by bound }

  bound := 2^k;
   if k ≥ 0   then B(x):= P(bound * x)
              else B(x):= (1/(bound*2^n))*P(bound * x);
  { call the isolation procedure which gives a list L'
    of isolation intervals of B }

  isolation_proc(B, 0, 1, 1, L');

  { call the procedure replace_L'_by_L to
    replace each interval  [a_i, b_i] by [bound*a_i, bound*b_i] }

  replace_L'_by_L(L, L');
end;
```

```
procedure isolation˙proc(B: polynomial; min˙int, max˙int: real; width: real;
                         var L: list˙of˙intervals);

var

  L1, L2: list˙of˙intervals;

  B*, B', B'': polynomial;
  I: interval;

  {min˙int and max˙int are respectively the smallest
    and largest endpoints of the current interval}

begin

  { transform the zeros of B in [0, 1] onto the zeros of B*
    in [0, ∞] }

  B*(x):= (x + 1)^n*B(1/(x + 1));

  { end of recursion }

if var(B*) = 0  then begin
                L := empty;
                return;
                end
  else if (var(B*) = 1)  and
                (width ¡= threshold)  then
           begin
          I := [min˙int, max˙int];
          insert˙in˙L(I);
end;


{ process the left-half subinterval by transforming
  the zeros of B in [0,1/2] on the zeros of B in [0, 1]}

B'(x) := 2^n*B(x/2);
isolation˙proc(B', min˙int, max˙int - width/2, width/2, L1);

{ process the right-half subinterval by transforming the
  zeros of B in [1/2,1] on the zeros of B in [0,1] }

B''(x) := B'(x + 1);
isolation˙proc(B'', min˙int + width/2, max˙int, width/2, L2);

{put the two lists L1 and L2 in L}

add˙list (L, L1);
add˙list (L, L2);

 end;
```

Bicubic Surfaces

Only bicubic patches are discussed. They are defined by:

$$Q(u,v) = [x(u,v), y(u,v), z(u,v)] = \sum_{i=0}^{3}\sum_{j=0}^{3} B_i(u).B_j(v).P_{ij} \tag{5.14}$$

where P_{ij} are the control points of the surface, and $B_i(u)$, $B_j(v)$ the blending functions which determine the type of surface (B-spline, Bézier, Beta-spline,...). These blending functions depend on the two parameters u and v which both range over [0, 1].

A ray may be considered as the intersection of two planes defined by:

$$[A_1, B_1, C_1] \cdot [x, y, z] = D_1 \tag{5.15}$$

$$[A_2, B_2, C_2] \cdot [x, y, z] = D_2 \tag{5.16}$$

If the ray equation is expressed as

$$[x, y, z] = [x_0, y_0, z_0] + t.[dx, dy, dz] \tag{5.17}$$

the two planes can be determined as follows:

$$[A_1, B_1, C_1] = [x_0, y_0, z_0] \times [dx, dy, dz] \tag{5.18}$$

$$[A_2, B_2, C_2] = [A_1, B_1, C_1] \times [dx, dy, dz] \tag{5.19}$$

$$D_1 = [A_1, B_1, C_1] \times [x_0, y_0, z_0] \tag{5.20}$$

$$D_2 = [A_2, B_2, C_2] \times [x_0, y_0, z_0] \tag{5.21}$$

If we substitute equation (5.14) in equation (5.15), we obtain the following system:

$$\sum_{i=0}^{3}\sum_{j=0}^{3}([A_1, B_1, C_1].P_{ij}).B_i(u).B_j(v) - D_1 = 0 \tag{5.22}$$

$$\sum_{i=0}^{3}\sum_{j=0}^{3}([A_2, B_2, C_2].P_{ij}).B_i(u).B_j(v) - D_2 = 0 \tag{5.23}$$

Once these equations have been stated, ray-surface intersection may be performed by means of one of the existing methods. At least three methods can be used:

- method which decomposes a patch into a set of planar polygons
- method which subdivides recursively a patch into four patches
- method which uses numerical techniques.

Decomposition into Planar Polygons

A patch is recursively subdivided into four subpatches. The recursion terminates when a subpatch satisfies a flatness criterion. A first criterion [38] consists in choosing three points among the 16 control points of the subpatch and determining the equation of the plane P containing these three points. Then, the maximum distance d, between P and the remaining control points, is computed. If d is smaller than a tolerance value, the subpatch can be approximated by a planar polygon.

Before the subdivision step, the original patch is transformed to a bézier patch because this latter has the smallest bounding volume. A resulting planar polygon is a quadrilateral whose vertices are the four corners of the corresponding subpatch.

A second criterion is to test if the four corners of the patch are coplanar and if its boundaries are linear [36], [37].

Once the original patch has been approximated by quadrilaterals, ray-patch intersections is transformed to ray-polygon intersection.

Subdivision into Subpatches

The original patch is first transformed into Bézier form and is subdivided recursively. The recursion stops when the bounding volumes of the subpatches satisfy a size criterion. One criterion is that the maximum length of the edges of a subpatch is smaller then a threshold value. Note that the faces of the bounding volumes are such that their faces are perpendicular to the eye coordinates system axes.

At the current level of subdivision, an intersection test is performed for each subpatch. If a subpatch does not satisfy this test, it is left out, else it is subdivided recursively until it satisfies the size criterion. We obtain as many terminal bounding volumes as points of intersection between the ray and the original patch. The closest point of the ray-terminal bounding volume intersection is also a point of the ray-original patch intersection.

Numerical Methods

This consists of solving system (5.23) with non linear programming techniques (Newton, Laguerre, Bairstow...). The starting values of u and v may be obtained either by the interval method [53] or by exploiting the objects spatial coherence [29] or by applying the Oslo algorithm [46] to add other control points to the original patch [50].

Kajiya's method [30] is different from the previous ones. Indeed, to solve system (5.23), Bezout's resultant is evaluated. This system can be rewritten as:

$$a(u,v) = \sum_{i=0}^{3}\sum_{j=0}^{3} a_{ij}.u^{3-i}.v^{3-j} \quad = \quad 0 \tag{5.24}$$

$$b(u,v) = \sum_{i=0}^{3}\sum_{j=0}^{3} b_{ij}.u^{3-i}.v^{3-j} \quad = \quad 0 \tag{5.25}$$

This new system can be written in a simple form:

$$a(u,v) \quad = \quad a(u)[v] \tag{5.26}$$

$$b(u,v) \quad = \quad b(u)[v] \tag{5.27}$$

Consequently, a and b may be considered as polynomials of a single variable v, whose coefficients depend on u. Bezout's resultant associated to a and b is:

$$R(a,b) = \begin{vmatrix} R_{11} & R_{12} & R_{13} \\ R_{21} & R_{22} & R_{23} \\ R_{31} & R_{32} & R_{33} \end{vmatrix}$$

where || represents the determinant and:

$$R_{11} = \begin{vmatrix} b_0 & b_1 \\ a_0 & a_1 \end{vmatrix} R_{12} = \begin{vmatrix} b_0 & b_2 \\ a_0 & a_2 \end{vmatrix} R_{13} = \begin{vmatrix} b_0 & b_3 \\ a_0 & a_3 \end{vmatrix}$$

$$R_{21} = \begin{vmatrix} b_0 & b_2 \\ a_0 & a_2 \end{vmatrix} R_{22} = \begin{vmatrix} b_0 & b_3 \\ a_0 & a_3 \end{vmatrix} + \begin{vmatrix} b_1 & b_2 \\ a_1 & a_2 \end{vmatrix} R_{23} = \begin{vmatrix} b_1 & b_3 \\ a_1 & a_3 \end{vmatrix}$$

$$R_{31} = \begin{vmatrix} b_0 & b_3 \\ a_0 & a_3 \end{vmatrix} R_{32} = \begin{vmatrix} b_1 & b_3 \\ a_1 & a_3 \end{vmatrix} R_{33} = \begin{vmatrix} b_2 & b_3 \\ a_2 & a_3 \end{vmatrix}$$

R(a, b) is an 18-th degree polynomial. Its roots are common roots of a and b. The ray-patch intersection is performed as follows:

- find the roots $u_1, u_2, \ldots, u_p, (p \leq 18)$ of $R(a,b)$ by using either the interval methods or a non linear programming technique (Laguerre, Newton, Bairstow,...)

- for each u_i, compute the GCD of $a(u_i)[v]$ and $b(u_i)[v]$
- evaluate v_i by substitution.

Fractals

In case of fractals generated by a recursive subdivision of a triangle, ray- fractal intersection is similar to ray-triangle intersection.

5.4 Accelerated Ray Tracing

5.4.1 Introduction

The use of a hierarchy of bounding volumes does not suffice to reduce the large amount of ray-scene intersections. Other methods have to be used in order to overcome the slowness of the ray tracing technique. Broadly, two types of methods have been proposed in the literature. The methods [35], [56], [25] of the first type consist in minimizing the extents of the objects within the scene. They reduce the computational cost of the ray-intersection test but do not exploit spatial coherence since intersection computation is performed for all those objects which lie on the ray path. As for methods of the second type, they use either 2D or 3D subdivision.

2D Subdivision

Bronswoort et al. [14] propose a method which refines the image by subdivision in macropixels, thereby avoiding explicit computation of the intensities of all the pixels of the image. This method presents the disadvantage of loosing small elements of the image. The algorithm of Sears et al. [48] consists in using the rectangular enclosures of the primitive objects (which result from the projection of their parallelepiped enclosures on the screen plane) to subdivide the screen into regions. A subtree is thereby associated with each region. This technique minimizes only primary ray intersections. Like Sears et al., Coquillart [19] proposes to project the parallelepipedic enclosures of primitives on a plane which may be different from the screen plane, and subdivides this plane into regions. The difference with the previous method is that the regions may overlap. Two graphs of connectivity and overlapping are deduced. This technique works for both primary and secondary rays. Since this method uses a plane projection, the drawback is that the spatial coherence is not fully exploited and intersection computation is performed for all the primitives whose rectangular enclosures lie along the ray path.

3D Subdivision

The parallelepipedic enclosure of the scene (whose faces are perpendicular to the view coordinate system axes) is subdivided into 3D regions containing a small number of objects. A primary or secondary ray which enters a region, intersects only those objects lying in this region. If no intersection is found or the objects intersected are all transparent (in the case of rays directed towards the light sources), a computation of the next region traversed by the ray, is performed. It is shown that space subdivision reduces considerably the amount of intersection computations.

Many authors have proposed different techniques of space subdivision to reduce the synthesis time. However all these methods lead us to the following remarks:

- Glassner [24] subdivides space in octree and uses a hashing table to access a leaf (voxel). The determination of the next region traversed by the ray, requires the traversal of all the octree structure and comparisons with voxel faces. This search is expensive in time

- Kaplan [33] uses the objects enclosures to subdivide regularly the parallelepipedic bounding volume of the scene by planes perpendicular to the coordinate system axes, according to the BSP technique (Binary Space Partitioning) [22]. The data structure obtained is a binary tree whose leaves are 3D regions. Each region contains a minimal number of objects. The traversal of all the tree structure is needed to determine the region containing a point. For a scene of many tens or hundreds of objects, this traversal becomes very time consuming. For a CSG model, Kaplan proposes to compute the intersection of a ray with the primitive objects lying in all regions traversed by the ray. It will be seen later that it is not necessary to do that

- Fujimoto et al. [23] use an octree structure or a spatial enumeration structure called SEADS and a 3D differential analyser called 3DDDA, to determine the next region in the direction of a ray. They give many details about the traversal of the octree structure by means of 3DDDA. The search of the next region along the ray path is fast since it uses only integer arithmetic operations. The drawback is that the SEADS structure requires a large amount of memory and a preprocessing time that cannot be neglected

- Snyder and Barr [49] subdivide the scene enclosure into a 3D grid. Each element of the grid has a list of triangles which result from a surface tessellation. The searching of the next region along the ray path is performed by determining the intersected face and by incrementing or decrementing the three indices of the grid

- Arnaldi et al. [3] perform an irregular subdivision of the bounding volume of the scene. The subdivision result is a set of 3D regions which enclose the objects more tightly. For each region, the authors use four pointers associated with four corners of the relevant region. The meaning of one pointer is to link regions by their corners. These pointers allow the searching of the next region along the ray path

- Wyvill and Kunii [58] choose a DAG (Direct Acyclic Graph) to model a scene. The result of their spatial subdivision is an octree. Finding a voxel containing a point, requires the traversal of the octree structure. The authors require that two primitive objects do not overlap, which limits the creation of realistic and complex objects.

In order to show the details of the implementation of an accelerated ray tracing based on spatial subdivision, only one method is described. It is due to Bouatouch et al. [10].

Bouatouch's method is a space tracing technique which uses a CSG model with no restriction about the overlapping of primitive objects. With each region resulting from the subdivision is associated a subtree. The access to a region does not require the traversal of a data structure like octree or BSP tree. This method is described as follows. First we give an overview of this method and then detail the subdivision process. After that, we see how one traces rays in a subdivided space. Finally, we present some results which are followed by a discussion.

5.4.2 Overview of the Algorithm

The main characteristics of the algorithm are the following:

- Scenes are modelled by a CSG tree. Primitive objects may be solids or surfaces and the boolean operators are union, intersection and difference [9]

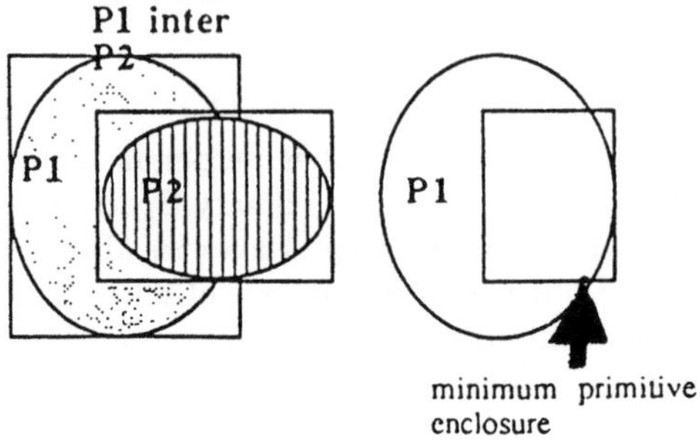

FIGURE 5.32. Minimum Enclosure of Primitive P1. Case of Two Primitives

- For each primitive, a minimal bounding volume is computed
- Like in Kaplan's method, the parallelepipedic enclosure of all the scene is subdivided recursively by planes aligned with the eye coordinates system axes. Each slicing plane divides a space into two equal subspaces (figure 5.34). The result is not a BSP tree butonly its leaves which represent 3D regions that we call boxes. A box is either empty or contains a subtree which is the projection of all the scene. A number is associated with each box
- A spatial index (that we call *SI*) which is a 3D grid, is used to determine rapidly the box containing a point in the subdivided space. Each component *SI[i,j,k]* of the grid is an integer representing the number of a box
- Given a ray, the box including its origin is found and intersection with the subtree associated is performed.
- If a ray fails to hit any primitive in a box, it must move to the next box lying in its direction. The algorithm for finding the next box is almost the same as Glassner's.

5.4.3 Space Subdivision

Subdivision Step

Before describing the spatial subdivision process, we explain the concept of minimum primitive enclosure used to find the list of primitives included in a box. A minimum enclosure bounds the part of the primitive which is effectively used. Let us consider figure 5.32 where two primitives P1 and P2 are combined by an intersection operator.

We see that it is not necessary to keep the whole enclosure of P1 since its minimum enclosure is sufficient to test if it intersects a box. Let us add a third primitive to the previous example, as it is shown in figure 5.33. Three primitives P1, P2 and P3 are combined by means of two intersection operators. We obtain a minimum enclosure of P1, smaller than that of figure 5.32.

The algorithm for computing the minimum enclosure of all the primitives of the CSG tree, is the following:

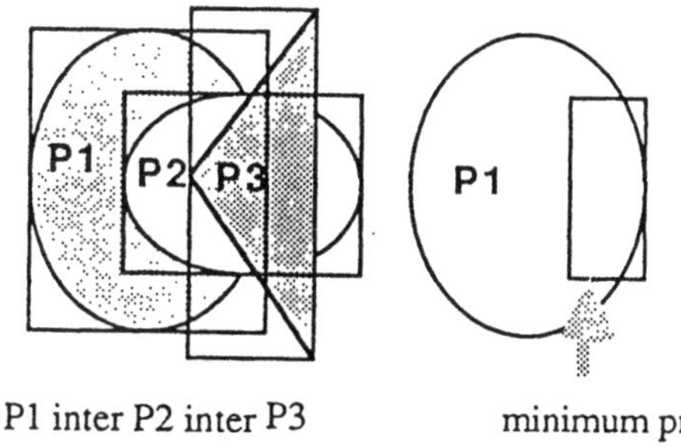

FIGURE 5.33. Minimum Enclosure of Primitive A: Case of Three Primitives

```
for each leaf P of the CSG tree do
begin
  B := P
  repeat
    if B enclosure is included in that of the parent node of P
      then do not change it
      else replace it by its intersection with the enclosure of the parent node of P
    replace P by its parent node
  until (P is the root)
  end { end of loops }
```

We obtain then a list of minimum primitive enclosures which are used for space subdivision. As mentioned previously, the space to subdivide is the enclosure of the scene which is a parallelepiped whose faces are aligned with the eye coordinate system axes. This space is sliced by planes perpendicular to the x, y and z axis. Each slicing plane divides a space (or box) into two subspaces (or boxes) of equal dimensions. The subdivision process stops either when a box is intersected by a minimum number of minimum primitive enclosures, or the maximum level of subdivision is reached with respect to each axis. These two criteria used to stop the subdivision are chosen by the user. This subdivision is similar to the one proposed in Kaplan. The difference is that the BSP tree structure is not saved. We keep only its leaves which constitute a linear array of boxes. With each box is associated a list of primitive objects whose minimum enclosures intersect it. Figure 5.34 illustrates a 2D example of subdivision where the scene contains four primitives A, B, C, and D.

Projection of the CSG Tree Representing all the Scene

Given a list of minimum enclosures, the goal is to project the scene CSG tree on each box, that is to restrict it to the primitive objects contained partly or totally in the box. The result of this projection is a subtree. This restriction is made by means of simple rules which are as follows.

Consider a node N and the two restricted subtrees NL and NR associated with N and combined by the boolean operator ope. NL and NR are respectively the left and right subtrees. Two cases are considered:

- the node is a primitive P (leaf): if P belongs to the box then the node projection is P, else it is null (null means that the box does not contain the object)

- otherwise:

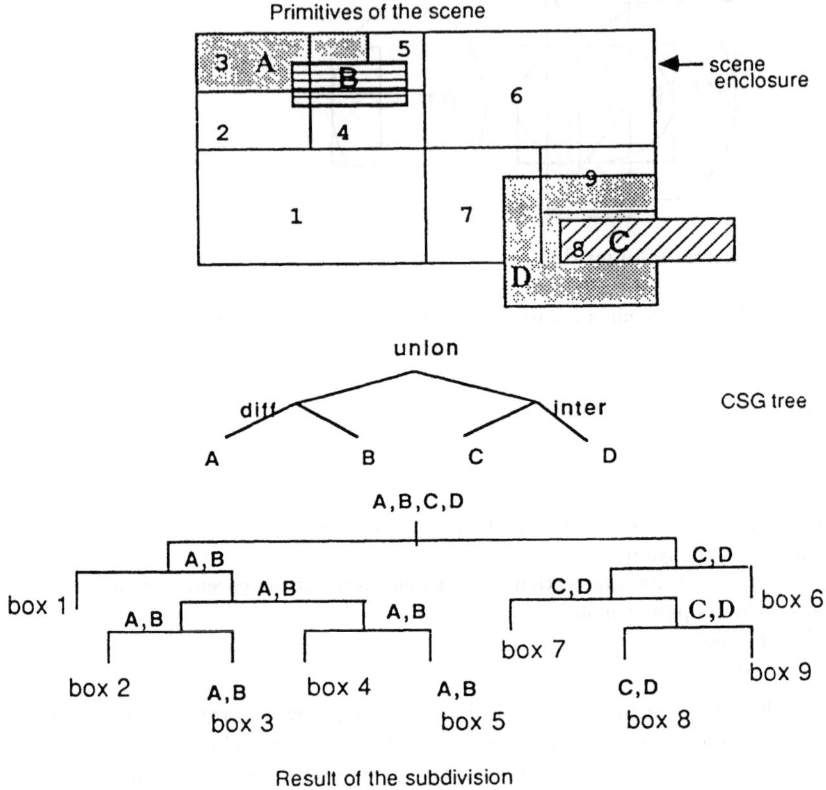

FIGURE 5.34. Subdivision Process, Nine Boxes are Found

- if NL and NR are not null then projection(node) = NL ope NR
- if NL and NR are null then projection(node) = null
- if NL is null and NR not null then there are three cases:
  * if ope = union then projection(node) = NR
  * if ope = difference then projection(node) = null
  * if ope = intersection then projection(node) = null
- if NL is not null and NR is null then there are three cases:
  * if ope = union then projection(node) = NL
  * if ope = difference then projection(node) = NL
  * if ope = intersection then projection(node) = null

A recursive procedure traverses the whole scene CSG tree to apply these restriction rules. A similar method was also described independently of Bouatouch's by Tilove [52] under the name "CSG tree pruning".

Data Structure of a Box

A box is represented by a structure constituted by the following fields:

address: a three integer components vector necessary to compute the 3D grid *SI*

level: a three integer components vector indicating the subdivision level reached according to the three coordinates axes

size: a six components vector representing the position of the box in space: xmax, xmin, ymax, ymin, zmax, zmin

ptr_list: a pointer to the list of primitive objects whose minimal enclosure intersects the box

ptr_tree: a pointer to a subtree

enclosure: parallelepipedic enclosure of this CSG tree

rec_enclosure: a 2D rectangular enclosure of this CSG tree which is used to filter the primary rays

empty: boolean taking the true value if there are no primitives in the box.

Before the subdivision process, the address of all the space to subdivide is [1, 1, 1]. Let C be the point of a box having the smallest coordinates. During the subdivision step, each address component named address.w corresponding to the slicing plane perpendicular to the w axis (where w is either x or y or z) progresses as follows:

- address.w of sub-box containing C = (address.w of parent box)*2
- address.w of the other sub-box = (address.w of parent box)*2+1

Construction of the 3D Grid *SI*

Once the space subdivision in boxes is accomplished, the construction of the 3D grid *SI* is performed according to the following algorithm written in Pseudo Pascal:

```
algorithm
{let i j and k be the three indices of the grid and B a box }

for each box B do
begin
if B.level = max { the box has a minimum size }
then begin
  i := B.address.x + 1
  j := B.address.y + 1
  k := B.address.z + 1
  SI[i,j,k] := box number associated with B
   end
else { the size of B is not minimum }
     subdivide B into minimum size boxes
     {each such box corresponds to an element of the grid }
     {compute the corresponding indices as follows}
      for xx := 0 to 2^(levelx - B.level.x)-1 do
        begin
      i := B.address.x*(2^(levelx-B.level.x))+xx+1
        for yy := 0 to 2^(levely-B.level.y)-1 do
           begin
          j := B.address.y*(2^(levely-B.level.y))+yy+1
```

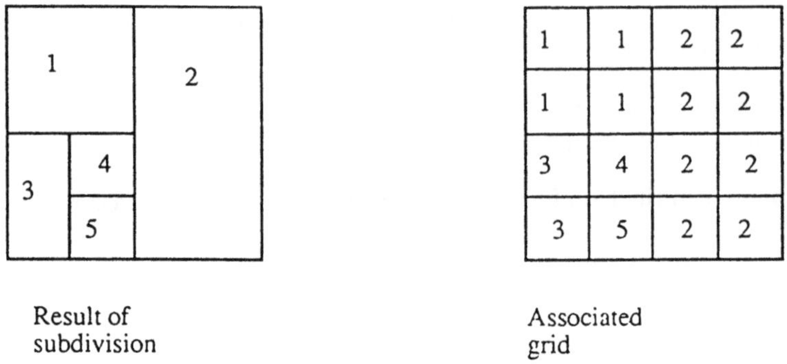

FIGURE 5.35. 2D Construction of a Grid

```
          for zz := 0  to 2^((levelz-B.level.z))-1  do
          begin
          k := B.address.z*(2^((levelz-B.level.z)))+zz+1
          SI[i,j,k] := box number associated with B
{levelx, levely and levelz are the maximum levels of
 subdivision with respect to x, y and z axis }
       end
     end
   end{ of loop xx}
end{ of loop box}
```

A 2D example of the construction of the grid is illustrated in figure 5.35.

The construction of the 3D grid is derived from the EXCELL method proposed by Tammimen [51] to minimize the computational cost of boolean operations on polyhedral objects.

5.4.4 Ray Tracing

Given a ray, the search of the box containing its origin is first performed. If the ray fails to intersect primitive objects of the box, it moves to the next box lying in its direction. This process is reiterated until no intersection is found or the ray leaves the scene.

Search of the Box Containing a Point $P\ (x,\ y,\ z)$

Let *vxmax, vxmin, vymax, vymin, vzmax, vzmin,* be the six planes bounding the scene. The contents of element grid $SI[i, j, k]$ represent the number of the box $P(x,\ y,\ z)$. The three indices i, j and k are computed as follows:

$$i = trunc((x - vxmin)/(vxmax - vxmin) * 2^{levelx}) + 1 \qquad (5.28)$$
$$j = trunc((y - vymin)/(vymax - vymin) * 2^{levely}) + 1 \qquad (5.29)$$
$$k = trunc((z - vzmin)/(vzmax - vzmin) * 2^{levelz}) + 1 \qquad (5.30)$$

If the observer does not belong to the subdivided space, it is easy to find the first box traversed by a primary ray. In this case the third index k of the 3D grid is zero.

Search of the Next Box

A ray is defined by the following parametric equations:

$$x = t * dx + x_0 \qquad (5.31)$$

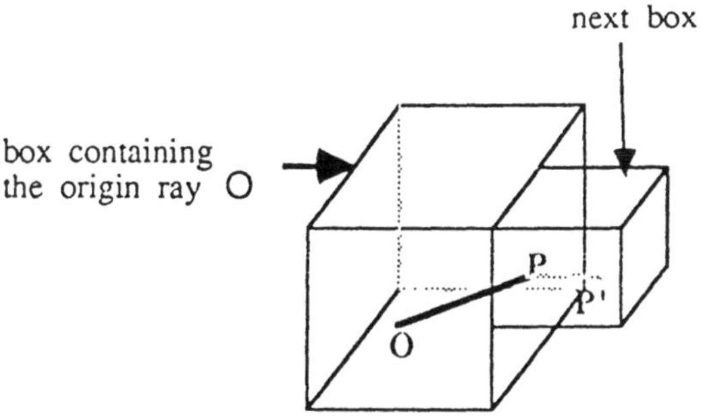

FIGURE 5.36. Search of the Next Box

$$y = t * dy + y_0 \tag{5.32}$$
$$z = t * dz + z_0 \tag{5.33}$$

t is the scalar parameter, $O(x_0, y_0, z_0)$ the ray origin and *(dx, dy, dz)* its direction vector.

The search algorithm is nearly the same as Glassner's except that only three ray-face intersections are needed instead of six:

- According to the sign of *dx*, *dy* and *dz*, we compute the intersection of the ray with the three faces of box B containing its origin: *xmin* or *xmax*, and *ymin* or *ymax*, and *zmin* or *zmax*. We then obtain three values of t
- The smallest value of t corresponds to the point P where the ray leaves box B. We then compute its coordinates X, Y, Z
- If P belongs to the face of B perpendicular to the x or y or z axis, we add or subtract to the x, y or z component of P, respectively a value *deltax*, *deltay* or *deltaz* which is equal to half the length of x, y or z side of a minimum size box. We then obtain a point P' and the box containing it. P' is only used to find the next box in the ray path. The mechanism consists in fact in pushing a point P in the direction of a face normal
- If P is on an edge or a vertex, we push it in the direction of the normals of faces sharing it. Figure 5.36 illustrates this search algorithm.

Intersection

Two bounding volumes are associated with each box: a rectangle for filtering primary rays and a parallelepiped to minimize secondary rays intersections. Note that the first one is the perspective projection of the second one. The parallelepipedic bounding volume encloses the subtree contained in the box. The intersection calculation is similar to the one proposed by Roth [47]. The difference is that the subtree is small and only its root has a bounding volume.

A primitive object may be shared by several boxes as shown in figure 5.37. To avoid computing repeatedly the same intersection of a ray with this kind of primitive, a mail box is associated with each primitive of the scene's CSG tree and a unique number is associated with each ray. When an intersection with a primitive is performed, we put the ray number and the intersection result in its mail box. Then if another intersection with this same primitive has to be performed, we test if the current ray has the same number as that in the mail box. If it is the case, we do not need to perform intersection since the

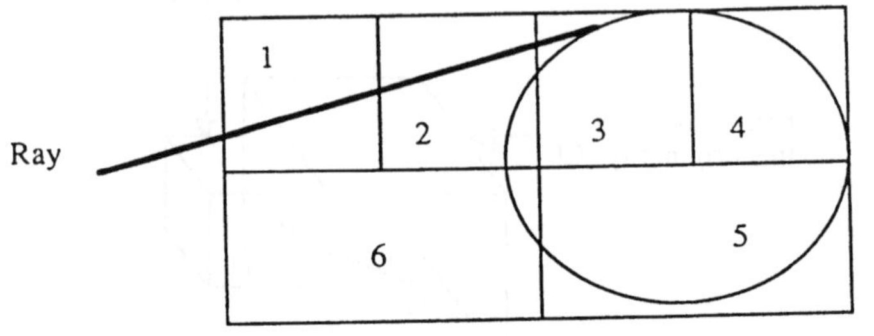

FIGURE 5.37. Example of a Ray Intersecting Boxes Containing the Same Primitive

Image	IMA1	IMA2	IMA3
Number of Primitives	32	72	12
Number of Light Sources	3	1	2
Depth of Rays Tree	1	1	5
Types of Object	sphere cylinder cube cone	sphere cylinder	sphere cylinder cube

TABLE 5.1. Characteristics of Test Images

result is already in the mail box. Otherwise intersection is performed and the result is put in the primitive's mail box.

5.4.5 Results and Discussion

The space tracing algorithm has been implemented in Pascal on a VAX 11/750 computer. Tests have been performed for three kinds of scenes called IMA1, IMA2 and IMA3. These scenes are described in table 5.1.

Table 5.2 compares Roth's algorithm to the new space tracing algorithm. The computation time ratio is 4 for a 12 primitives scene and 25 for a 72 primitives scene. This gain may be more important for more complex images. Another interesting result given by table 5.2 is that preprocessing time (space subdivision) is very small.

Table 5.3 presents the results of statistical tests performed on the three kinds of images. The mail boxes seem very efficient since they contain 40 to 50 per cent of the intersection results. Thereby a lot of intersection computations are avoided. The mean number of calls to the next box procedure seems small for both primary and secondary rays but a little larger for rays directed towards light sources, in the case of scenes containing transparent objects. In order to prove that with this algorithm, synthesis time does not depend on primitive objects number but on the type of primitives, a variable number of primitives (36, 48, 60 and 64) have been removed from IMA2 to get scenes having a variable number of primitives of the same kind.

Table 5.4 shows that for these images, the mean time synthesis per ray is relatively constant for space tracing but grows rapidly for Roth's algorithm. Also, it shows that the gain factor in synthesis time grows with primitive objects number. Other tests have shown that sometimes, increasing subdivision level is not necessary since it increases synthesis time instead of decreasing it. This can be explained by the increase in the number of calls to the next box procedure which is due to a larger number of boxes.

Image	IMA1		IMA2		IMA3	
Resolution	512*512		512*512		256*256	
Spatial Subdivision	no	yes	no	yes	no	yes
Maximum Subdivision level		4		4		3
Maximum number of Primitives per box		2		2		3
Number of Boxes		251		549		34
Subdivision Time		6s		17s		1s
Synthesis Time	13h 54m	47m	13h 44m	47m 43	7h 38m	2h
Gain Factor		17.74		25.65		3.78

TABLE 5.2. Comparison of Roth's and Bouatouch's algorithm; (h=hours, m=minutes, s=seconds)

Image	IMA1		IMA2		IMA3	
Resolution	256*256		256*256		256*256	
Subdivision Level	3	4	3	4	3	4
Number of Primitives per box	2	2	2	2	2	2
Number of Boxes	123	251	279	549	86	260
Mean number of calls to next box procedure:						
— primary rays	2.49	2.57	3.11	3.21	2.92	3.28
— secondary rays					2.5	2.98
— rays towards light sources	3.14	3.39	2.62	2.8	4.19	5.25
Efficiency of mail boxes (%)	39.6	41.7	39.8	39.97	4.19	50.2
Synthesis time	14m 27s	13m 24s	18m 26s	16m		
Subdivision time	3s 50	6s 58		19s		

TABLE 5.3. Statistical Results; (h=hours, m=minutes, s=seconds)

5.4.6 Conclusion

With this space tracing algorithm, computation time is decreased by a factor of 25 for a scene having few primitives. For more complex scenes of many hundreds of primitives, this factor would be more important. Scenes are represented by a CSG tree which is a simple and efficient way to model objects exactly. In this algorithm, the searching time of the next box lying along the ray path, is constant whatever the image complexity, whereas it is variable for space subdivision methods needing the traversal of a data structure and thereby it may be very important for complex scenes. Minimum primitive enclosures and mail boxes contribute efficiently to the occupation due to boxes array and 3D grid creation. A technique to minimize this occupation is to reduce the grid size by using sub-grids as proposed by Tammimen at al.[51] for performing boolean operations on polyhedral objects. These sub-grids are determined as follows:

- choose small values for the maximum subdivision levels associated with the three axes

Image	IMA1	IMA2	IMA3	IMA4	IMA5
Primitives number	72	36	24	12	8
Mean time synthesis per ray, case of traditional ray tracing	177	93	75	57	52
Mean time synthesis per ray, case of space tracing	10.1	7.68	8.1	8.5	8
Gain factor	18.7	12.1	9.3	6.75	6.5

TABLE 5.4. Influence of Primitives Number

- during the spatial subdivision step, when maximum subdivision levels are reached, a sub-grid is associated with each box containing many primitives.

Thereby, two grid levels are created. An element of the first level grid may point to either a box or a subgrid which points in its turn to a box.

5.5 Deformation

The ways to deform a surface and to render it by ray tracing, are presented.

5.5.1 Method

Let $\mathbf{P}(x, y, z)$ be a point on the surface and let $\mathbf{F}(x, y, z)$ be the deformation applied to this surface and defined by

$$\mathbf{P}'(X, Y, Z) = \mathbf{F}(\mathbf{P}(x, y, z)) = \mathbf{F}(x, y, z).$$

Let $\mathbf{J}$ be the jacobian of $\mathbf{F}$(x, y, z):

$$\mathbf{J} = [F_x, F_y, F_z]$$

where F_x, F_y, F_z are the derivatives of F with respect to each coordinate. Barr [6] has shown that the normal $\mathbf{N}(\mathbf{P}')$ at a point P' is equal to

$$\mathbf{N}(\mathbf{P}') = \mid \mathbf{J} \mid .\mathbf{N}(\mathbf{P}).\mathbf{J}^{-1}$$

where $\mathbf{N}(\mathbf{P})$ is the normal at point $\mathbf{P}$ and $\mid \mathbf{J} \mid$ represents the determinant of J. In addition, the tangents at $\mathbf{P}'$ (with respect to each parameter u and v) to the parameterised deformed surface $\mathbf{P}'(u, v)$, are given by

$$\mathbf{T}_u = \mathbf{P}'_u(u, v) = \mathbf{P}_u(u, v).\mathbf{J}^T$$

$$\mathbf{T}_v = \mathbf{P}'_v(u, v) = \mathbf{P}_v(u, v).\mathbf{J}^T$$

Consequently, we have $\mathbf{N}(P') = \mathbf{T}_u \times \mathbf{T}_v$

The intersection between a ray and a deformed surface may be easily performed, first by deforming the ray, then by computing the intersection between the deformed ray and the non-deformed surface. Note that the deformed ray may become a curve. If so, the intersection is performed analytically.

5.5.2 Examples of Deformation

Let **F** be a scaling deformation given by

$$\mathbf{P}'(X,Y,Z) = \mathbf{F}(x,y,z) \quad (5.34)$$
$$= [a.x, b.y, c.z] \quad (5.35)$$

where a, b and c are constants. Let the following be the ray equation expressed in the eye coordinate system:

$$\mathbf{P} = \mathbf{P}_0 + t.\mathbf{D} \qquad \text{or} \qquad [x,y,z] = [x_0, y_0, z_0] + t.[dx, dy, dz]$$

To perform the ray-surface intersection, first the ray is expressed in the local coordinate system of the surface (transformation matrix **Mel**), then it is deformed by **F**. The equation of the deformed ray then becomes:

$$\mathbf{P}' = \mathbf{F}(\mathbf{P}_0.\mathbf{Mel}) + t.\mathbf{F}(\mathbf{D}.\mathbf{Mel}).$$

Note that the deformed ray remains a straight line. After that, the non deformed surface is intersected by the deformed ray to give the value t_i corresponding to the closest intersection point $\mathbf{P}'_i$ so that:

$$\mathbf{P}'_i = \mathbf{P}_0 + t_i.\mathbf{D} = [X_i, Y_i, Z_i].$$

The point $\mathbf{P}_i$, on the non-deformed surface, which corresponds to $\mathbf{P}'_i$, is given by:

$$\mathbf{P}_i = [X_i/a, Y_i/b, Z_i/c]$$

Once $\mathbf{P}_i$ is found, we compute the normal $\mathbf{N}(\mathbf{P}_i)$ at point $\mathbf{P}_i$, then we deduce $\mathbf{N}(\mathbf{P}'_i)$ which is the normal at $\mathbf{P}'_i$:

$$\mathbf{N}(\mathbf{P}'_i) = |\,\mathbf{J}\,|.\mathbf{N}(\mathbf{P}_i).\mathbf{J}^{-1} \quad (5.36)$$
$$= \mathbf{N}(P_i).\begin{bmatrix} b.c & 0 & 0 \\ 0 & a.c & 0 \\ 0 & 0 & a.b \end{bmatrix} \quad (5.37)$$

Deformation Along the Z Axis

Let us consider the following deformation where *f(z)* is a differentiable function:

$$[X,Y,Z] = \mathbf{F}(x,y,z) = [f(z).x, f(z).y, z]$$

After applying the transformations described in the previous example, we determine point $\mathbf{P}_i$ and its normal $\mathbf{N}(\mathbf{P}_i)$ as:

$$z_i = Z_i \quad (5.38)$$
$$\mathbf{P}_i = [X_i/f(z_i), Y_i/f(z_i), Z_i] \quad (5.39)$$

$$\mathbf{N}(\mathbf{P}_i) = f^2(z_i).\mathbf{N}(\mathbf{P}_i).\mathbf{J}^{-1} \quad (5.40)$$
$$= \mathbf{N}(\mathbf{P}_i) = \begin{bmatrix} f(z_i) & 0 & -f(z_i).f'(z_i).x_i \\ 0 & f(z_i) & -f(z_i).f'(z_i).y_i \\ 0 & 0 & f(z_i,).f'(z_i) \end{bmatrix} \quad (5.41)$$

The jacobian **J** is given by:

$$\begin{pmatrix} f(z_i) & 0 & f'(z_i).x_i \\ 0 & f(z_i) & f'(z_i).y_i \\ 0 & 0 & 1 \end{pmatrix}$$

Note that the matrix **J** is evaluated for each ray.

5.6 Conclusion

On the whole, ray tracing is by no means a closed subject and many research work on the topics of both speed improvement and realism are in progress. As for the first topic, an important issue is the searching mechanism implied in the ray-object intersection procedure. This problem becomes highly critical when the scene involves large collections of primitives. The work in this field has led to impressive results owing to the space tracing approach. As Snyder and Barr have shown, the rendering of scenes involving billions of triangles is possible with this approach [49]. However, the speed improvements brought by space tracing remain limited by the computation time of the ray-primitive intersection. An important issue that remains is the k.N2 dependancy of computation time as a function of image definition (for an *NxN* array of pixels). Low sampling densities are possible with the use of stochastic sampling techniques; however, as D Mitchel points out, additional assumptions are necessary to obtain acceptable results [42]. If scattering phenomena are not accounted for, the pencil tracing approach described in [55] yields interesting results. Pencil tracing makes use of paraxial approximation theory for efficiently tracing a pencil of rays.

As far as realism is concerned, distributed ray tracing makes it possible to obtain not only "nice pictures" but also accurate representations of true scenes, i.e., photosimulation. However, as we have seen above, ray tracing techniques are effective for dealing with highly directional phenomena and their extension to global diffuse illumination is still questionable. The association of ray tracing with the radiosity approach seems to be an interesting alternative when all light transport mechanisms need to be accounted for. However, the efficiency of the proposed methods still needs to be improved and further investigations on this subject are necessary.

5.7 References

[1] J Amantides. Ray-Tracing with Cones. *Computer Graphics (Proc. Siggraph 84)*, 18(3):129–135, July 1984.

[2] A Appel. Some Techniques for Shading Machine Rendering of Solids. *SJCC*, pages 37–45, 1968.

[3] B Arnaldi, T Priol, and K Bouatouch. A New Space Subdivision Method for Ray Tracing CSG Modelled Scenes. *The Visual Computer*, 3:98–108, 1987.

[4] J Arvo. Backward Ray Tracing. Siggraph '86 Course Notes 12 - (Developments in Ray-tracing), 1986.

[5] D H Ballard. Strip Trees: A Hierarchical Representation of Curves. *Communications of the ACM*, 24, May 1984.

[6] A H Barr. Global and L Deformations of Solid Primitives. *Computer Graphics (Proc. Siggraph 84)*, 18(13), July 1984.

[7] E P Berlin Jr. Efficiency Considerations in Image Synthesis. Siggraph 85 Course Notes 11, July 1985.

[8] J F Blinn. Methods of Light Reflection for Computer Synthesized Pictures. *Computer Graphics (Proc. Siggraph 77)*, 11(2):192–198, July 1977.

[9] K Bouatouch, B Arnaldi, and T Priol. LGRC: Un Langage pour la Synthèse d'Image par Lancer de Rayon. *TSI*, December 1986.

[10] K Bouatouch, M Madani, T Priol, and B Arnaldi. A New Algorithm of Space Tracing Using a CSG Model. In *Proceedings of Eurographics-88*, 1988.

[11] C Bouville. Bounding Ellipsoids for Ray Fractal Intersection. *Computer Graphics (Proc. Siggraph 85)*, 19(3):45–52, July 1985.

[12] C Bouville, R Brusq, L Dubois, and I Marchal. Generating High Quality Pictures by Ray Tracing. *Computer Graphics Forum*, 4:87–99, 1985.

[13] C Bouville, J L Dubois, I Marchal, and M L Viaud. Monte Carlo Integration Applied to an Illumination Model. In *Proceedings Eurographics-88*, 1988.

[14] F Bronsvoort, J V Jarke, and F W Jansen. Two Methods for Improving the Efficiency of Ray Casting in Solid Modelling. *Computer Aided Design*, 16(1), 1984.

[15] M F Cohen and D P Greenberg. The Hemi-cube, a Radiosity Solution for Complex Environments. *Computer Graphics (Proc. Siggraph 85)*, 19(3):31–40, July 1985.

[16] G E Collins and R Loos. Real Zeros of Polynomials. *Computing Supplement*, 4:83–94, 1982. Also in Computer Graphics and Image Processing, 1972.

[17] L C Cook. Stochastic Sampling in Computer Graphics. *ACM Transaction on Graphics*, 5(1):51–72, January 1986.

[18] R L Cook and K E Torrance. A Reflectance Model for Computer Graphics. *ACM Transaction on Graphics*, 1(1):7–24, January 1982.

[19] S Coquillart. An Improvement of the Ray Tracing Algorithm. *In Proceedings of Eurographics - 85*, 1985.

[20] M A Z Dippe and E H Wolf. Antialising Through Stochastic Sampling. *Computer Graphics (Proc. Siggraph 84)*, 19(3):69–78, July 1985.

[21] A Fournier, D Fussel, and L Carpenter. Computer Rendering of Stochasting Models. *Communications of the ACM*, 25(6), June 1982.

[22] H Fuchs, Z M Redem, and B F Naylor. On Visible Surface Generation by a Priori Tree Structure. *Communications of the ACM*, 1980.

[23] A Fujimoto, T Tanaka, and K Iwata. ARTS: Accelerated Ray Tracing System. *IEEE Computer Graphics and Applications*, April 1986.

[24] A Glassner. Space Subdivision for Fast Ray Tracing. *IEEE Computer Graphics and Applications*, October 1984.

[25] J Goldsmith and J Salmon. Automatic Creation of Object Hierarchies for Ray Tracing. *IEEE Computer Graphics and Applications*, pages 14–20, March 1987.

[26] E Goldsteins and R Nagel. 3D Visual Simulation. *Simulation*, pages 25–31, January 1971.

[27] E Haines. Essential Ray Tracing Algorithms. Siggraph 87 Tutorial Notes, 1987.

[28] K A Hall and L P Greenberg. A Testbed for Realistic Image Synthesis. *IEEE Computer Graphics and Applications*, pages 10–20, November 1983.

[29] K I Joy. Ray Tracing Parametric Surface Patches Utilizing Numerical Techniques and Ray Coherence. *Computer Graphics (Proc. Siggraph 86)*, 20(3), 1986.

[30] J T Kajiya. Ray Tracing Parametric Patches. *Computer Graphics (Proc. Siggraph 82)*, 16(3):245–254, July 1982.

[31] J T Kajiya. New Techniques for Ray Tracing Procedurally Defined Objects. *Computer Graphics (Proc. Siggraph 83)*, 17(3):91–102, July 1983.

[32] J T Kajiya. The Rendering Equation. *Computer Graphics (Proc. Siggraph 86)*, 20(4):143–150, July 1986.

[33] M R Kaplan. Space Tracing, a Constant Time Ray Tracer. Siggraph 85, Tutorial on the Uses of Spatial Coherence in Ray Tracing, 1985.

[34] D S Kay and D Greenberg. Transparency for Computer Synthesized Images. Master's thesis, Cornel University, Ithaca, N.Y., January 1979.

[35] T L Kay and J T Kajiya. Ray Tracing Complex Scenes. *Computer Graphics (Proc. Siggraph 86)*, 20(3), August 1986.

[36] P A Koparkar and S P Mudur. A New Class of Algorithm for the Processing of Parametric Curves. *Computer Aided Design*, 15(1), January 1983.

[37] P A Koparkar and S P Mudur. Computational Techniques for Processing of Parametric Surfaces;. *Computer Vision, Graphics and Image Processing*, 28:303–322, 1984.

[38] J Lane and L Carpenter. A Generalised Scan Line Algorithm for the Computer Display of Parametrically Defined Surfaces. *Computer Graphics and Image Processing*, 11:290–297, 1979.

[39] M E Lee, R A Redner, and S P Uselton. Statistically Optimized Sampling for Distributed Ray Tracing. *Computer Graphics (Proc. Siggraph 85)*, 19(3):61–67, July 1985.

[40] I Marchal. Private Communication.

[41] G W Meyer. Wavelength Selection for Synthetic Image Generation. *Computer Vision, Graphics and Image Processing*, 41:57–59, 1988.

[42] D P Mitchell. Generating Antialised Images at Low Sampling Densities. *Computer Graphics (Proc. Siggraph 87)*, 21(4):65–69, July 1987.

[43] R E Moore. *Interval Analysis.* Prentice Hall, Englewood Cliffs, N.J., 1986.

[44] B T Phong. Illumination for Computer Generated Images. *Communications of the ACM*, 18(6):311–317, June 1975.

[45] W Purgathofer. A Statistical Method for Adaptive Stochastic Sampling. In *Proceedings of Eurographics-86*, pages 145–152, 1986.

[46] R F Riesenfield, E Cohen, and T Lyche. Discrete B-splines and Subdivision Techniques in Computer Aided Geometric Design and Computer Graphics. *Computer Graphics and Image Processing*, 14(2):87–111, October 1980.

[47] S D Roth. Ray Casting for Modelling Solids. *Computer Graphics and Image Processing*, 18:109–144, 1982.

[48] K H Sears and A E Middleditch. Set-Theoretic Volume Model Evaluation and Picture-Plane Coherence. *IEEE Computer Graphics and Applications*, March 1984.

[49] J M Snyder and A H Barr. Ray Tracing Complex Models Containing Surface Tesselations. *Computer Graphics (Proc. Siggraph 87)*, 21(4):119–128, July 1987.

[50] M A J Sweeney and R H Bartels. Ray Tracing Free Form B-Spline Surfaces. *IEEE Computer Graphics and Applications*, pages 41–49, February 1986.

[51] M Tammimen, O Karonen, and M Mantyla. Ray Casting and Block Model Conversion Using Spatial Index. *Computer Aided Design*, 16(4):203–208, July 1984.

[52] R B Tilove. *Exploiting Spatial and Structural Locality in Geometric Modeling.* PhD thesis, University of Rochester, October 1981.

[53] D L Toth. On Ray Tracing Parametric Patches. *Computer Graphics (Proc. Siggraph 85)*, 19(3), 1985.

[54] M Vincent. Sur la Resolution des Equations Numeriques. *Journal de Mathématiques, Pures et Appliquées*, 1:341–372, 1836.

[55] J R Wallace, M F Cohen, and D P Greenberg. A Two-Pass Solution to the Rendering Equation: A Synthesis of Ray Tracing and Radiosity Methods. *Computer Graphics (Proc. Siggraph 87)*, 21(4):311–320, July 1987.

[56] H Weghorst, G Hooper, and D Greenberg. Improved Computational Methods for Ray Tracing. *ACM Transactions on Graphics*, 3(1):52–69, January 1984.

[57] T Whitted. An Improved Illumination Model for Shaded Display. *Communications of the ACM*, pages 343–349, June 1980.

[58] G Wyvill and T L Kunii. A Functional Model for Constructive Solid Geometry. *The Visual Computer*, 1:3–14, 1985.

6 Rendering Techniques

Tom Nadas and Armand Fellous

ABSTRACT
This tutorial will provide a general overview of the traditional rendering process and both the theory and application of the most common techniques in use today.
We will describe the basic stages of traditional rendering along with techniques commonly used to obtain more realistic images. Topics covered will include geometric transformations, visible surface determination, sampling and filtering, shading, mapping techniques, shadow generation, motion blur and atmospheric and other special effects. In addition, a number of examples and case studies will be used to illustrate specific problems and their solutions. Finally, methods and systems which allow the user to obtain maximal rendering capabilities will be discussed and illustrated.

6.1 Introduction

The creation of a computer generated image is essentially a two stage process. First the image is defined, or *modelled*, and then, a computer generates, or *renders*, it.

The definition of the image, or images in the case of animation, is often a long and involved process. A geometric description of each object must first be created. Also, an appearance or material description of each object must be defined. Then, the relative size and position of all objects in the scene must be chosen, along with some special objects such as lights. Sometimes, special rules, which define how objects visually interact, are also added. This results in the description of an imaginary world. viewing this world is via a *camera* built especially for this purpose. This camera takes the world description and generates an image.

A *renderer*, or rendering program, may therefore be thought of as a camera. However, unlike real cameras for which the laws of light are predefined, renderers internally define their own physical laws of appearance. It is, in fact, these laws that govern the way in which the created world is originally described. This close relationship is why interactive computer tools are used during the modelling stage, since that will ensure that the world description is one that can be understood by the renderer.

Although renderers come in many types, most follow some of the same basic algorithms. The following pages will describe the basics of traditional rendering programs and the techniques that they use.

6.1.1 Modelling

Before creating an image, the *world* must first be created in a particular fashion. The first step is to create a *model* of the objects in the world. A model is a description of an object consisting of sufficient information so that it may be suitably displayed. Although it is not often stressed, a single object model may be divided into a number of sub-models.

The most obvious is the *geometric* or *physical model* which defines the physical shape or geometric properties of the object. This is the type of model usually referred to in the computer graphics literature when the term *modelling* is used.

Next, is the model that determines the object's appearance, which we shall refer to as the *visual model*. Basically, it describes what the object looks like, excluding its geometry.

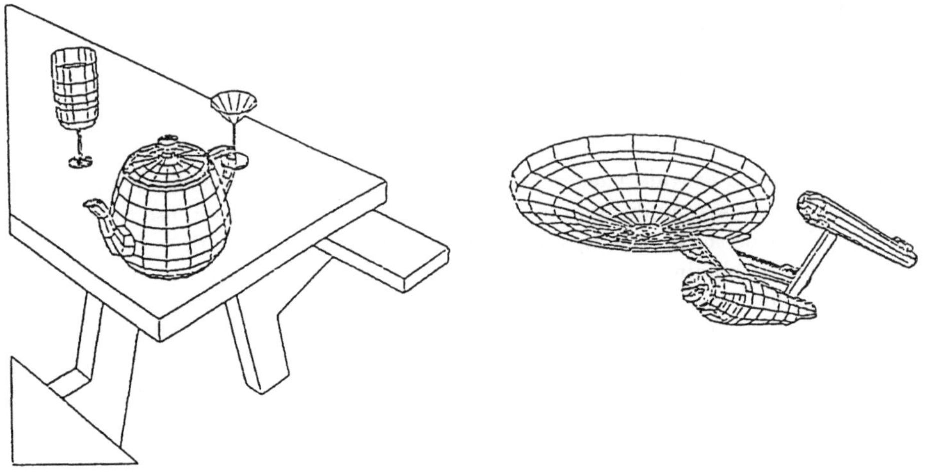

FIGURE 6.1. Here the versatility of polygons is illustrated. Curved surfaces may be approximated by a large number of small polygons in a mesh

A visual model usually consists of all the material and surface properties that affect the object's interaction with light, such as the material of which the object is made and the smoothness of the surface finish. The boundary between physical and visual models is a fuzzy one, since some visual models may be used to create the illusion of a geometric model. The distinction will become evident later, since models can be categorized in this way according to the stage of the rendering process in which they are used.

Finally, *temporal models* are used in applications such as simulation and animation. They describe the interactions between objects, and the environment. Properties such as the mass of the object, its attractive force, its elasticity, the strength of its joints, and any forces which act upon it may all determine an object's geometry and appearance with respect to time. However, for any given instant in time, the object may be sufficiently described by the (potentially altered) physical and visual models.

Physical Modelling

The three dimensional representation of 3D objects is accomplished using *modelling primitives*. These are simple entities that may be combined to produce complex entities. By analogy, the atomic elements are the primitives of all matter. Although there are many types of primitives in use in computer graphics, one of the most common is *polygons*: a planar collection of vertices and edges in 3D space that define a closed boundary. By connecting together a number of polygons into a so-called *polygon mesh*, one may create almost any shape. Polygon meshes are ideal for modelling objects composed of a number of flat surfaces, however, curved surfaces can only be approximated. By increasing the number of polygons used in the mesh, the approximation can be made as close as desired. However, a fine mesh with a high polygon count puts larger data storage demands on the system, increasing total polygon — and therefore image — processing time.

The common solution is to use a mathematical description of a curved surface known as a *parametric bi-cubic patch*, which are usually based on Bezier or B-spline surfaces [3]. In this way, only 16 *control points* are required to define a single patch and a mesh of patches can describe rather complex smooth surfaces, with far less information than a polygon mesh of comparable detail.

Both polygon meshes and parametric patches provide a boundary or surface representation of solid 3D objects. Another method is to use a solid representation, as in *solid modelling.* Its primitives are simple solids such as cubes, spheres and cones which may be combined using a number of Boolean operations such as union, intersection and difference. Solid modelling is mostly used in Computer Aided Design systems.

Although these modelling techniques may be used to model a large class of objects, they are not suitable for modelling rather complex natural objects such as rocks and trees.

Stochastic modelling uses a controlled stochastic, or random, process to create random, irregular features in objects [24]. Simple primitives, such as triangles, are automatically generated and stochastically arranged under loose control. Using this process a number of natural terrain models have been produced.

A variation of stochastic modelling is the use of *particle systems* [41]. Here a large number of particles are stochastically assigned values such as position, velocity, colour, and so on. In this way, natural objects such as trees and grass, as well as phenomena such as fire, may be more easily modelled.

Visual Modelling

Visual modelling defines the surface properties or *appearance parameters* of an object. In general, these properties define the material from which the object is made, and are used to determine the surface's interaction with light.

Appearance parameters include colour, shininess, surface roughness, reflectance properties, and anything else of a purely visual nature other than geometry.

At least one set of these parameters is required per object to define it appearance, however this will give an object a homogeneous look. Visually complex objects may require a larger number of parameter sets per primitive, which is more difficult to handle.

Texture mapping solves this problem by providing a method of applying a range of parameters onto a modelling primitive. When mapping 2D texture to 3D objects, an object is *wall-papered* with an image or pattern, for example, wood grain [6]. 3D texture mapping is similar except that the texture is now defined in 3D space and the object is *sculpted* from it [28, 35, 36].

Another aspect to visual modelling is related not to the object but rather to the environment. The types of lights and their locations play an important role in the appearance of the objects. In fact, lights should be considered a special type of object. Although they are usually defined as a point or infinite source, lights are sometimes given physical properties such as is the case with area or spherical sources. Other light parameters include colour, intensity, direction, and spread functions.

Other models, non-object-related, include atmospheric effects such as fog and smoke depend on the relative position of an object and observer, while those such as rainbows also depend on the relative position of the lights. Since these affect the appearance of the final image, as well as the objects within the image, they are considered a part of visual modelling.

Finally, camera effects should also be considered part of visual modelling since they also affect the appearance of the final image. Most computer graphics have been done using a pin-hole camera model, although models that take the aperture size and exposure time into account exist.

Although visual modelling is usually considered secondary to physical modelling, it is just as important. In some cases a good visual model can substitute for a more elaborate physical model. For example, a marble floor may be modelled by a large number of mono-coloured polygons or more simply with a single texture mapped polygon.

This illustrates an important point. In many cases the same result may be obtained by using either physical-modelling techniques or visual-modelling techniques. If physical modelling is used, the number of modelling primitives is always greater. This means that modelling primitive computations are increased while the individual pixel computations remain relatively simple. When visual modelling is used, the modelling primitives computations are simplified but at the expense of the individual pixel computations. In general, the incremental expense of a pixel computation is more than the incremental expense of a modelling primitive. However, if the model is complex enough, which is often the case, the total computational complexity is less using visual-modelling techniques.

6.1.2 Display

Once the basic objects have been modelled, they become the primitives of the *world*. They are placed into their final positions with *geometric transformations*, which in animations, generally change from frame to frame.

At this point, the modelling of the objects and the scene have been set and are ready to be rendered, which means reducing all the information into an array of pixel colours.

This process includes two major steps. The first is the *visible surface determination* stage, which decides which objects or parts of objects are actually seen. Typically something in an image occuldes something else; at the very least, the front of an object occludes its own back. The goal is to produce a set of appearance parameters for each pixel containing sufficient information from which to determine the pixel colour.

The second stage is *pixel colour determination*: transforming the appearance parameters into the final pixel colour. Computations performed include *shading*, which determines how much light is reflected off the surface, as well as any texture mapping required.

6.2 Visible Surface Determination

Visible-surface determination uses the information from the physical model (which, for our purposes, includes the transformations used to arrange the scene), while pixel-colour determination mainly uses the information from the visual model.

Visible surface determination algorithms vary, and may be categorized by their *process flow*, that is, the order in which they process information. The *traditional* process flow proceeds from modelling to geometric to display operations, as illustrated in figure 6.2. Most general rendering systems make use of this type of process flow, transforming object data into an image in a step-by-step process. The most pervasive of these are either *depth-buffer* or *scan-line* based. These traditional renderers can process the basic types of physical and visual models, and are flexible enough to be used in a number of applications.

6.2.1 Model Operations

Usually each object is defined in its own unique coordinate system. These must be combined to form one over-all coordinate system. These transformations are described during the scene modelling process, and are usually applied during the rendering stage. They are known as *instance transformations* since they take the master object described in model coordinates and create one or more instances of it in the *world coordinate system*.

In the world coordinate system the entire scene is described and the objects are positioned and scaled relative to each other. Also, it is usually the world coordinate system in which special objects such as lights and the camera are described.

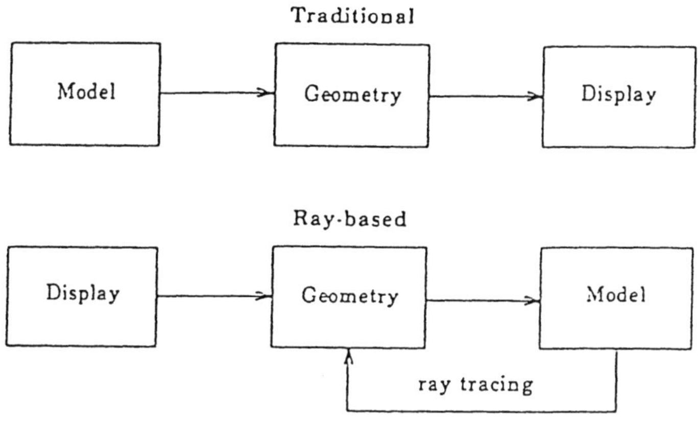

FIGURE 6.2. A comparison of various process flows

Any number of instances may be created from a particular master, each of which may have a unique mapping into world space. Internally, an instance is simply represented by a transformation matrix, and it is the master that contains most of the geometric information as well as all of the appearance parameters associated with that object, such as the texture coordinates and the shading parameters.

During instance transformations, the modelling primitives are converted to the *rendering primitives*, which are in a form that may be directly displayed. While rendering systems that can process many more than one type of rendering primitive exist, they are usually elaborate and constrained. However, if all the different kinds of modelling primitives are reduced to the same rendering primitive then they may all be treated in the same way, thereby reducing some of the complexity of the software.

Due to the relative, simplicity, most systems use general polygons as their rendering primitive. In fact, some systems further reduce complexity by limiting these polygons to be triangles. For most modelling primitives, this is not too much of a constraint since most can be broken down, or *subdivided* into a polygonal or polygon-mesh representation.

The *subdivision level* of a curved surface determines the number of polygons used to approximate it. The simplistic approach is to set the subdivision level at some user-defined constant. The problem lies in the fact that highly curved surfaces require such high subdivision levels that the number of polygons to be rendered can be quite large for even simple objects. Also, surfaces of lower curvature in the image would require a much smaller number of polygons to provide the same appearance.

Adaptive subdivision is a method used to reduce the polygon count by varying the subdivision level automatically according to surface curvature measurements. An example of this process for a patch is to recursively apply the procedure: if the patch is planar enough, then transform it into a polygon else subdivide it into four sub-patches.

In addition, by performing the subdivision after the viewing transformation (see below), the size of the surface in the final image may also be taken into account. For example, a patch of high curvature spanning only a few pixels on the screen need only be subdivided into a small number of polygons. In this way, the number of polygons required to represent a surface adequately may be minimized.

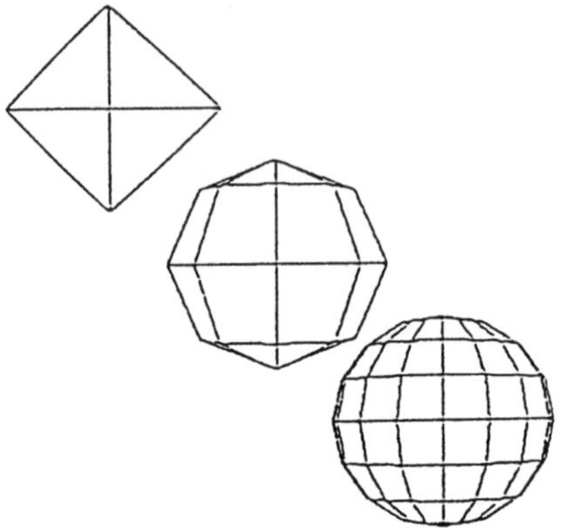

FIGURE 6.3. The objects above are all the same procedural model of a sphere. The only difference between them is the subdivision level

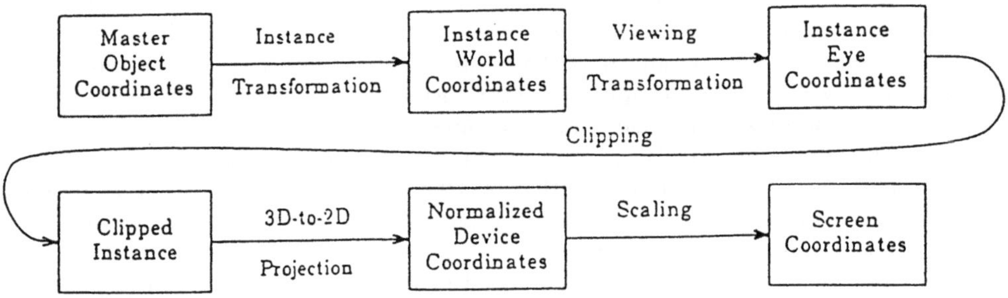

FIGURE 6.4. The transformation pipeline

6.2.2 Geometric Operations

Once all the primitives have been established in the world coordinate system, the rendering process continues to apply a number of geometric transformations which collectively are referred to as the *transformation pipeline* (figure 6.4). Usually, the instance transformation is considered the first stage to the transformation pipeline. However, it is really a modelling operation, which unlike the rest of the pipeline, must be performed regardless of the rendering algorithm. For simplicity, the rest of our discussion will assume that the instances have already been converted into a polygonal representation.

The geometric-transformation pipeline converts the 3D world-coordinate polygons into 2D device-coordinate screen polygons in confined and defined screen space.

The world polygons are first transformed by the *viewing transformation* so that they are described in the *eye-coordinate system.* The *eye*, or camera, is usually defined by 1) a position in world space, 2) a *look-at-point* which, along with the position, determines the line-of-sight direction, and 3) a vector that specifies an up direction to determine the rotation along the line of sight (figure 6.5). In the eye-coordinate system, the eye is at the origin and the line of sight is along an axis, usually the z axis.

At this point, the *clipping transformation* and *3D-to-2D projection* are performed. The clipping transformation removes those polygons and parts of polygons that cannot be seen through the specified *window*, the 2D surface onto which the control points are to be projected. The window is usually a rectangular plane restricted to be centred along,

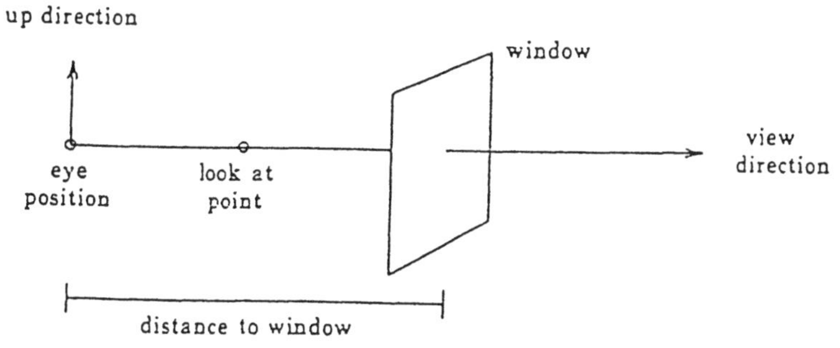

FIGURE 6.5. An example of how the eye and window may be defined

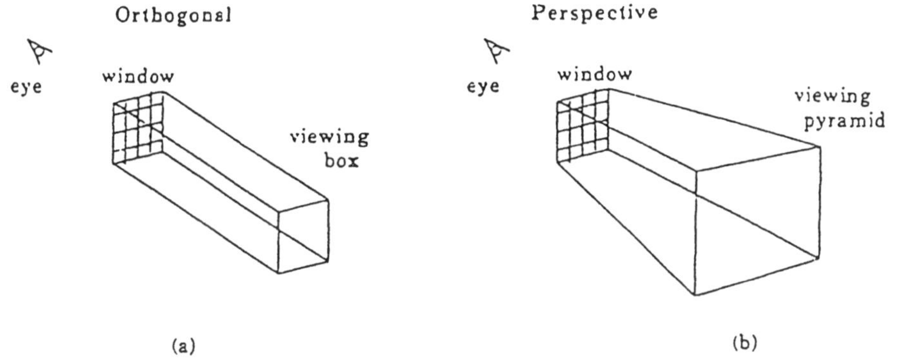

FIGURE 6.6. An orthogonal and perspective projection indicating the view volumes

and perpendicular to, the line of sight. In the eye-coordinate system the window has sides parallel to the x and y axes centred along the z axis.

The 3D-to-2D projection is generally one of two forms: *orthogonal* or *perspective*. In the orthogonal projection, each vertex is mapped onto the window with the projection parallel to the z axis. Those polygons lying outside the viewing volume or box, an extrusion of the window (figure 6.6), are rejected, or clipped, since they will not be part of the final image. In the cases where only part of a polygon can be seen, a new sub-polygon, which consists of the part within the viewing volume, is created to replace it.

The extent of the volume is determined by hither and yon clipping distances. These may be used to provide cross-sections of objects or limit the number of objects visible, however, their main purpose is to define a depth resolution for the viewing volume. Vertices within the viewing volume are normalized so that they are defined within a unit cube.

Perspective projections are more common. Here each vertex is mapped onto the window by a projection centred at the eye (the origin). A viewing pyramid is formed between the origin and the lines extending through the corners of the window (figure 6.6), and the mapping depends on the distance of the window, d, and the z value of the vertex, and is given as $x_p = x(d/z)$, and $y_p = y(d/z)$. Along with the original z, this mapping has the effect of converting the viewing volume from a pyramidal to rectangular prism shape. The polygons may now be clipped and normalized as above.

At this point polygons are defined in *normalized device coordinates*, and a simple scaling operation converts them to physical *screen coordinates*. Note that the precision of the screen coordinates must be maintained to assure accuracy in subsequent operations. Hardware display systems often use integer values, thus detracting the image quality. For the visible surface problem, the precision of the z values is of particular concern, and for this reason, care must be taken in choosing the hither and yon clipping planes.

6.2.3 Display Operations

The previous operations have now simplified the task of finding the visible surfaces. We are left with only those polygons completely confined within the screen boundary and the perspective distortions already taken into account. Therefore, the viewing volume need only be projected on the screen with a depth priority given by the z values.

Although this may be accomplished using a variety of techniques, the *z-buffer* and the *scan conversion*, or scan-line, algorithms are among the most common.

The Z-Buffer Algorithm

The z-buffer (or depth-buffer) algorithm is basically the simplest method of determining the visible surface. It stores the depth information in a raster array, a z-buffer, so that the final pixel with consist of a colour and a z value.

In any order, each polygon is *scanned* out onto the raster array. This process involves drawing the polygon onto the raster buffer one pixel at a time, calculating the pixel depth by an interpolation of the vertex depth. However, before actually writing the pixel into the frame buffer, the depth value of the destination pixel is read. If the new pixel is closer, it overwrites the pixel in the frame buffer, otherwise, it is discarded. If the frame buffer is initialized with the farthest possible depth value, and all the polygons are processed in this way, the result will be a raster image that displays only the visible surfaces.

As each polygon is passed to the z-buffer algorithm, it is usually scanned out in *spans*. Spans are solid segments of polygons that are parallel to the scan lines on the screen; they are polygon segments bounded by the polygon edges. The scanning out process involves an interpolation of the vertex depth values along the polygon edges from scanline to scanline, as well as between the two ends of each span. If one assumes that the polygon is planar, then these interpolations may be linear. With a triangular polygon limitation, this assumption is always true. A more complicated case deals with the colour assignments of each pixel during the scan-out phase. However, this will be discussed later in more detail. The z-buffer algorithm is used extensively due to its simplicity in implementation and speed of execution. Due to the nature of the z-buffer algorithm, since the polygons are scanned out in the order in which they are passed, the output raster device must be a random-access frame buffer. Therefore the z-buffer algorithm must either uses a physical frame buffer or a virtual frame buffer in memory. The disadvantage of using a physical frame buffer is the fact that the bit resolution may be sufficient to distinguish all possible depth values which may result in incorrect visible-surface determination. To overcome this, the colour value may be stored in the frame buffer, and the depth values may be recorded in a virtual frame buffer.

Since the polygons are scanned out in random order, there will be many cases of calculating the pixel colours of surface that will be hidden later. This is a major efficiency problem, since pixel colour determination is usually computationally more expensive than the visible surface determination [44]. It may be solved by either sorting the polygons from front to back ahead of time, or creating a multiple pass z-buffer in which the first pass determines only the visibility, deferring pixel colour determinations to successive passes.

It should be noted that the z-buffer is basically a point-sampling algorithm. Since there is one-to-one correspondence between pixels and samples, the basic algorithm may be executed on the internal processor associated with the frame buffer. This allows the z-buffer algorithm to run quickly.

However, a major problem with point-sampled images is that they suffer from defects often referred to as *jaggies*. Within a pixel, polygon segments might not span the entire area. As described later, to more correctly determine the final pixel colour, the pixel *coverage* of all visible polygon segments within the pixel must be used. However, due to the random polygon scan-out order, such total pixel information is not available. For this reason, z-buffer based algorithms are often not used where high picture quality is required.

Scan Conversions

Due to the inadequacies of the z-buffer, many have turned to the *scan-line algorithm* as the solution to the visible-surface-determination problem. Although there are a number of different scan-line algorithms, they all share a few basic properties, and their differences lie in the way information is internally stored [45]. The premise is to order the pixel determination process; that is, the screen is processed in scan-line order, as opposed to polygon order.

The following is a description of the Watkins scan-line algorithm [50], which is simpler and quicker than some others since it assumes that all input primitives are simple planar polygons. The basic algorithm uses polygon edge lists, as opposed to the polygon lists of the z-buffer algorithm. Each edge is made up of the vertex information on its two ends, and is tagged to identify the originating polygon. These edges are first sorted by the minimum screen y values of there vertices. Then, scan-line processing proceeds in order from the bottom to the top scan line. Note that the edges may alternatively be sorted by the minimum y values, in which case, the scan conversion proceeds from top to bottom. Initially, all the edges are tagged as active if they intersect the present y scan line. By processing from bottom to top, edges become active once, remain so for one or more scan lines, and then become inactive and may be discarded. For polygons, since each vertex bounds two edges, edges become active and inactive in pairs. Each pair of active edges with the same polygon tag bounds a span, the end points of which are interpolated vertex values of the corresponding edge. So that these values may be calculated incrementally, the edge information also includes delta values used to determine the next scan line's edge values.

For each scan line, a list of spans is created, one for every pair of active edges. By cutting the scan line at active span boundaries, and comparing span depths at these points, one can create a list of visible spans. For each pair of adjacent span boundaries, if the closest span is the same at both ends then it is added to the visible span list. If they are different, then the two spans intersect, and these spans are subdivided at the intersection point. Virtual edges are created, one for each span, to keep the edge count in pairs and to minimize intersection calculations for subsequent scanlines. The resulting visible span list is guaranteed not to overlap, and is the solution to the visible surface determination problem for the scan line. Finally, at the pixel level, the visible spans are simply the interpolation, along the x-axis, of the span end points, and may be done within the pixel-colour-determination stage.

There are inherent properties of scan conversion that allow for increased efficiency. The most obvious is that depth comparisons are only made at span boundaries — as opposed to at every pixel in the span — and spans are generally many pixels in length. However the greatest performance savings can be made by taking advantage of any coherence between consecutive scan-lines. In a typical image, the visible-span lists of any two adjacent scan lines may be very similar. In these cases most of the work may be carried over from the previous line, and the visible surface computation is reduced to only calculating the differences between scan lines and inserting and deleting spans as required. Unfortunately, the efficiency of such methods is much reduced for images containing many small polygons at near-pixel size or smaller.

A clear advantage over the z-buffer algorithm is that since the result of the scan-line algorithm is a complete visible-span list, once can minimize the pixel-colour-determination calculations, since they will not be done for any hidden surfaces. The z-buffer algorithm can be better where pixel colour determination calculations are minimal, since visibility determination is faster, or where memory constraints prohibit the use of scan conversions, since the information must be sorted. In general, however, this is not the case.

Finally, there are a number of techniques to improve image quality or to add certain effects, such as transparency, that require complete pixel information (i.e., information from all visible polygon segments that effect that pixel) to correctly calculate the pixel colour. Such algorithms are generally more easily adaptable to scan-line algorithms than z-buffer algorithms, because processing order is predictable for scan conversion.

Some of the major difficulties with scan-line based algorithms originate from the fact that the entire data base must be sorted. For large data bases, the amount of time and dynamic memory required for this task may become significant. As the demands for more complex objects are placed on graphics systems, the number of raw primitives required to be rendered per image may increase at dramatic rates. Images of rocky landscapes, forests of trees, and beaches with crashing ocean waves can contain primitives that number in the hundreds of thousands. For such images, since each primitive becomes very small, the advantages of pixel and scanline coherence are lost.

Ray-based Systems

In the quest for image realism, algorithms have been developed to treat the special visual cases of shadows, and reflection and refraction effects. Although computationally expensive, the base algorithm, known as *ray casting*, is beginning to be more commonly used to solve the visible surface problem.

Ray Casting

Ray casting is based on a process flow which is almost opposite to that of the traditional algorithms (figure 6.2). Instead of starting with the raw model data and scene description, ray-casting methods start at the display.

A *ray* is a straight line in 3D space which represents the path that a ray of light might take. The whole idea of ray-based algorithms is to back trace the virtual light rays that enter the eye. Once their origin has been determined, their colour may also be determined.

Ray casting takes place in the world coordinate system, so instance transformations must first be performed. Processing continues for each pixel on the display screen, as opposed to each primitive in the world.

The screen coordinates are transformed to the window in the world coordinate system. Rays are then *fired* from the eye position through these virtual pixels. There is no need for a perspective projection transformation of the objects in the scene, since it is this variation of ray angles that provides the perspective. If all the rays were perpendicular to the window, and therefore all parallel, this would provide an orthogonal projection. Clipping is achieved without additional work since one simply does not fire rays outside the window. Therefore, the geometry in ray casting has been reduced to determining which primitives are intersected by the ray and which of these is closest to the eye.

First, by the use of bounding boxes around object and other techniques, a list of primitives that the ray potentially intersects is made. Each of these primitives is tested for intersection, and the depth of any intersections are recorded. A sort of these depths yields the solution to visible determination.

The corresponding model is examined to retrieve the appearance parameters for this point and the final pixel colour can be determined by the traditional techniques described below. Since the process is done on a pixel basis, one can acquire complete pixel information to provide most of the same advantages of the scan-line algorithm over the z-buffer.

Although ray casting is intuitively simple, the overall processing time can be quite long. Although more suited to objects defined by constructive solid geometry (i.e. solid modelling), the process is usually expensive because general intersection calculations are rather intensive.

Also, ray casting does not easily lend itself to coherence techniques. One intuitive reason for this is that each step in the traditional process described above simplifies the geometric information available for the next step. However, in ray casting the original geometry of the world scene is retained for each ray cast.

Finally, it should be noted that where the number of computations in traditional visible surface algorithms are mostly dependent on the number of primitives, ray casting's computations are mostly dependent on the number of rays cast (i.e. the number of pixels in the image). This means that ray casting rendering time approaches that of scan-line for the rendering of highly complex images.

Ray Tracing

An extension of ray casting is *ray tracing*, which takes advantage of the fact that the original geometry is retained by applying it to other phenomena, most notably shadows, reflections and refractions. As with ray casting, the process flow of ray tracing goes from display to geometry to model, however, a feedback loop is added from the model to the geometry to allow a continuous process flow (figure 6.2).

In ray tracing, the visible surface determination is only the first step. At this point, by casting a ray from the intersection point towards the light source, one can determine whether or not that surface is shadowed from that particular source. If the ray intersects a primitive before it intersects the light, then it is in shadow, otherwise it is not. Reflections and refractions are simulated by recursively casting rays in the reflection and refraction directions and blending the resultant colours from these rays accordingly.

Although the resulting images tend to be strikingly realistic, the complexity of ray tracing over ray casting is greatly increased. This is because the number of rays grow linearly with the number of light sources and grows exponentially with the levels of recursion.

6.2.4 Special Cases in Visibility

There are a number of cases that complicate the visible surface determination. These may be either due to visual effects desired or to the limitations of a model.

Transparency is a visual effect that causes more than one surface to be seen at a given point. Finding the closest polygon to the eye does not completely satisfy visibility when the polygon is not totally opaque. Instead, the visible surface determination phase must find *all* surfaces up to and including the closest opaque surface in order to determine the final pixel colour.

The colour of a point on a transparent surface is first calculated as if the surface were opaque. Then, the transparency level[1] is used as a factor in blending the colour with that

[1]The transparency level is usually calculated as a function of the overall object transparency and the angle between the surface normal and the eye viewing direction. In this way, the angular dependencies of reflection vs. transmission and attenuation due to thickness may be simulated.

of the next closest surface [30]. This means that the effects of each visible surface on a pixel must be computed in order from back to front.

This requirement causes the most problems for the z-buffer algorithm. However, the visible surface and colour determination phases are more clearly separated for scan-line and ray-based algorithms. In the scan-line algorithm, all non-opaque spans in front of the closest opaque spans must be added to the visible span list in depth order. Ray casters must also keep a similar list of intersection points. In this way, transparent surfaces may be included in object models.

It should be noted that these are *ideal* transparent surfaces which cannot cause any refraction, since it would require more than just the depth information at the one pixel. Ray tracing is the only way to retrieve proper refractions.

Some other visual effects require additional information which is not normally calculated. For example, a gas volumetric object may be created by using the visible object *depth*, not surface, information to determine the objects transparency level. Such information must be specially calculated for those objects which require them, and once again, this proves to be the most difficult to do with the z-buffer.

Finally, there exists a number of types of objects that cannot be rendered using the above methods. It would either be completely impractical or even impossible to break them down into polygons. Examples of such objects are those made with particle systems and blobby spheres. Such objects require renderers which have been tailored especially for them. This means that these objects must be rendered separately and added, or *composited*, into the final image either as a post process [37], or during rendering in a common z-buffer [19, 20].

6.3 Pixel Colour Determination

After the visible surface determination, the next most important task is that of determining the pixel colour. In fact, in almost all cases, it is the pixel colour determination phase of a renderer in which most of the computation time is spent.

Most of the final pixel colour depends on the *surface colour calculations* which transform a surface's appearance parameters into a surface colour. Depending on the algorithm used, they are performed before, during, or after visible determination. The surface colour depends mostly on the *shading* calculations which takes into the effect the presence and relative positions of the light sources. During these calculations, *texture mapping* techniques may be used to alter any number of appearance parameters.

The actual *pixel* colour depends on the colours of *all* surfaces within the pixel. For simplicity, the visible surface algorithms above have been presented as if there were only one visible surface per pixel, however, there may be a number of surfaces visible within a given pixel. This may be due to the presence of transparent surfaces, but more often it is due to a surface's partial coverage of a pixel's area. In order to include such information when determining the final pixel colour, *anti-aliasing* techniques are performed.

6.3.1 Aliasing

When a rendering system produces an image, what is actually happening is that a scene in the continuous world coordinate system is transformed to a discrete pixel array. This transformation is achieved by point sampling the continuous space at discrete intervals. As a result, obvious defects in the image are produced, and are most pronounced along the edges of objects, which appear to be jagged (figure 6.7). These defects are due to a sampling problem known as *aliasing*. Dealing with aliasing is of primary concern since

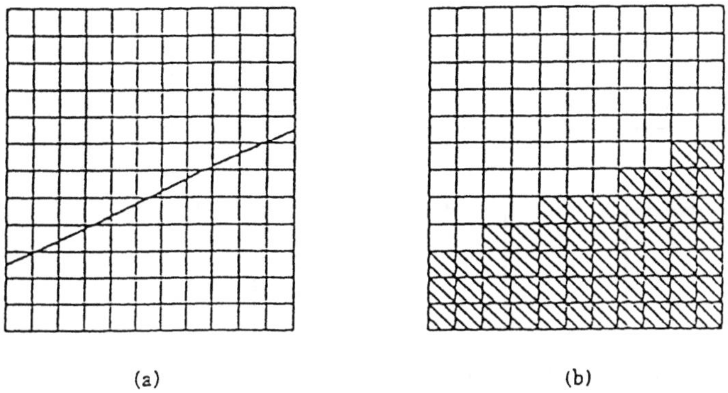

FIGURE 6.7. An edge is overlayed over a pixel grid in (a). In (b), the resulting jagged edge is illustrated for point samples taken at pixel centres

these *jaggies* can be quite disturbing, especially in animations where they produce scintillation effects. Adding more advanced features such as shadows and reflections are not very useful in image synthesis unless aliasing is addressed.

To fully understand aliasing and its solutions we must first examine sampling theory. When an analogue signal is digitized it is point-sampled at regular intervals. The sampling frequency is determined by the size of these intervals, which in this case is determined by the distance between pixels. The goal is to sample the signal so that the original signal may be reconstructed from the samples alone.

Shannon's sampling theorem — the basis of sampling theory — states that a signal which is strictly band-limited can be recovered from samples spaced at intervals taken at more than twice the highest frequency in the signal spectrum, the *Nyquist* sampling rate [40]. This may be understood by first noting that any periodic function can be represented by a *Fourier series*, a sum of sinusoidal functions of various phases, amplitudes and frequencies. For non-periodic functions, an extension of the Fourier series, a *Fourier transform* function, is used. This function is a representation of the same function in the *frequency domain*, that is, it is a function of frequency. Shannon's theory states that to properly reconstruct a function one must sample it such that there is at least one sample for each peak and valley of each sinusiod in the Fourier series of the function, so that the potential effect of each sinusoid is recorded in the samples. This makes the minimum sampling rate twice the highest frequency. Of course, the assumption is that the original function must have a Fourier series consisting of a finite number of sinusoidals, so that the highest frequency is known. This is what is meant by *band-limited*.

If a signal is sampled at a frequency lower than the Nyquist rate, then aliasing problems occur since the potential effects of the higher frequencies will not be interpolated correctly. Therefore, aliasing problems are most pronounced when sampling signals with significant amplitudes at high frequencies, since it is these frequencies which may potentially be aliased as lower ones (figure 6.8).

An image may be defined as an intensity function in two dimensions $I(x, y)$. For a colour image, there is a separate intensity function for each component of the colour space. The rate at which the intensity function changes at any point is proportional to the frequency at that point. Therefore the areas of high frequencies in an image appear as rapid colour changes. This means that aliasing problems are always expected along those polygon edges which cause colour discontinuity, such as those representing object silhouettes.

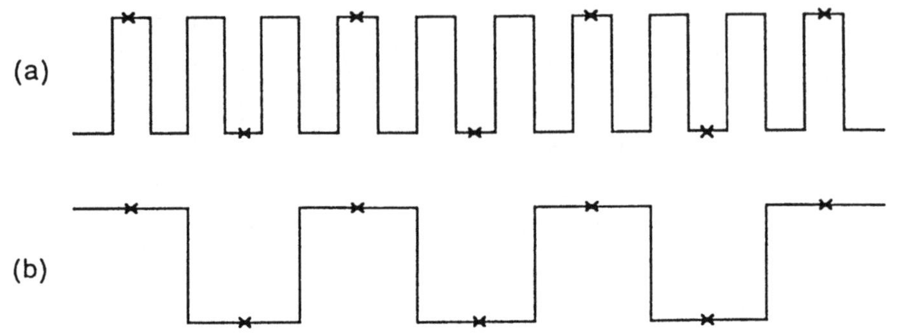

FIGURE 6.8. The samples, x, of both signals are identical and will produce the same results. Therefore, signal (b) is said to be an alias of signal (a)

Anti-Aliasing

Anti-aliasing is a term given to the techniques that are used to reduce or remove aliasing artifacts.

As a direct result from Shannon's theorem, the most obvious solution would be to sample at the Nyquist rate, however, this is not possible. Actually measuring the frequency limits contained in an image is not a trivial task. Even if such measurements were done the resolution requirements could increase the costs of pixel computations to unreasonable levels, or more likely, they would simply be beyond the limits of the display technology. Remember that an ideal edge, such as sharp polygon boundaries, are of infinite frequency.

A more practical solution is to filter out the high frequencies of the image before sampling, so that, in effect, the sampling rate **is** the Nyquist rate. This means that the actual original function would be altered by a filter that removes all high frequency components of the Fourier series. Then, if this *pre-filtered* function is sampled, Shannon's theorem says it can be reconstructed exactly. Of course, the reconstruction will be that of the filtered function but it would be the closest approximation to the original function given the limited resolution.

The filtering required can be accomplished through the convolution[2] of the function with an *ideal low-pass filter* function, defined as

$$H(x) = \frac{\sin(\pi\omega_0 x)}{\pi\omega_0 x} = sinc(\pi\omega_0 x)$$

where ω_0 is the high frequency cutoff, half the Nyquist frequency. The reason for this result is mathematical, namely that $H(x)$ is the Fourier transform pair of

$$h(\omega) = \begin{cases} 1, & \text{if} \mid \omega \mid \leq \omega_0 \\ 0 & \text{otherwise} \end{cases}$$

the ideal low-pass filter as described in the frequency domain.

[2]The convolution of f_1 and f_2 (commonly used in sample theory) is defined as the function

$$C_{f_1 f_2}(t) = \int_{-\infty}^{\infty} f_1(t-s) f_2(s) ds, \qquad -\infty < t < \infty$$

Therefore, the pre-filtered function of the image $I(x, y)$ with high frequencies removed, is defined as

$$f(x,y) = \int_{-\infty}^{\infty} \int_{-\infty}^{\infty} I(x,y) sinc(\omega_0(x-s)) sinc(\omega_0(y-t)) ds dt$$

Unfortunately, it is impossible to accurately compute this convolution. The fact that the lobes of the *sinc* function continue to infinity is not a concern, since their contributions become insignificant after a few periods. The main problems are that both the *sinc* function and the actual integration with the usually complex image intensity function are difficult and expensive to calculate. Therefore, approximations are generally used for both the *sinc* function and the integration.

Theoretically, the convolution must be performed before the point sampling is done. However, in practice, these are usually done concurrently, since the filtering convolution need only be calculated at the sample points. Therefore, anti-aliasing implementations use a low-pass filter convolution (or an approximations thereof) as the sampling function.

Instead of using an actual *sinc* function, many approaches use approximations of it, the most common of which are the truncated Gaussian, triangular, the box filters (figure 6.9). Each has the property that the integral over the range in which they are defined is one, thus meeting that criterion of the *sinc* function. While none are even close approximations in practice, they are all superior to not filtering at all. Ironically, the worse the approximation, the more commonly it is used.

The use of any *sinc* approximation — as any approximation in computer graphics — is dependent solely on the basis of computational efficiency and speed, for which there is a trade-off with image quality. The truncated Gaussian seems to be the closest approximation. Its advantage is that it has no negative lobes, and approximates the unit area integral condition in a smaller range. However, its use still requires quite complex evaluations and lookup table is sometimes used to speed up the process [21]. The triangular function has the advantage that its direct evaluation time is much quicker, requiring only one addition and multiplication at any point. However, it extends past the pixel's boundaries thereby requiring information from an array of pixels. The box filter is the simplest to implement and the fastest to execute since it consists of a constant value and its integral is generally limited to one pixel.

Box filters often give acceptable results, and triangular filters produce good quality results. This is because the convolution approximation is more dependent on the approximation of the integral than on the approximation of the actual *sinc* function.

The simplest way to approximate the convolution is by using a number of discrete point samples within the integration range, and using the filter function to assign weights for a summation of these samples. In general, the more samples taken, the better the approximation and final image quality. This technique is known as *super sampling*, since the image is sampled at a resolution higher than required with sets of samples then being filtered (averaged). A straightforward implementation is extremely expensive since the computation costs increase linearly with the number of samples.

Adaptive filtering can be used to reduce the number of complex filtering operations to those areas which are most susceptible to aliasing problems, such as object silhouettes [51]. With the assumption that the colour of a single surface changes only at low frequencies, the expensive shading computations may be greatly reduced, since the colour of a polygon need only be determined once within any given pixel. In this way, *coverage* information, which consists of the area and position that each polygon occupies within the pixel, is used along with the filter function to determine the percentage of each polygon's contribution to the final pixel colour.

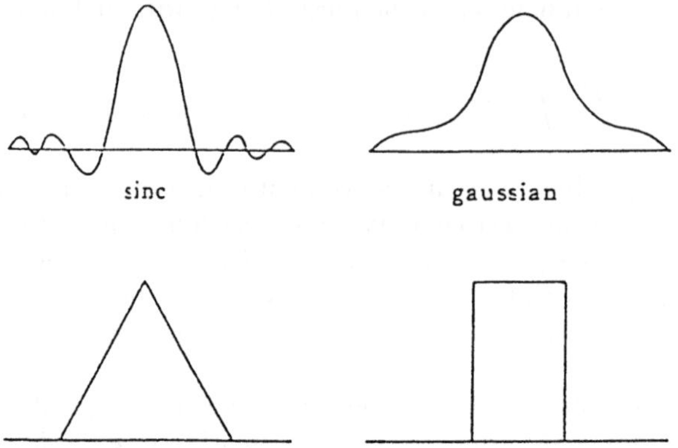

FIGURE 6.9. A *sinc* function along with common filters that are used to approximate it

In some implementations, coverage information is stored in the form of coverage masks calculated during visibility determination [22]. By keeping lists of adjacent polygons, evaluations may be further reduced by combining coverage masks of adjacent segments before the pixel computations [8].

More recently, stochastic sampling techniques have been developed to scatter high-frequency information into broadband noise [18]. Instead of sampling at regular intervals irregularly spaced samples are used. In this way, jaggies are less regular and become noise. The above methods of increasing the sampling rate and filtering may be applied to reduce this noise. The main advantage comes because the human eye is far less sensitive to noise than to the aliasing caused by regular samples. This means that errors due to all the approximations are less noticeable in stochastically sampled images.

6.3.2 Shading

Whether or not an image is anti-aliased, it is almost always shaded in some way. One of the first things an art student learns is that shading brings out the 3D nature of an object; without shading, only the silhouette shape may be determined.

Shading may be defined as the process of determining the result of the interaction between a surface and light. In particular, it is the calculation of the intensity of the reflected and re-radiated light off a surface from all light sources. Shading calculations take into account several appearance parameters including surface and lighting properties, as well as the relative orientation and position of the surface, light source, and camera (or eye). Since it determines the intensity of the reflected light, it makes use of the *colour* surface property. This appearance parameter indicates the fraction of light that is *not* absorbed by the surface, and is used as a weighting factor for the reflected intensity.

Before discussing shading in detail, it should be noted that a colour space (as seen by the human eye) may be defined as a three-dimensional vector space, to correspond with the responses of the tree types of colour receptors or *cones* in the human eye. Therefore, if a basis[3] for the colour space is known, then the effect of each basis component may be calculated separately. Such a basis for colour, most commonly used in computer graphics, is red, green, and blue. Shading calculations may be performed for each of these separately,

[3] A basis for a vector space is defined as a set of linearly independent vectors that spans the space.

only if linear operations are applied. It should be noted that this limitation is often ignored in shading computations in computer graphics. In addition, since the light sources act independently, the shading contributions of each may be computed separately and summed to the total result. Therefore, all the following discussions will be restricted to intensity calculations for one basis component from a single light source.

The goal of shading algorithms, as with most other aspects of computer graphics, is not necessary to model the physical phenomena that exist in reality, but to simulate the effects of such phenomena only to the point where the images produced seem reasonably realistic from a perceptual point of view. Therefore, although some algorithms are based on the physics of a situation, others are based purely on the empirical observations of it, and others are based on computational conveniences.

Lighting Models

Before one may calculate the interaction of light with the surface, one must know the direction and intensity of light that falls on the surface. Such information is provided by the *lighting model.* The simplest lighting model is that of *ambient lighting*, which represents light that is uniformly incident from the environment. Ambient light is usually defined as light reflected from all objects in the world space, and depends on a great number of factors (e.g. the number and types of light sources, the number and arrangement of objects, and the reflective properties of the objects). This is extremely difficult to evaluate and a simple approximation of a global constant intensity is usually used. Ambient lighting is mainly used as a convenience to remove harsh shading effects by ensuring that surface points not directly lit by a light source will still be seen, ensuring that there is still a minimum amount of light falling upon all surfaces.

Most other lighting models are based on modeling of the actual light sources. The purpose of a light source is to define a method of determining a light direction $\mathbf{L}$ for any point $\mathbf{P}_0$ in world space, as well as an intensity I_l that reaches that point from that source.

The simplest and most common light source is known as a *directional* source. It simply consists of a constant light direction $\mathbf{L}$ and a constant intensity I_l for all points. This fairly accurately models a light at a relatively large distance, and requires the least amount of calculation of any light-source model.

Another common light source (it demands only a little more calculation) is the *point* source which models an ideal light bulb in that it emits light equally in all directions from a given point. It is defined as a point $\mathbf{P}_l$, from which the light direction for $\mathbf{P}_0$ is determined by the normalized difference of $\mathbf{P}_0$ and $\mathbf{P}_l$. As with directional sources, I_l is usually defined as constant even though this does not accurately model the nature of light.

If the power emitted by the light at $\mathbf{P}_l$ is p then the actual intensity at $\mathbf{P}_0$ would be $p/(4\pi d^2)$, where d is the distance from $\mathbf{P}_l$ to $\mathbf{P}_0$. It is assumed here that the energy is spread homogeneously over a sphere of radius d, and therefore the light energy per unit area falls off as the inverse square of the distance it travels from the source. In practice, however, point sources placed close to objects create a wide range of intensities that reach various surfaces, and often produce undesirable effects. Also, across most surfaces the distance to the source does not vary much, so that unnoticeable effects are produced at a relatively high computational cost. For these reasons, this inverse square law is often not used in any stage of the shading process, and even when it is, it is often approximated as a linear drop based on the distance between surface point and the eye position [23].

Many other types of light sources may be designed as extensions of point sources by defining the light intensity as a function of the light direction [49]. For example, for a spot light effect, the light intensity may be defined as

$$I_l = I_0(\mathbf{D}.\mathbf{E}')^n$$

where I_0 is a constant, $\mathbf{D}$ is the direction of the spotlight, $\mathbf{E}' = -\mathbf{E}$ (the direction from the spotlight), and n is the concentration factor[4]. I_l is the maximum of I_0 when the angle between $\mathbf{E}'$ and $\mathbf{D}$, θ, is zero, and n determines the rate of fall as this angle increases. For $n = 0$, the spotlight acts as a point source; for $n = 1$, it appears more as a flood light; as n increases so does the spot concentration. Note that if θ is greater then $\pi/2$ the I_l is clipped to zero, or alternatively, I_l may be clipped to zero at smaller angles to produce a sharply-delineated spotlight.

The intensity value may also be a function of other factors. For example, photographers' flaps used to restrict the path of light may be modelled by clipping the intensity function to zero outside a bounding plane defined in world coordinates. Therefore, the intensity function may be formed to model almost any type of point-based light sources.

Point-based light sources are presently the basis of most lighting models, since they have been found to be versatile enough to simulate most desired effects. The simulations of linear and area light sources, fluorescent light fixtures for example, have been accomplished by using a collection of point-light-source models [48]. One main reason that point light sources have been found sufficient in most applications is that some specialized lighting conditions may be more easily specified by shading calculations.

Shading Models

Traditional shading algorithms are relatively simple, quick and effective enough to be used in most applications. In the following discussions, there are a number of unit vector and angles that will be used. They are illustrated and defined in figure 6.10. Note that all of these values may be derived from the position of the light $\mathbf{P}_l$, the position of the surface point $\mathbf{P}_0$, and the unit surface normal $\mathbf{N}$ at $\mathbf{P}_0$.

If ambient lighting is used, it is defined as a global constant, I_0. Since ambient light is equally incident from all directions, some of it will be absorbed, and the rest will be equally re-radiated in all directions. Therefore, ambient shading may be defined by

$$I_a = ck_aI_0$$

where c is the fraction of light not absorbed by the surface (i.e. colour), and k_a is a factor that determines the effective fraction of the ambient lighting.

Unlike ambient lighting, directional and point-source lighting emanate from a particular direction and affect the way in which a surface is shaded. The light reflected and re-radiated from a surface due to such a source may be divided into two components, *diffuse* and *specular*. The diffuse component (or diffuse reflection) represents the light that is absorbed and re-radiated off the surface equally in all directions, while the specular component (or specular reflection) represents the light that is directly reflected off the surface and not absorbed.

An ideal diffuse surface, such as chalk, is dull and matt. It absorbs light, and re-radiates a percentage of it in all directions. Therefore, its intensity is independent of the viewing direction, $\mathbf{E}$. The main difference is that in this case the original source light emanates from a particular direction $\mathbf{L}$. Therefore, the light energy distributed over the surface is governed by *Lambert's cosine law*, which relates this energy to the cosine of the angle of

[4]The operation $\mathbf{A} \cdot \mathbf{B}$ is a *dot product* in which the vectors are assumed to be normalized so that it is simply the cosine of the angle, θ, between the vectors

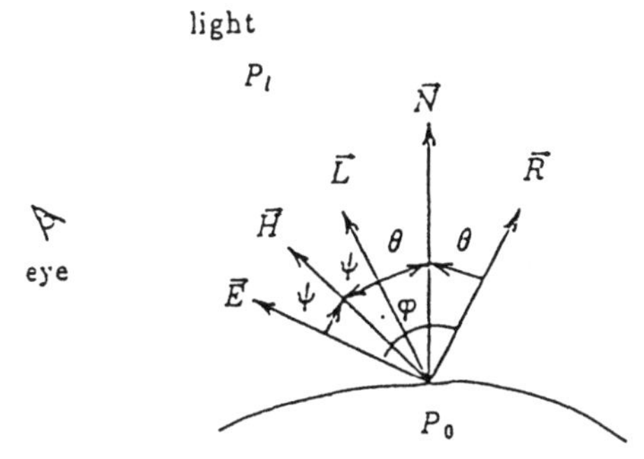

FIGURE 6.10. A point on a surface ($\mathbf{P_0}$), and the vectors associated with shading calculations. All vectors are assumed to be normalized (i.e., the vectors are all one unit in length)

L is the direction to the light source
N is the surface normal
E is the direction to the eye or camera
R is the ideal direction of reflection
H is the direction of the angular bisector direction of **E** and **L**
θ is the angle between **N** and **L** = angle between **N** and **R**
ϕ is the angle between **E** and **R**
ψ is the angle between **E** and **H** = angle between **L** and **H**

incidence, the angle between the surface normal, **N** and **L**. Lambert's law indicates that the *effective* light intensity I_e is a fraction of the *total* light intensity reaching the surface I_l, namely

$$I_e = I_l \cos(\theta) = I_l(\mathbf{L.N})$$

where θ is the angle of incidence. Therefore, the diffuse component may be defined as

$$I_d = ck_d I_e = ck_d I_l(\mathbf{L.N})$$

where k_d indicates the fraction of the incident light that undergoes diffuse reflection and varies from one material to another. The rest of the light undergoes specular reflection, an effect which is somewhat more difficult to model correctly.

An ideal specular surface, such as a mirror, reflects light in only one direction, the direction of reflectance **R**, where the angles of incidence and reflection are equal. However, most objects are not ideal specular reflectors, but nevertheless directly reflect light observed as highlights on the surface. This is related to the shininess of the surface.

Specular reflection differs from diffuse reflection in that it is dependent on the viewing direction **E**, and in particular, the angle between **E** and **R**, that is φ. For the ideal specular surface, specular reflected light may be seen only when φ is zero. For near specular or shiny surfaces, the intensity of the specular reflection falls rapidly as φ increases; the less shiny the surface, the less rapidly it drops off.

Based on such empirical observations, Phong approximated this drop in specular reflectance by $\cos^n(\varphi)$, where n is an indication of the shininess of the surface [7]. This cosine is maximum when φ is zero, and n increases, the function drops off more rapidly. For dull surfaces with large diffuse highlights, n is small, but for shiny surfaces with concentrated highlights, n is large. In fact, for an ideal specular surface, n would be infinite.

Similar to the diffuse component, the specularly reflected light is a fraction of the incident light and is a function of θ. Given that this function is $W(\theta)$, the Phong model for the specular reflection is defined as

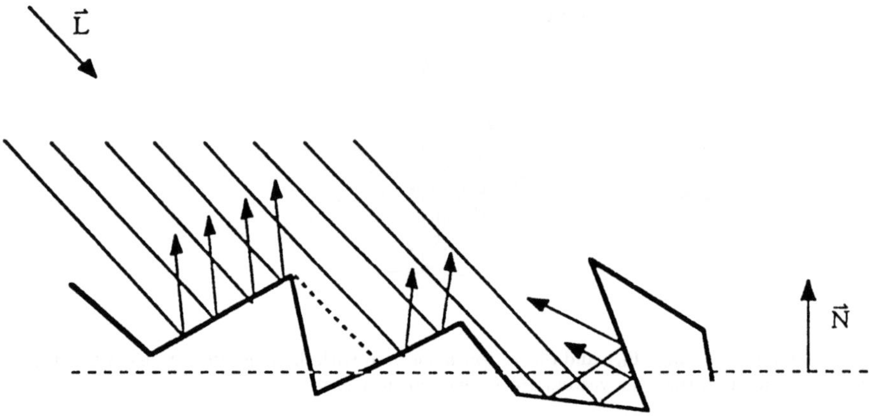

FIGURE 6.11. The Torrance-Sparrow model of a specular surface. Each micro-facet is oriented at a perturbation of the surface normal.

$$I_s = k_s I_l W(\theta) \cos^n(\varphi)$$

where k_s indicates the fraction of incident light that undergoes specular reflection. However, in practice, $W(\theta)$ is often set to a constant that is absorbed into k_s. Therefore, if **R** and **E** are normalized, then the Phong model, as generally used, becomes

$$I_s = k_s I_l (\mathbf{R}.\mathbf{E})^n$$

It is important to note that the equation does not include the factor c. This is because, unlike the diffuse component, the specular component is completely composed of reflected light and none of it is absorbed and re-radiated. Therefore, the specular component is independent of the colour of the surface and only depends on the colour of the light.

Since the Phong model is simple, relatively quick to calculate, and generally produces aesthetically pleasing results, it is the most widely used model of specular reflectance. It is nevertheless based on an empirically-derived model and not on a theoretically-based model.

The generally accepted model of specular surfaces is the *Torrance-Sparrow* model [46]. The surface is assumed to be a collection of microscopic facets, each of which is an ideal specular surface. The orientation of each of these *micro-facets* is based on a random probability distribution function centred about the surface normal (figure 6.11).

The specular reflection, in terms of the Torrance-Sparrow model, is made up of that fraction of light which is directly reflected towards the eye. Only those micro-facets whose normals are **H**, the bisector direction of **E** and **L**, contribute to this specular reflection. The diffuse component is due to multiple reflections and internal scattering, where light penetrates beneath the surface. It should therefore be noted that the diffuse component, in the Torrance-Sparrow model, still obeys Lambert's law and remains as previously defined. However, the specular component becomes

$$I_s = k_s I_l \frac{DGF}{\mathbf{N}.\mathbf{E}}$$

where D is the micro-facet distribution function, G is the micro-facet shadowing and masking function, and F is the Fresnel function, each of which is to be elaborated on shortly.

The specular component is related to the relative surface area of micro-facets with normals in the **H** direction, which is most influenced by the angle between **N** and **E** and the micro-facet orientation distribution function, D.

The distribution is a function of the angle between **H** and **N** with the spread a function of the surface roughness; the smaller the spread the shinier the surface. The Phong model in effect uses the cosine raised to a power as the distribution function. Torrance and Sparrow proposed a Gaussian distribution, but other distribution functions may be used either alone or in combinations, to model various classes of materials, such as metals and polymers.

In addition, the geometrical attenuation factor G accounts for other micro-facets whose normals are not **H**, which shadow some of the incident light and mask some of the reflected light (figure 6.11). This factor is the minimum fraction of light that is not shadowed or masked. The final factor F that influences the specular intensity is the Fresnel reflection law function. This function gives the fraction of the incident light that is actually reflected, rather than absorbed. Therefore, it is similar to the $W(\theta)$ term in the Phong model. However, F is not only a function of the angle of incidence but it is also a function of the wavelength of the incident light. The Fresnel function is therefore, highly dependent on the reflectance spectrum of the surface material and predicts colour shifts in the reflectance direction.

There are two major differences between the Torrance-Sparrow model and the empirical Phong model. The location and absolute intensity of the highlights produced by both models appear to be essentially the same for small angles of incidence, but are quite different for large angles of incidence. The main departure is the fact that for large angles, the peak of the specular highlights does not lie in the reflectance direction, but at a slightly larger angle and at a greater intensity. This overall effect is most notable at grazing light angles or on edge-lit objects.

The other major difference lies with the colour of the highlight. In the Phong model, the specular component was assumed to be the same colour as the incident light. However, the Fresnel function predicts that the surface material may influence the highlight colour and in fact, this colour changes with the angle of incidence. These two differences can be quite significant, depending on the surface materials that are being used. The Torrance-Sparrow model is ideal for metallic surfaces, while the overall effect of Phong shading is an inescapable *plastic* look.

The major problem with using the Torrance-Sparrow model is the fact that a full implementation is computationally expensive [13]. By making some approximations, the costs may be made more reasonable [4], however, the simplifications provide values that are not a function of the wavelength of the incident light.

It has been proposed that the colour shift may be incorporated into this value by interpolating the reflected colour from angles of incidence at 0 and 90 degrees, with respect to the variation of the Fresnel value as defined above [13]. This requires looking up these respective colour-shift values for various light source and material combinations in standard illumination tables [10, 47]. As an alternative, an approximation of the colour shift may be accomplished by using a constant colour for the specular component that is a user-specified mixture of the surface and light source colours. For metals this colour should be predominantly the surface colour and for non-metals it should be predominantly the colour of the light source. Although this is a crude approximation, the results are a striking improvement over simply using the colour of the light surface when simulating metallic surfaces. Finally, the greatest advantage of this method is that since it does not rely on the Fresnel value, it may also be applied to the computationally-faster Phong model.

Therefore, by avoiding lighting conditions that produce edge lighting, and by specifying a constant highlight colour that is different from the surface or light source colour, one is able to simulate the Torrance-Sparrow model quickly and sufficiently with a simpler model such as that of Phong. It is for this reason that most applications today use the Phong model.

Specialized Shading

So far, all the lighting and shading models presented make some assumptions that are unsuitable for particular applications. To overcome some of these problems, specialized shading and lighting models have been used. The problems with these models is that some may only be applied under particular conditions and are inappropriate in the general case, and other provide solutions to rare conditions and only add high cost and negligible influence in the general case.

For example, to correctly evaluate the global illumination effects, as opposed to the ambient lighting approximation discussed above, one must model the effects of object-to-object reflection between diffuse surfaces [26, 11]. In this way, a white object placed within a room with red walls will display red reflections even when the illuminating light source is white. This is known as *radiosity*, and the algorithms involve modeling objects with polygons, finding the relative visibility of each pair of polygons, and then solving a system of n equations for n unknowns, where n is the number of polygons in the scene. Although the results are striking, the cost of such computations make the algorithm impractical for use unless such effects are significant.

Another example of shading, where traditional techniques have proven to be inadequate, is in the shading of trees modelled with particle systems [42]. In this case, to take into account the shadowing and diffuse illumination effects of other particles in the tree, such parameters as the location of shading point within the tree and the distance into the tree from the light source are included in the shading calculations. Obviously, such a shading model cannot be used out of this particular context.

Finally, it should be noted that, unlike in the real world, even the standard light sources may be applied selectively so that each object in the scene may be lit individually without effecting other objects. Although this technique is used quite often, care must be taken to maintain a consistent global light effect, especially in animations.

Aliasing Problems in Shading

Fortunately, shading often occurs in images at sufficiently low frequency as to not contain any aliasing effects. However, there may be aliasing problems in areas of high curvature, or when the specular spread is small (as with sharp highlights).

The most obvious solution would be to use super-sampling techniques. However, this is undesirable since it would involve multiple shading calculations per pixel for all pixels, whereas shade aliasing may only occur in a small percentage of pixels. Adaptive super-sampling would solve this problem by essentially super-sampling only for pixels which contain surfaces of high curvature. This, of course, requires a curvature measurement to be done for each object per pixel, usually performed by comparing the normals at the pixel corners.

A more elegant solution is to clamp the spread of the specular component so that the spread is reduced for high curvature [1]. For example, in the Phong model, the cosine power n would actually be a function of the curvature measurement. In this way, the number of shading calculations required to produce an image can still be minimized.

Finally, the shade-aliasing problem may be converted to a texture-map-aliasing problem through the use of *reflectance mapping*, explained in the next section. Shade aliasing does

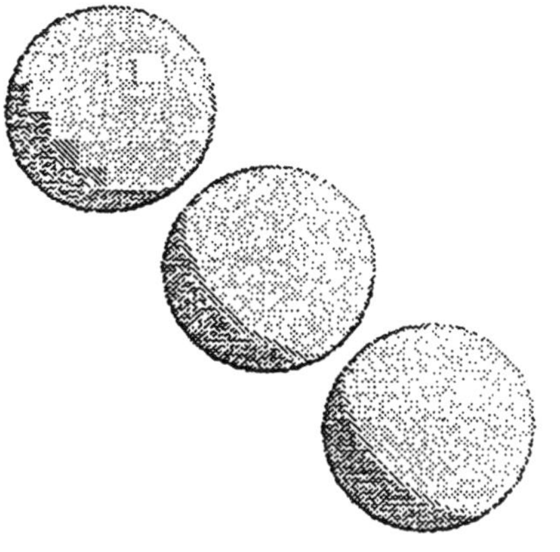

FIGURE 6.12. Three spheres, under the same lighting model, are shaded with flat, Gouraud, and Phong shading respectively from top to bottom

not occur often in most applications and, therefore, it is often overlooked or ignored in many graphic systems.

Shading Application

The shading calculations may be performed in the world-coordinate system, where the light sources are usually defined, however, the eye-coordinate system is slightly more convenient, since it simplifies the eye direction calculation. In fact, in the case where the objects are not too near the eye, one can approximate the eye direction to be a constant equal to the z-axis. In many cases, this can significantly speed up the computations without much apparent difference.

In either case, all shading calculations must be performed using positions and vectors defined *before* the perspective transformation. This is because the distortions created by this transformation would result in unrealistic shading effects, especially in the normalized device coordinate system.

This leads to the question of exactly when and how the shading are performed in relation to the visible surface determination. Basically, the shading calculation is applied to a polygon during the scan out process, and this is usually done in one of three ways.

The simplest is referred to as *flat shading*. Since each polygon is flat, if the light source is at a sufficient distance form the surface, then the entire surface will be shaded with the same colour. This means that only one shading calculation need be performed with respect to the surface normal of the polygon, and this colour can be used for each pixel in the scan out process. The resulting image appears (as one would expect) to be made out of a series of flat polygons (figure 6.12). However, objects modelled with curved surface patches will also appear this way, and the result is often undesirable.

To solve this problem, a method known as *Gouraud shading* uses the shading calculation to determine a colour at each polygon vertex, and then interpolates the resulting colours across the polygon during the scan-out phase [27]. For surface patches, calculating a normal at each vertex is simply a matter of taking the cross product of the two parametric vectors that define the surface. For objects composed of a polygonal mesh, a normal may be assigned to each vertex, which may be computed by averaging the normals of all adjacent polygons. The polygon boundaries in the resulting image disappear since shading discontinuities are no longer present and the polygonal representation is only present in the silhouette (figure 6.12). Although this works well for diffuse shading, problems result with relatively higher frequency highlights, due to the low shade-sampling rate. In fact, the highlight may be missed completely if it appears completely with a polygon.

Phong's solution to this problem provides a unique shading calculation for each pixel [7]. This method is known as *Phong shading* but should not be confused with the Phong shading-model described above. Phong shading is similar to Gouraud shading in that it requires a unique normal for each vertex, however, it is the *normal* that is interpolated across the polygon during scan out, by interpolating each vector component. Although not totally accurate, the resulting image is an improvement in that the highlights are present without defects, and the shading is not only continuous but varies smoothly (figure 6.12). To be totally correct, the normals used for surface patches should be calculated from the surface patch itself. However, polygons originating from surface patches are indistinguishable from any other type of polygon by the scan-out stage.

Finally: note that if flat or Gouraud shading is to be used, then all surface colour computations must be done for each vertex *before* the visible-surface-determination stage.

6.3.3 Texture Mapping

The shading calculation is the basis of the visual appearance of all rendered objects. Since each point that is to be shaded requires a set of the appropriate surface-property parameters, the visual complexity of any object is directly proportional to the number of *unique* surface property-parameter-sets defined for it. Physical models restrict the number of such sets, from one per object to a small number per primitive (e.g. one per vertex for polygons), and these restrictions are often too limiting to simulate such complex surfaces as wood or marble unless the number of primitives are increased substantially. Such problems were the motivating factors in the design of visual-object-modelling tools.

There are two approaches to increase the complexity of a surface through shading, besides using physical modelling tools. One may either create individual shading functions for each type of surface that is to be simulated, or one may devise a way to adjust the shading parameters according to a visual model. Obviously, the latter is the better solution, since it provides greater flexibility to non-technical users and programmers. The method incorporated to do this is known as *texture mapping*.

Texture mapping is a simple technique of mapping a *texture* onto an appearance parameter. The texture is a function that describes the variations and fluctuations desired for the range of the appearance parameters, and the values thereof may be used as the actual appearance parameters themselves or as one of the arguments for functions that will determine the appearance parameters. The texture function itself exists in *texture space* and is therefore only a function of the texture coordinate system. So by creating a mapping of these texture coordinates on to the surface of an object, one is mapping the texture function onto the desired appearance parameter.

In this way, the quantity of information in the physical model need not increase by a significant amount, if it increases at all, to accommodate complex pattern for an appearance parameter. For example, a polygon may include a set of texture coordinates for each vertex, in place of colour. The texture space may then define a complex colour pattern, say that of marble. By interpolating the texture coordinate across the polygon, the complex colour pattern may be mapped onto the colour parameter, thus producing a *marble* coloured polygon.

Therefore, texture mapping provides an illusion of complexity without adding complexity to the physical model, and without adding complexity to the geometric transformations and the visible-surface-determination algorithm. By applying such a procedure to the var-

FIGURE 6.13. Here, a 2D texture (left) is mapped onto a sphere (right). The mapping function projects the texture's Cartesian coordinates onto the sphere's latitude and longitude coordinate system, and thus has the effect of *squeezing* the texture at the poles

ious appearance parameters and by using a variety of texture functions, a large range of effects is possible.

Texture Spaces

Although texture space may be of any dimension, it is commonly defined as a two-dimensional space. This is intuitively the easiest to understand since one may associate a direct mapping of a 2D texture onto the 2D surface of an object, as a type of *wall papering* of the object with the texture.

This 2D mapping had been initially applied to bivariate surface patches, since each patch was conveniently described as a surface of two parameters which could also be used directly as texture coordinates [9]. The technique was easily transferred to other surfaces by simply providing a 2D coordinate system that spanned across the surface. Unfortunately, there are inherent problems that complicate this.

In most cases, texture coordinates are a direct mapping from the surface of the object to the texture space. However, many applications may require an indirect mapping to texture coordinate, since the texture may need to be adjusted on the surface by shifting, scaling or rotating it. These require a mapping function from the original texture coordinates to the actual texture coordinates, which must be pre-defined by a user. Although such transformations are often required, the high computational costs of textured rendering greatly restrict their interactive assignment, thereby complicating the visual-modelling process.

Another potential source of difficulties lies in the fact that the 2D texture is usually defined on a plane but is usually mapped onto a curved surface. This may result in distortions of the original image (figure 6.13), which in many applications is an undesirable effect. Using a different coordinate system to define the texture space, such as the angles of spherical coordinates, is only a solution for a particular class of objects, in this case, spheres. The general solution involves either creating a complex mapping function or individually adjusting the originally assigned coordinates. The latter is the simpler and more general of the two solutions. However, graphics systems may not allow arbitrary assignment of these coordinates, such as those for parametric surfaces, or even those that do, require complex tools for such assignments. In an attempt to avoid such problems, some visual modeling tools facilitate users with the means of pre-processing the texture image with an inverse deformation.

Recently, it has been shown that 3D texture-space definitions, sometimes referred to as *solid textures*, provide more simple solutions to these problems, while creating a number of other problems, described below [35, 36]. In this case, the actual coordinates of the surface point, usually in object space, is used as the texture-space coordinate, *carving* the object out of the texture space. The only mapping functions that may have to be applied will usually consist of the basic transformations of translation, scale, and rotation applied uniformly. Since the mapping in this case is from 3D space to 3D space (i.e. the same form of space), a simple mapping will not contain any deformations of the original texture space. The major disadvantage with the use of 3D as opposed to 2D texture spaces is in its greater time and space complexities.

In contrast, one-dimensional texture coordinates are relatively simple to use. However, due to the 1D constraints as applied to 3D models, its use is limited, but may be well-suited for particular applications.

Finally, texture spaces may be defined in a number of ways. The simplest way is to provide a *texture function* which, when given a texture coordinate, returns the value at that point. Regardless of the dimensionality, texture functions may be defined in various ways ranging from a simple mathematical equation to a complex stochastic functions. The complexity may be increased by using a combination of functions and varying the functions used with the texture coordinates.

Although such functions are well-suited to a number of applications, they are rather limited. A texture composed of a simple periodic pattern such as a checker-board may be coded without difficulty. In fact highly complex textures such as a 3D marble texture have been achieved analytically be using combinations of *noise* and *turbulence* functions [36]. However, other common texture spaces are almost impossible to define (consider the case of text or a logo that is required to be mapped onto a surface).

For highly complex surfaces which cannot be described analytically, the texture is often supplied in the form of a look-up table. Although these look-up tables may be generated with the use of analytic functions — often the case for 1D textures — to reduce the amount of computation during the rendering stage, they are usually created through other means. For 2D textures, once it is realized that a 2D look-up table may be equated with a raster image, all the tools used to create raster images may be used to create textures. These include, but are not limited to, paint systems, digitizing cameras, and other rendering systems. In this way, the door is opened to an almost unlimited range of texture patterns.

However, 3D texture spaces are not suitably described by look-up tables due to memory constraints, and are therefore usually described analytically. Nevertheless, some forms of 3D textures may make use of 2D textures. For example, a 2D texture may be considered to be extruded to 3D so that the 3D texture is only defined by a 2D look-up [25]. Another technique is to define a 3D texture as the intersection of two or more extruded 2D textures [35]. This has the advantage of providing unique plane cuts throughout the texture space.

FIGURE 6.14. A square pixel from the screen is mapped onto the perspective view of a surface. Shown here is an example of its corresponding mapping into a 2D texture space. Note that although the edges of the mapped region are shown to be linear, they may in fact be non-linear

Aliasing Problems in Texture Mapping

As with any sampling procedure, the sampling of texture space is vulnerable to aliasing artifacts. Of course, the amount of aliasing problems depends on the frequency spectrum of the texture itself. However, in general, textures are often rich in high frequencies[5].

Unlike the sampling previously described where the samples are taken at even spatial intervals, the sampling of texture is often uneven die to both the mapping function of texture space onto the object and the projection transformation of the object onto the screen. The result is that there are variable amounts of aliasing defect throughout the object. This means the anti-aliasing solutions must be variable depending on the sampling rate, which therefore becomes another parameter in the texture mapping function. If the sampling rate did not vary, then the only solution would be to anti-alias at the lowest texture sampling frequency. For objects that contain a wide range of such frequencies, areas of high sampling frequency will appear to have blurred texture, since the low-pass cutoff will be too low.

As with the other aliasing problems previously described, the anti-aliasing techniques to overcome texture aliasing are either to super-sample or to use a pre-sampling low-pass filter. Again, super-sampling is not a recommended solution since the potentially complex mapping to texture coordinates must be done many times per pixel. However, due to the fact that the texture-space sampling rate varies depending on the coordinate mapping, correctly applying a low-pass filter is more difficult than in the cases previously mentioned.

The problem lies in the fact that the integration range in the convolution must be mapped onto the texture coordinates, and therefore undergo all deformations associated with that mapping (figure 6.14), making this integration difficult.

The common solution is to approximate the integration area or volume with a standard one which is easier to work with. For example, in the 2D case a square integration area that encloses the actual mapped area is used. If the precise area cannot be used it is better to over-estimate the area, since under-estimating may not filter out all the high frequencies. On the other hand, over estimating may filter out desired frequencies and thus result in an over-blurred texture. If the estimated area is less then twice the actual area, this over-blurring is not very noticeable. Problems occur only when the area is greatly over-estimated (such that the square region cannot be used to approximate the mapped region) as in the case of a long thin region. However, these cases are not common.

[5]It should be noted that it is for this reason that texture mapping is generally not used in conjunction with flat or Gouraud shading.

All that is left is to perform the convolution. For analytically defined textures, this convolution may be done analytically. The complexity of such a procedure is highly dependent on the type of filter used and the texture function itself. In fact, if a Fourier-series representation of the texture function is pre-computed, then the convolution would simply involve clamping the high frequencies [33].

For textures defined by raster images, the convolution would involve the weighted sum of the texture pixels, or *texels*, within the integration region [6]. The weighting function would depend on the filter that is being used. However, in some applications, such a process is computationally expensive. Therefore, techniques have been developed to speed up this computation by pre-processing the texture.

A popular method make use of *pyramidal parametrics* [52], which involves filtering the texture at a number of resolutions. These textures form a 3D pyramidal structure from which values are computed though tri-linear interpolation. The area of the convolution region defines the level of access and the sample point in texture space defines where on the level to sample. The advantage of this procedure is that only one sample is required per pixel. Also, since the pre-filtering need only be done once per texture and may be used for all rendering that requires that texture, the filter that is used may be more accurate since computation time is less important. However, the filter shape is also pre-computed and cannot be changed.

Another disadvantage is that the pre-filtered texture requires more space. However, this space is not too great compared to the time savings, being only a 30% increase if the texture resolutions are powers of two apart.

Another pre-processing method involves using *summed-area tables* [17]. This table stores the total integrated sum of the rectangular area defined by the texel corresponding to the table entry and the texel at some texture corner. In this way, the area of any rectangular region may be determined by three arithmetic operations. Although this method is faster than the one previously described, the memory requirements are much larger, by a factor of two to four.

3D textures have the greatest need for such pre-filtering techniques, and they may be pre-filtered using the same techniques as for 2D textures. The problem is that most 3D textures are analytical, for which convolutions are either extremely expensive or difficult to define. In many cases, 3D textures are simply point-sampled and it is up to the user to adjust the parameters such that the frequencies present are low enough to avoid aliasing problems.

Finally, it should be noted that pre-filtering of 1D textures is usually not required due to the simplicity of the convolution involved.

Texture Applications

Texture-mapping techniques have been used in a variety of applications. In fact, texture mapping has been generalized to the point where it is the major tool in the visual-modelling process and textures have been used either alone or in combination to model some of the most visually complex images with elegant simplicity.

The actual information recorded in the texture space is that returned by a texture-space sampling function. Colour is the most common, mainly because it allow textures to be created and examined visually. But the format of the texture values are unrestricted, and may range from a simple logical unit (i.e., a bit) to a complex array of types and structures. The unique generality of texture mapping lies in the fact that there are no restrictions on the mapping into texture space or on the mapping of texture values.

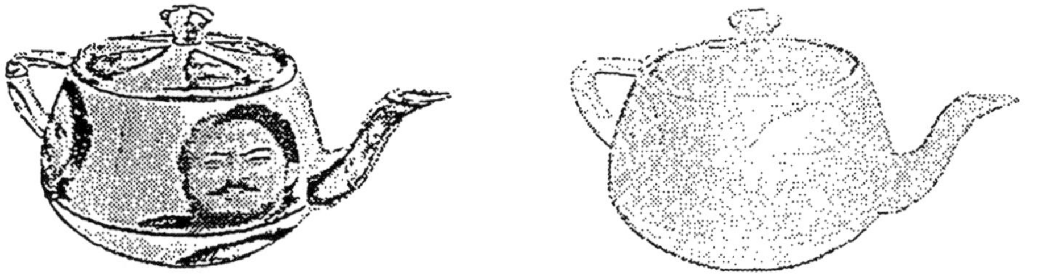

FIGURE 6.15. An example of standard texture mapping (left), and bump mapping using the same texture (right)

In its simplest form, texture mapping permits a colour pattern to be mapped onto a surface [9]. In this case, values associated with the location on the surface, such as the parametric values of a patch, are used to generate the texture coordinates. These are then used to sample a texture which consists of the colour pattern desired, and the sampled value is used as the surface colour for that point (figure 6.15). In this way, the colour pattern has been affixed to the surface and remains static regardless of the angle of view. All available texture-mapping techniques are variations of this standard method.

The function that generates the actual texture coordinates may be arbitrarily complex. Its input may be any parameter or group of parameters of any type[6], and its only restriction is that it must produce a valid input to the texture-space sampling function.

A simple example of such a variation is in *reflectance* or *environment mapping* [6]. Here the reflectance direction R is used to access the texture space. The polar and azimuthal angles of R are mapped onto the axes of a 2D texture space. In this way, the texture may be thought of as an environmental sphere that surrounds the object, and since the reflectance direction is used, the colour sampled is the environmental colour that would be seen if the surface were a perfect reflector, assuming no self occlusion. Therefore, if the texture pattern consists of the environment of the object (i.e., other objects and light sources), then by blending the resulting colour with the surface colour, specular reflections may be simulated without using ray-tracing techniques. Unlike standard texture mapping, the mapped pattern is not static on the object surface, but varies depending on the orientation of the object and the viewing position. Environment mapping may also be used to replace multiple shading calculations when there are many light sources, and as a solution to shade aliasing since texture anti-aliasing will take care of it [52].

In the above examples, the texture values were interpreted as colour. However, the way in which these values are interpreted is totally arbitrary. They may be used to replace or alter any one or more appearance parameters, and each interpretation results in a different effect. For example, the texture values may be used to assign surface transparency factors. Such *transparency mapping* techniques have been used to adequately model clouds and trees with simple ellipsoids [25].

A more complex example is the simulation of wrinkled surfaces, which uses a technique often referred to as *bump mapping* [5]. The texture values are used to perform small perturbations on the direction of the surface normal. This affects the shading calculation by shading the surface as if it were oriented in the perturbed direction. In this way, surface irregularities such as surface roughness, depressions and bumps may be modeled

[6]These parameters are not restricted to be traditional appearance parameters. However, since they are being used as input to a texture-mapping function they are appearance parameters by definition.

(figure 6.15). Note the process is a *visual* modelling tool that only affects the shading calculation and not the physical shape of the surface. Therefore, the effects is limited in degree but is nevertheless quite convincing.

Such a process requires some more major adjustments to the standard texture mapping process. This is due to the fact that the intuitive approach to creating a bump texture would be to represent texture values as the perpendicular perturbation of the surface point. This means that the perturbation required for the surface normal depends on the derivative of the texture value and not on the value itself. Also, the normal must be perturbed in the same direction with respect to the object surface at all times. For these reasons, the orientation of the texture on the object must be taken into account, and standard texture mapping does not do this.

With standard texture mapping, all that need be known to sample the texture are the texture coordinates, and possibly the sample size for anti-aliasing. Even if the texture contains the derivative of the height of the surface, this derivative will be given in a particular orientation. Therefore, depending on the orientation of the projection of the texture onto the window pixel, a rotation will most likely have to be performed to ensure that the normal is perturbed correctly.

Finally, it should be noted that all of the above approaches (along with others) may be used in combinations with each other. In fact, texture values may even be used to determine which texture mapping techniques to use [12]. With such flexibility, the visual complexity that may be achieved through the use of texture mapping is virtually unbounded.

6.3.4 Towards Realism

The techniques described above can be used to create a wide variety of surface descriptions. Even so, computer generated images can still lack a totally realistic look. The many variations of texture mapping greatly help to visually enrich our synthetic images, but they may still lack some important features.

Shadows

Shadows can enhance image realism and provide some important depth queues. If a point on a surface is blocked from a light source by another surface, than a shadow is created by not shading the point for that light (i.e.,) only the ambient component is used). Therefore, one saves the time of shading computations, and the major problem is determining where the shadows lie.

Shadow algorithms are therefore identical to visible surface determination algorithms. Any point in the scene that is *visible* to the light is *not* in shadow. By solving the visible surface problem with the light as the eye, one can generate enough information to locate the shadows. For a directional source, the eye is placed outside the scene, and an orthogonal projection is used. For a spot light, the eye is at the light, and a perspective projection is used. In both cases, the look-at direction is the same as that of the light source.

Some methods for generating shadows add a number of polygons to the geometric data base. These polygons either represent projected polygon shadows [2] or, in sets, they form polygon *shadow volumes* [15]. During this visible surface determination, they are used to decide whether or not a point is shadowed. The disadvantage of such methods, other than greatly increasing the polygon count, is that they only apply to polygonal objects.

A more general approach involves the creation of a shadow *image* in which the pixels are used to record a tag of the closest surface, rather than a colour [52, 29]. During surface colour determination, this image is accessed much like a texture. A point on a surface is in shadow if the sampled tag does not belong to the same surface. This method suffers from aliasing problems more than other texture techniques since the tags cannot be filtered, so super and stochastic sampling are applied [43].

In general, synthetic images use shadows quite sparingly since they are computationally expensive. The visible surface problem must be solved for each light source. If the relative positions of the objects and the light sources do not change in an animation, then the shadow generation need only be done once, however, this is usually not the case. To reduce some of the computation time, the number of objects and lights involved in the shadow calculations are sometimes limited. It should be noted that these methods all assume point light sources so that the shadows generated are without any penumbras.

Atmospheric Effects

Computer generated images are often sterile in appearance; textures may be used to apply scratches and dust to the too perfect surfaces, but at a relatively high cost. However, the addition of some simple atmospheric effects adds to the realism of an image.

Fog is a simple effect which may be added to outdoor scenes. It is simply created by a desaturation of a surface point's colour as a function of its distance from the eye. A more complicated model would also make use of the height above a ground plane.

The scattering of light near a light source and the creation of light beams are difficult to calculate directly. In may cases, they are simulated with the use of transparent cones projecting from the source. By assuming a constant air particle density, the transparency level at a given point on the cone surface is a function of its distance from the light. Gaseous objects such as clouds and smoke are also simulated using transparency techniques, but often less convincingly. This is because the assumption of uniform particle density within these volumes is incorrect.

Camera Effects

Computer generated images are also associated with the fact that the entire scene is always in perfect focus. This is because the visible surface algorithms are based on ideal pin-hole camera models. Some people have attempted to use more elaborate camera models that include *depth-of-field* [38], which is due to increased aperture size, and *motion-blur* [31, 39], which is due to a finite exposure time (as opposed to instantaneous exposure).

If the depth value for each pixel is retained, depth-of-field may be added to an image as a post-process. The blurring effect is achieved by expanding a pixel's influence to neighbouring pixels to the extent of a *circle of confusion* whose size is a function of both the focal length of the camera and the depth information at the pixel.

The simulation of motion blur is an attempt to solve the problem of temporal aliasing in animation. Along similar lines to the depth-of-field solution, a velocity vector may be recorded with each pixel. These would than be used to post process the image by dragging the pixel colour in a direction opposite to that of the velocity vector, by an amount proportional to its length. However, this method can produce incorrect results since it does not take the depth information into account. Proper motion blur can only be calculated within the visible surface algorithm. As with spatial anti-aliasing, each pixel is basically super-sampled, however, the samples are taken at different *times*. In addition, the time samples must also include the surface colour calculations since the shading will change with position. The net effect is computing a number of instantaneous images to produce a single blurred image. This makes motion blur one of the most expensive rendering effects.

6.3.5 Rendering Trends

The rendering techniques described above (along with their variations) have been used in the creation of almost all of the 3D computer generated images that exist to date. The potential visual complexity of images that can be produced is quite high, however, due to the costs involved, most images generated today are relatively simple.

In general, the use of more shading and texture techniques leads to a richer quality image. Unfortunately, the memory and processing requirements also rise, and efficiency in computation time is usually one of most important considerations when designing images (or renderers). This is due to the fact that each millisecond of pixel computation time results in over two hours of rendering time for each second of animation at video resolution.

Therefore, in practical terms, image complexity and quality may be increased by either reducing the computation requirements of the algorithms used, or by using more powerful computers. Much research has concentrated on finding faster methods by using approximations that still produce acceptable results. Some algorithms are faster due to larger memory requirements, however, as memory is becoming less expensive, this is not much of a constraint. In fact, the recent trend of falling prices with more processing power will permit the use of techniques that were once thought to be too slow to use in graphic production.

Recently, new generations of algorithms are being developed that take advantage of various multi-processor architectures [14]. In fact, architectures specialized for graphics applications are rapidly emerging. Some workstations are now capable of generating visible surface images of fairly complex objects, phong shaded with multiple local light sources, in real time. The only problem with this trend is that it reinforces one of the major problems of rendering systems, that of flexibility.

There are a number of *special* effects which are sometimes required to obtain just the right look for a particular object or scene. Often, users have the proper insight to know what special combination of functions could achieve the desired effect. Unfortunately, the rendering systems are usually restricted in the way in which they may be used. For example, although a renderer might permit textures to apply to any shading parameter, they most likely do not allow one texture to be used to access another. The development of systems which provide such freedom and flexibility had recently become an important research topic [16, 12, 34, 32]. One of the most important considerations is to provide tools which do not require programming skills. The conflict between more restrictive, faster, hardware renderers and the more flexible, slower, software environments must be addressed in the near future.

The decision of which techniques to use is basically application dependent. The goal of every graphics application is to produce acceptable-looking images in the shortest time possible. In the end, the choice mainly depends on how one defines *acceptable* as a function of how long one is willing to wait for it.

6.4 References

[1] J Amanatides. Highlight Anti-Aliasing. Unpublished, 1986.

[2] P Atherton, K Weiler, and D Greenberg. Polygon Shadow Generation. *Computer Graphics (Proc. Siggraph 78)*, 12(3), July 1978.

[3] R E Barnhill and R F Riesenfeld, editors. *Computer Aided Geometric Design*. Academic Press, 1974.

[4] J F Blinn. Models of Light Reflection for Computer Synthesized Pictures. *Computer Graphics (Proc. Siggraph 77)*, 11(3):192–198, July 1977.

[5] J F Blinn. Simulation of Wrinkled Surfaces. *Computer Graphics (Proc. Siggraph 78)*, 12(3):192–198, July 1978.

[6] J F Blinn and M E Newell. Texture and Reflection in Computer Generated Images. *Communications of the ACM*, 19(10):542–547, October 1976.

[7] Phong Bui-Toung. Illumination for Computer Generated Pictures. *Communications of the ACM*, 18(6):311–317, June 1975.

[8] L Carpenter. The A–buffer, an Antialiased Hidden Surface Method. *Computer Graphics (Proc. Siggraph 84)*, 18(3):103–108, July 1984.

[9] E Catmull. *A Subdivision Algorithm for Computer Display of Curved Surfaces*. PhD thesis, University of Utah, 1974.

[10] CIE, International Commission on Illumination, Publ. CIE 15, Bureau Central de la CIE, Paris. *Official Recommendations of the International Commission on Illumination,* Colorimetry *(E-1.3.1)*, 1970.

[11] M F Cohen and D F Greenberg. The Hemi-cube: a Radiosity Solution for Complex Environments. *Computer Graphics (Proc. Siggraph 85)*, 19(3):31–40, July 1985.

[12] R L Cook. Shade Trees. *Computer Graphics (Proc. Siggraph 84)*, 18(3):223–231, July 1984.

[13] R L Cook and K L Torrance. A Reflectance Model for Computer Graphics. *ACM Transactions on Graphics*, 1(1):41–44, January 1982.

[14] F Crow, G Demos, J Hardy, J McLaughlin, and K Sims. 3D Image Synthesis on the Connection Machine, 1988. Leeds.

[15] F C Crow. Shadow Algorithms for Computer Graphics. *Computer Graphics (Proc. Siggraph 77)*, 11(3):242–248, July 1977.

[16] F C Crow. A More Flexible Image Generation Environment. *Computer Graphics (Proc. Siggraph 82)*, 16(3):9–18, July 1982.

[17] F C Crow. Summed-Area Tables for Texture Mapping. *Computer Graphics (Proc. Siggraph 84)*, 18(3):207–212, July 1984.

[18] M A Z Dippé and E H Wold. Antialiasing Through Stochastic Sampling. *Computer Graphics (Proc. Siggraph 85)*, 19(3):69–78, July 1985.

[19] T Duff. *The Soid and Roid Manual.* NYIT Computer Graphics Laboratory internal memorandum, 1980.

[20] T Duff. Compositing 3-D Rendered Images. *Computer Graphics (Proc. Siggraph 85)*, 19(3):41–44, July 1985.

[21] E A Feibush, M Levoy, and R L Cook. Synthetic Texturing Using Digital Filters. *Computer Graphics (Proc. Siggraph 80)*, 14(3):294–301, July 1980.

[22] E Fiume, A Fournier, and L Rudolph. A Parallel Scan Conversion Algorithm with Anti-Aliasing for a General-purpose Ultracomputer. *Computer Graphics (Proc. Siggraph 83)*, 17(3):141–150, July 1983.

[23] J D Foley and A Van Dam. *Fundamentals of Interactive Computer Graphics.* Addison-Wesley, 1982.

[24] A Fournier, D Fussell, and L Carpenter. Computer Rendering and Stochastic Models. *Communications of the ACM*, 25(6):371–384, June 1982.

[25] G Y Gardner. Simulation of Natural Scenes using Textured Quadratic Surfaces. *Computer Graphics (Proc. Siggraph 84)*, 18(3):11–20, July 1984.

[26] C M Goral, K E Torrence, D P Greenberg, and B Battaile. Modelling the Interaction of Light Between Diffuse Surfaces. *Computer Graphics (Proc. Siggraph 84)*, 18(3):213–222, July 1984.

[27] H Gouraud. Continuous Shading of Curved Surfaces. *IEE Transactions on Computers*, 20(6):623–629, June 1971.

[28] D A Grindal. The Stochastic Creation of Tree Images. Master's thesis, Department of Computer Science, University of Toronto, June 1984.

[29] J C Horcade and A Nicolas. Algorithms for Antialisased Cast Shadows. *Computers and Graphics*, 9(3):259–265, 1985.

[30] D S Kay and D Greenberg. Transparency for Computer Synthesized Images. *Computer Graphics (Proc. Siggraph 79)*, 13(2):158–164, August 1979.

[31] J Korein and N Badler. Temporal Anti-Aliasing in Computer Generated Animation. *Computer Graphics (Proc. Siggraph 83)*, 17(3):377–388, July 1983.

[32] T Nadas and A Fournier. GRAPE: An Environment to Build Display Processes. *Unkown*, 1987.

[33] A Norton, A P Rockwood, and P T Skolmoski. Clamping: A Method of Antialiasing Textured Surfaces by Bandwidth Limiting in Object Space. *Computer Graphics (Proc. Siggraph 82)*, 16(3):1–8, July 1982.

[34] A W Paeth and K S Booth. Design and Experience with a Generalized Raster Toolkit. *Graphics Interface 1986 Proceedings*, pages 91–97, May 1986.

[35] D R Peachey. Solid Texturing of Complex Surfaces. *Computer Graphics (Proc. Siggraph 85)*, 19(3):279–286, July 1985.

[36] K Perlin. An Image Synthesizer. *Computer Graphics (Proc. Siggraph 85)*, 19(3):287–296, July 1985.

[37] T Porter and T Duff. Compositing Digital Images. *Computer Graphics (Proc. Siggraph 84)*, 18(3):253–259, July 1984.

[38] M Potmesil and I Chakravarty. Synthetic Image Generation with a Lens and Aperture Camera Model. *ACM Transactions on Graphics*, 1(2):85–101, April 1982.

[39] M Potmesil and I Chakravarty. Modelling Motion Blur in Computer Generated Images. *Computer Graphics (Proc. Siggraph 83)*, 17(3):389–399, July 1983.

[40] W K Pratt. *Digital Image Processing.* John Wiley & Sons Ltd, 1978.

[41] W T Reeves. Particle Systems - A Technique for Modelling a Class of Fuzzy Objects. *Computer Graphics (Proc. Siggraph 83)*, 17(3):359–376, July 1983.

[42] W T Reeves and R Blau. Approximate and Probabilistic Algorithms for Shading and Rendering Structured Particle Systems. *Computer Graphics (Proc. Siggraph 85)*, 19(3):313–322, July 1985.

[43] W T Reeves, D F Salesin, and R L Cook. Rendering Antialiased Shadows with Depth Maps. *Computer Graphics (Proc. Siggraph 87)*, 21(4):283–291, July 1987.

[44] P Schoeler and A Fournier. Profiling Graphic Display Systems. *Graphics Interface 1986 Proceedings*, pages 49–55, May 1986.

[45] I E Sutherland, R F Sproull, and R A Schumacker. A Characterization of Ten Hidden-Surface Algorithms. *Computing Surveys*, 6(1):1–55, March 1974.

[46] K E Torrance and E M Sparrow. Theory for off-specular reflection from roughened surfaces. *J Opt Soc Am 57*, pages 1105–1114, September 1967.

[47] Purdue University. *Thermophysical Properties of Mattter*, volume 7, 8, and 9. Plenum, New York, 1970.

[48] C P Verbeck and D P Greenberg. A Comprehensive Light-Source Description for Computer Graphics. *IEEE Computer Graphics and Applications*, 4(7):66–75, July 1984.

[49] D R Warn. Lighting Controls for Synthetic Images. *Computer Graphics (Proc. Siggraph 83)*, 17(3):13–21, July 1983.

[50] G S Watkins. *A Real-Time Visible Surface Algorithm.* PhD thesis, Department of Computer Science, University of Utah, June 1970.

[51] T Whitted. An Improved Illumination Model for Shaded Display. *Communications of the ACM*, 23(6):343–349, June 1980.

[52] L Williams. Pyramidal Parametrics. *Computer Graphics (Proc. Siggraph 83)*, 17(3):1–11, July 1983.

List of Authors

K. Bouatouch
IRISA/IFSIC
Rennes, Frances

C. Bouville
CCETT
Rue du clos Courtel, BP 59, 35512 Cesson, Sevigne, France

A. Fellous
Institut National de l'Audiovisuel
Bry-sur-Marne, France

E. Fiume
Computer Systems Research Institute, University of Toronto
10 King's College Road, Toronto, Canada M5S 1A4

A. Gagalowicz
INRIA
Domaine de Voluceau, Rocquencourt
78153 Le Chesnay, France

I. Herman
Centre for Mathematics and Computer Science (CWI)
Kruislaan 413, 1098 SJ Amsterdam, The Netherlands

W.T. Hewitt
Department of Computer Science, University of Manchester
Oxford Road, Manchester M13 9PL, UK

R.J. Hubbold
Department of Computer Science, University of Manchester
Oxford Road, Manchester M13 9PL, UK

T. Nadas
Thomson Digital Image,
Paris, France

EurographicSeminars

Tutorials and Perspectives in Computer Graphics

Eurographics Tutorials '83. Edited by P. J. W. ten Hagen.
XI, 425 pages, 164 figs., 1984

User Interface Management Systems. Edited by G. E. Pfaff.
XII, 224 pages, 65 figs., 1985

Methodology of Window Management. Edited by F. R. A. Hopgood, D. A. Duce,
E. V. C. Fielding, K. Robinson, A. S. Williams.
XV, 250 pages, 41 figs., 1985

Data Structures for Raster Graphics. Edited by L. R. A. Kessener, F. J. Peters,
M. L.P. van Lierop. VII, 201 pages, 80 figs., 1986

Advances in Computer Graphics I. Edited by G. Enderle, M. Grave, F. Lillehagen.
XII, 512 pages, 168 figs., 1986

Advances in Computer Graphics II. Edited by F. R. A. Hopgood, R. J. Hubbold,
D. A. Duce. X, 186 pages, 96 figs., 1986

Advances in Computer Graphics Hardware I. Edited by W. Straßer.
X, 147 pages, 76 figs., 1987

GKS Theory and Practice. Edited by P. R. Bono, I. Herman.
X, 316 pages, 92 figs., 1987

Intelligent CAD Systems I. Theoretical and Methodological Aspects.
Edited by P. J. W. ten Hagen, T. Tomiyama. XIV, 360 pages, 119 figs., 1987

Advances in Computer Graphics III. Edited by M. M. de Ruiter.
IX, 323 pages, 247 figs., 1988

Advances in Computer Graphics Hardware II. Edited by A. A. M. Kuijk, W. Straßer.
VIII, 258 pages, 99 figs., 1988

CGM in the Real World. Edited by A. M. Mumford, M. W. Skall.
VIII, 288 pages, 23 figs., 1988

Intelligent CAD Systems II. Implementational Issues. Edited by V. Akman,
P. J. W. ten Hagen, P. J. Veerkamp. X, 324 pages, 114 figs., 1989

Advances in Computer Graphics IV. Edited by W. T. Hewitt, M. Grave, M. Roch.
XVI, 248 pages, 138 figs., 1991

Advances in Computer Graphics V. Edited by W. Purgathofer, J. Schönhut.
VIII, 223 pages, 101 figs., 1989

User Interface Management and Design. Edited by D. A. Duce, M. R. Gomes, F. R. A. Hopgood, J. R. Lee. VIII, 324 pages, 117 figs., 1991

Advances in Computer Graphics Hardware III. Edited by A. A. M. Kuijk. VIII, 214 pages, 88 figs., 1991

Advances in Object-Oriented Graphics I. Edited by E. H. Blake, P. Wisskirchen. X, 218 pages, 74 figs., 1991

In preparation:

Advances in Computer Graphics VI. Edited by G. Garcia, I. Herman. Approx. 465 pages, 1991

Advances in Computer Graphics Hardware IV. Edited by R. L. Grimsdale, W. Straßer. Approx. 290 pages, 1991

Intelligent CAD Systems III. Practical Experience and Evaluation. Edited by P. J. W. ten Hagen, P. J. Veerkamp. Approx. 280 pages, 1991

MIX
Papier aus verantwortungsvollen Quellen
Paper from responsible sources
FSC® C105338

If you have any concerns about our products,
you can contact us on
ProductSafety@springernature.com

In case Publisher is established outside the EU,
the EU authorized representative is:
Springer Nature Customer Service Center GmbH
Europaplatz 3, 69115 Heidelberg, Germany

Printed by Libri Plureos GmbH
in Hamburg, Germany